Energy Efficient Spectrum Resources Usage in WPANs

IEEE 802.15.4 MAC Sub-layer Protocols

RIVER PUBLISHERS SERIES IN COMMUNICATIONS AND NETWORKING

Series Editors:

ABBAS JAMALIPOUR
The University of Sydney, Australia

MARINA RUGGIERI
University of Rome Tor Vergata, Italy

The "River Publishers Series in Communications and Networking" is a series of comprehensive academic and professional books which focus on communication and network systems. Topics range from the theory and use of systems involving all terminals, computers, and information processors to wired and wireless networks and network layouts, protocols, architectures, and implementations. Also covered are developments stemming from new market demands in systems, products, and technologies such as personal communications services, multimedia systems, enterprise networks, and optical communications.

The series includes research monographs, edited volumes, handbooks and textbooks, providing professionals, researchers, educators, and advanced students in the field with an invaluable insight into the latest research and developments.

Topics included in this series include:-

- Communication theory
- Multimedia systems
- Network architecture
- Optical communications
- Personal communication services
- Telecoms networks
- Wifi network protocols

For a list of other books in this series, visit www.riverpublishers.com

Energy Efficient Spectrum Resources Usage in WPANs

IEEE 802.15.4 MAC Sub-layer Protocols

Fernando José Velez

Instituto de Telecomunicações and Universidade da Beira Interior,
DEM, Portugal

Luís Miguel Borges

Instituto de Telecomunicações and Universidade da Beira Interior,
DEM, Portugal

Norberto Barroca

Instituto de Telecomunicações and Universidade da Beira Interior,
DEM, Portugal

Periklis Chatzimisios

International Hellenic University, Greece

River Publishers

Routledge
Taylor & Francis Group

NEW YORK AND LONDON

Published 2023 by River Publishers
River Publishers
Alsbjergvej 10, 9260 Gistrup, Denmark
www.riverpublishers.com

Distributed exclusively by Routledge
605 Third Avenue, New York, NY 10017, USA
4 Park Square, Milton Park, Abingdon, Oxon OX14 4RN

Energy Efficient Spectrum Resources Usage in WPANs / by Fernando José Velez, Luís Miguel Borges, Norberto Barroca, Periklis Chatzimisios.

Routledge is an imprint of the Taylor & Francis Group, an informa business

ISBN 978-87-7022-214-3 (print)
ISBN 978-10-0079-473-1 (online)
ISBN 978-10-0333-811-6 (ebook master)

While every effort is made to provide dependable information, the publisher, authors, and editors cannot be held responsible for any errors or omissions.

Contents

Preface

Wireless Sensor Networks (WSNs) and the Internet of Things are facing tremendous advances both in terms of energy-efficiency as well as in the number of available applications. Consequently, there are open challenges that need to be tackled for the future generation of WSNs, and this book gives answers to some of the important questions that need to be answered.

The seed for the project of writing the book was launched in a meeting of COST TERRA (Techno-Economic Regulatory Framework for Radio Spectrum Access for Cognitive Radio Software Defined Radio) in November 2012 in Brussels, when the Covilhã team, from Instituto de Telecomunicações (IT) and University of Beira Interior (UBI), started to interact with the Thessaloniki team, in the context of the discussion of a work on *Energy Harvesting for Wireless Body Area Networks with Cognitive Radio Capabilities*. The willingness to explore new paths of research evolved during the COST-TERRA meeting held in Thessaloniki in April 2013, where the block acknowledgement and piggyback mechanisms for the optimization of channel use was further analyzed. A Short-term scientific mission of Norberto Barroca, from UBI/IT, to conceptualize the proposed energy-efficient protocols and implement them in MiXim was then planned for the summer of 2013.

Resulting from these joint efforts, a contribution to the IEEE P802.15 Working Group for Wireless Personal Area Networks (WPANs), IEEE 802 Meeting, held in Dallas, Texas, USA, on 14th Nov. 2013, was prepared. Also in November 2013, Fernando J. Velez presented the results from the joint work on *Two innovative energy efficient IEEE 802.15.4 MAC sub-layer protocols: employing RTS/CTS and multi-channel scheduled channel polling with packet concatenation*, co-authored with Norberto Barroca, Luís Miguel Borges and Periklis Chatzimisios, in the 8th Meeting of the Management Committee of COST TERRA that was held in Biel, Switzerland.

As the resources that had been created are rich, further collaboration was sought in the context of the International Conference on Telecommunications, held in Lisbon in April 2014, when Periklis Chatzimisios and Athanassios

Iossifides, from the International Hellenic University (IHU), gave a tutorial on aspects of physical layer (PHY) and medium access control (MAC) sub-layer of WLANs, followed by the preparation work during the sabbatical leave from Fernando J. Velez in the University of Aalborg and Carnegie Mellon University, in 2015/16, where the initial book proposal was drafted and discussed with Rajeev Prasad while feedback was being obtained from the external reviewers of River Publishers. Next collaboration steps between the IT/UBI and IHU teams were the IEEE 5G Lisbon Summit and the meetings for the creation of an MSCA Innovative Training Network (ITN) proposal that allowed for the teams to further develop research in IoT towards 5G and actively contribute to standardization bodies. This MSCA/ITN proposal is TeamUp5G, which started in 2019, and allowed for the collaboration that enabled the successful conclusion of the book.

After giving an overview of the WSN protocols and IEEE 802.15.4 standard, aspects of its physical layer, and MAC sub-layer protocols (including a complete survey on a broad range of protocol options), the book proposes IEEE 802.15.4 Medium Access Control (MAC) sub-layer performance enhancements by employing not only RTS/CTS combined with packet concatenation but also scheduled channel poling (SCP) and its derived version with multi-channel access (MC-SCP). It addresses the importance of such an appropriate design for the MAC sub-layer protocol for the desired WSN application. Depending on the mission of the WSN application, different protocols are required. Therefore, the overall performance of a WSN application certainly depends on the development and application of suitable MAC and network layer protocols. We hope that the book will be a source of inspiration for a new generation of young researchers starting their research career path, creating their models and preparing the code for the simulation of their novel protocols.

Acknowledgements

This work was partially supported and funded by the National Funding from the FCT - Fundação para a Ciência e a Tecnologia through the PhD grant FRH/BD/66803/2009, UIDB/EEA/50008/2020, SFRH/BSAB/113798/2015, 3221/BMOB/16 CMU Portugal Faculty Exchange Programme grant, and CONQUEST (CMU/ECE/030/2017), and by COST IC0905 TERRA, COST CA15104 IRACON, COST CA20120 INTERACT, SNF Scientific Exchange - AISpectrum (project 205842), TeamUp5G and ORCIP. TeamUp5G project has received funding from the European Union's Horizon 2020 research and innovation programme under the Marie Skodowska-Curie project number 813391.

List of Figures

List of Tables

List of Acronyms

ACK	Acknowledgement
Act	Active period
Adapt	Adaptive
AFH	Adaptive frequency hopping
AP	Access point
App	Application demand
ASK	Amplitude shift keying
AWGN	Additive white gaussian noise
BACK	Block acknowledgment
BE	Backoff Exponent
Be	Beacon mode
BER	Bit error rate
BI	Beacon interval
BO	Beacon order
BP	Backoff period
BPSK	Band binary phase-shift keyin
BSN	Beacon sequence number
Busy	Busy tone
CA	Channel assessment
CAP	Contention access period
CCA	Clear channel assessment
CD	Channel detection
Cent	Centralized scheduling
CFP	Contention free period
CL	Cross-layer
Co	Reducing collisions

Conv	Converge cast
CPU	Central processing unit
CR	Cognitive radio
CRC	Cyclic redundancy check
CS	Carrier sense
CSMA-CA	Carrier sense multiple access with collision avoidance
CSS	Chirp spread spectrum
CTS	Clear-to-send
CW	Contention window
CW_1	First contention window
CW_2	Second contention window
DaP	Data prediction
Db and Fi	Double CW & fixed size
Db and Var	Double CW & variable size
Dis	Dynamic size
Dist	Distributed scheduling
DLL	Data link layer
DpR	Data frame preamble
DPSR	Data packet success rate
DSN	Data sequence number
DSSS	Direct sequence spread spectrum
DtS	Data packet sampling
Dual	Dual-channel
EC	European commission
ER	Extra resolution
ETSI	European telecommunications standards institute
FCC	Federal communications commission
FCS	Frame check sequence
FFD	Full function device
FFT	Fast fourier transform
FHSS	Frequency hopping spread spectrum
Fi	Fixed

FIFO	First in, first out
FrD	Framelets division
FrH	Frequency hopping
Glob	Global schedule
GTS	Guaranteed time slot
Hi	High traffic
IFS	Interframe space
Il	Reducing idle listening
IoT	Internet of things
IR	Influential range
ISM	Industrial, scientific and medical
LCG	Linear congruential generator
LiB	Link activation based
LIFS	Long interframe spacing
Lo	Low traffic
Local	Localized collision free based
LPL	Low power listening
LQI	Link quality indicator
LR-WPAN	Low-rate wireless personal area network
LS	Least square
MAC	Medium access control
MC-SCP-MAC	Multi channel scheduled channel polling MAC
Me	Medium traffic
Media	Multimedia
MFR	MAC footer
MHR	MAC header
MHWN	Multi-hop wireless network
Micro	Micro-frame
MMSN	Multi-frequency MAC for wireless sensor networks
Mob	Mobility adjust
MPDU	MAC protocol data unit
MSB	Most significant bit

MSDU	MAC service data unit
MSK	Minimum shift keying
Multi	Multi-channel
Multi	Multi-frequency
NaI	Narrow band interference
NAV	Network allocation vector
NB	Number of Backoffs
N-Be	Non-beacon mode
ND	Node-contention
Nghb	Neighbour-based
NoB	Node activation based
Oh	Reducing overhearing
On	On-demand based
O-QPSK	Offset quadrature phase-shift keying
Ov	Reducing overhead
Pack	Packetization
PAN	Personal area network
PER	Packet error rate
PHR	PHY header
PHY	Physical
Pigg	Piggybacking synchronization information
PM	Piggyback mechanism
PN	Pseudo-random noise
PPDU	PHY protocol data unit
Pre	Pre-defined sets of duty cycles
PrS	Preamble sampling
PSDU	PHY service data unit
Pseudo-RNG	Pseudo-random number generator
PSR	Packet success rate
QoS	Quality-of-service
QPSK	Quadrature phase-shift keying
Radio	Wake-up radio

Rang	Range
Rc	Receiver-based
RF	Radio frequency
RFD	Reduced-function device
RFH	Random frequency hopping
RSC	Radio spectrum committee
RSSI	Received signal strength indicator
Rt	Rotative
RTS/CTS	Request-to-send/Clear-to-send
RTS	Request-to-send
RX	Receive
Samp	Sampling duration
SBACK	Sensor block acknowledgement
SBACK-MAC	Sensor block acknowledgment MAC
Schd	Wake-up schedule
Sched	Schedules period
ScP	Schedule channel polling
SCP-MAC	Scheduled channel polling MAC
SCWNG	Wireless next generation standing committee
SD	Superframe duration
Sensor	Sensor
SER	Symbol error rate
SFD	Start of frame delimiter
Sg and Fi	Single CW & fixed size
Sg and Var	Single CW & variable size
SG	Spreading gain
SHR	Synchronisation header
SIFS	Short interframe spacing
SIG	Special interest group
Single	Single-channel
Single	Single-frequency
SIR	Signal-to-interference ratio

Sleep	Sleep delay
S-MAC	Sensor-MAC
SNIR	Signal-to-interference-plus-noise ratio
SNR	Signal-to-noise ratio
SO	Superframe order
St	Static
Stag	Staggered based
Stat	Statistical function
Synch	Synchronization period
Tf	Traffic load
Tone	Wake-up tone
Uni	Uniform
Util	Utilization function
UWB	Ultra-wideband
Var	Variable
VAT	Value added tax
WBAN	Wireless body area network
WLAN	Wireless local area network
WPAN	Wireless personal area network
WSANs	Wireless sensor actuator networks
WSN	Wireless sensor network

List of Symbols

A_e	Exclusive area of two circles with equal radius
a_i	Representation of the slot i
$A_f(m)$	Number of possibilities for the lowest order slot to be chosen by more than one node
$A_s(m)$	Number of possibilities that exactly one node chooses the i-th lowest order slot and the remaining ones choose the subsequent slots
$aTurnaroundTime$	RX/TX or TX/RX maximum turnaround length
$aUnitBackoffPeriod$	Backoff period length
BE	Backoff Exponent
BER	Bit error rate
BI	Beacon Interval
BO	MAC Beacon Order
C	Capacity of the medium (i.e., aggregated throughput)
C_{batt}	Capacity of the battery in ampere hour
$ccaTime$	Time delay due to Clear channel assessment (CCA)
ch	channel order
CW	Contention Window
c_+	Increment of the LCG
d	Distance separation between the transmitter and receiver
D_{ch}	Slot channel degradation threshold
D_{min}	Minimum Average Delay
D_{min_CCA}	Minimum delay due to CCA
$D_{min\_CCA\_RTS}$	Minimum delay due to CCA before each RTS/CTS
$D_{min\_Data\_Ret}$	Minimum delay due to packet retransmissions
$D_{min\_Data\_Ret\_RTS}$	Minimum delay due to packet retransmissions and ACK is not received

$D_{min\_RTS\_CTS}$	Minimum average delay due to the channel state and packet retransmissions
D_T	Random Deferral Time Period
E_b/N_0	SNR to noise density
$\overline{E_{cons\_lw}^{SCP}}$	Lower bound for the energy consumption with piggybacked synchronization
$\overline{\overline{E_{cons\_lw}^{SCP}}}$	Lower bound for the energy consumption without piggybacked synchronization
$\overline{E_{cons}^{SCP}}$	Total energy consumed by the node
E_{dat}	Energy associated with the sampled data
e_i	Expected theoretical frequency
E_{listen}	Energy consumed during the node listening process
$\overline{E_{listen}^{SCP}}$	energy consumption in the node listening process
$E_{multi\_lw}^{SCP}$	Lower bound for the energy consumption with piggybacked synchronization in a linear chain topology
E_s/N_0	Symbol to noise density
$E[S]$	Expected number of contending nodes in CW_2
$E[m]$	Nearest integer value of $E[S]$
E_{poll}	Energy consumption during the node polling process
E_{poll}^{SCP}	energy consumption in the node polling process
$\overline{E_{poll}^{SCP}}$	Energy consumption in node polling process
E_{rx}	Energy consumed during the node reception process
$\overline{E_{rx}^{SCP}}$	Energy consumption in node reception process
E_{sensor}^{SCP}	Energy associated with sampled data
E_{sleep}	Energy consumed during the sleeping time of a sensor node
$\overline{E_{sleep}^{SCP}}$	Energy consumption in node sleeping process
E_{tx}	Energy consumed during the node transmission process
$\overline{E_{tx}^{SCP}}$	Energy consumption in node transmission process
f	Clock skew
f_b	Carrier Frequency

f_c	Center Frequency
G_r	Receiver gain
G_t	Transmitter gain
I	Discharge current in ampere
i_{dat}	Sample sensors current, in mA
I_{listen}	Current in listening
I_{poll}	Current in polling
i_{rx}	Current to receive a Byte
I_{rx}	Current in receiving
i_{sleep}	Current in sleeping mode, in μA
I_{sleep}	Current in sleep
i_{tx}	Current to send a byte [mA]
I_{tx}	Current in transmitting
L	Side length
L_{ACK}	ACK frame length
L_{DATA}	DATA payload
L_{data}	data frame length, in Bytes
L_{FL}	DATA frame length
L_{H_MAC}	MAC overhead
L_{H_PHY}	PHY length overhead
L_{SHR}	PHY SHR length
M	Codebook
m_{lcg}	Modulus of the LCG
n	Number of nodes in the network
N	Number of needed slot channels
n^*	Inactive period
N_{agg}	Number of aggregated packets
N_{ch}	Available channels
NB	Number of Backoffs
$nBits$	Length of the frame
n_{frames}	Number of TX frames (referred as n DATA frames/packets in some Figures)
n_{hd}	Number of hidden terminals
n_{max}	Maximum number of nodes
n_{ngb}	Number of neighbours
n_peuk	Peukert's exponent for a particular battery type
n_{tx}	Total number of nodes per slot
n_{tx_F}	Number of nodes transmitting a packet involved in a collision

n_{tx_S}	Number of nodes per slot that transmit a data packet and experienced a successfully reception
P_b	Probability of Bit Error
P_{c1}	Probability of collision in the first contention period, CW_1 (saturated condition)
$\widehat{P_{c1}}$	Probability of collision in the CW_1 (unsaturated condition)
P_{cF}	CW_2 collision probability (saturated condition)
$\widehat{P_{cF}}$	Collision probability in CW_2 (unsaturated regime)
$\overline{P_{cons}^{SCP}}$)	Power consumed in SCP-MAC
$phySymbolsPerOctet$	Symbols per octet for the current PHY
$P_{eF}(m)$	Probability of a collision occurring in CW_2 by combinatorics approach (saturated regime)
$\widehat{P_{eF}}(m)$	Probability of a collision occurring in CW_2 by combinatorics approach (unsaturated regime)
PER	Packet error rate
$phySymbolsPerOctet$	Symbols per octet for the current PHY
p_i	Probability of choosing a slot i
P_I	Power consumption in the idle state
P_L	Path loss
$\overline{P_{listen}}$	Power in listening
$\overline{P_{multi\_hg}^{SCP}}$	Higher bound for the power in SCP-MAC with piggybacked synchronization in a linear multi-hop chain
$\overline{P_{multi\_lw}^{SCP}}$	Lower bound for the power in SCP-MAC with piggybacked synchronization in a linear multi-hop chain
P_{poll}	Power during poll interval
P_r	Received Power
P_{rx}	Power during reception interval (or power in receiving)
P_{RX}	Power Consumption in the Receiving state
P_{tx}	Power during transmission interval
P_s	Probability of Symbol Error
P_{S_C}	Probability of nodes choosing the same slot channel

P_{Scol}	Probability of choosing same slot in CW_2 (saturated regime)
$\widehat{P_{Scol}}$	Probability of choosing same slot in CW_2 (unsaturated regime)
P_{sF}	Probability of the event of a successful data packet transmission after i-th transmission attempts after $i-1$ consecutive transmission failures, for $i = 1, ..., \zeta, \ \forall \ \zeta \to \infty$
P_{Sidle}	Probability of a slot being idle (saturated regime)
$\widehat{P_{Sidle}}$	Probability of a slot being idle (unsaturated regime)
P_{Sleep}	Power Consumption in the Sleep state
PSR	Probability that the frame has no errors
P_{Ssuc}	Probability of not choosing the same slot in CW_2
P_t	Transmitter Power
P_{tx}	Power in transmitting
P_{TX}	Power Consumption in the Transmitting state
P_{u_G}	Uniformly distributed packet generation probability / probability of a given node to choose one of the G intervals unsaturated
Q	Capacity
Q^2	Chi-square statistic
Q_d	Drained Capacity
Q_r	Remain Capacity
R	Data rate
R_{batt}	Battery hour rating
R_c	Chip rate
rdm_{nb}	Random backoff time
$RSSI$	Receiver Signal Strength Indicator
$rxSetupTime$	Setup radio to RX or TX states
S	Number of contending nodes in CW_2
S_{11}	Return loss
S_{c11}	SYNC and control channel
SD	Superframe duration
S_{max}	Maximum Average Throughput
$S_{max\_RTS\_CTS}$	Maximum average throughput by employing RTS/CTS combined with packet concatenation
S_{min}	Sensitivity

SO	MAC superframe order
T	Battery Lifetime
$T_{1,n}$	Current time of the parent node (coordinator)
$T_{2,n}$	Current time of the child node
T_{ACK}	ACK transmission time
T_{AW}	ACK wait duration time
t_B	time to transmit/receive a Byte
T_{BO}	Backoff period duration
$T_{BRequest}$	BACK Request transmission time
$T_{BResponse}$	BACK Response transmission time
T_{CCA}	CCA detection time
t_{CS}	Average carrier sense time (or expected time a node spends in carrier sense)
t_{CS1}	average carrier sense time
T_{CTS}	CTS transmission time
T_{CTS_ADDBA}	CTS ADDBA transmission time
t_{dat}	Time to sample sensors, in s
T_{DATA}	Transmission time for DATA
t_F	Duration of the Frame
t_{guard}	Time out timer to receive last data packet
T_h	Time in hours
t_{idle}	Time on IDLE state
$t_{init_{CS}}$	Time to start performing carrier sense
T_{LIFS}	LIFS time
t_{node}	lifetime of a sensor node
T_p	channel polling period
t_{p1}	average time to poll the channel
T_{RTS}	RTS transmission time
T_{RTS_ADDBA}	RTS ADDBA transmission time
$t_{RX\_SYNC\_pkt}$	Time interval to receive SYNC packets
T_{SHR}	PHY SHR duration
T_{SIFS}	SIFS time
$t_{simulation}$	Simulation time
t_{sleep}	duration of the sleeping period
T_{subs_slots}	Duration of the remaining time slots
T_{symbol}	Symbol time
t_{sync_max}	Maximum time to disseminate SYNC packet
t_{sync}	Synchronization period
T_{TA}	TX/RX or RX/TX switching time

T_{tR}	Inter-arrival time value
t_{tone}	Duration of the wake-up tone
T_{tone_min}	Minimun duration of the wake-up tone
t_{tx}	Time on TX state
t_{tx_sync}	Time instant when the SYNC packet was sent
$t_{wake-up}$	Time for wake-up
t_{θ_i}	Time instant value for the first data packet generation
$t_{\theta_{max}}$	Maximum time of the interval for the packet generation
t_{rx}	Time on RX state
$t_{RX\_SYNC\_pkt}$	Duration for the reception of a SYNC packet
t_{sleep}	Time on SLEEP state
$\overline{T_{sv}}$	Average MAC service time
$t_{\varphi}^{CW_1}$	Time instant to initiate the CS
$t_{\mu Pact}$	Active time microcontroller
$t_{\mu Psleep}$	Sleep time microcontroller
V_{cc}	Voltage, in V
X	Geometric random variable
x_i	Throughput of the node i
X_i	Random variable which corresponds to the number of nodes that choose the slot i (to start the transmission of the wake-up tone)
X_n'	Variable of the transmission delays
X_n	Current seed of the LCG
X_{n+1}	LCG generated value
X_σ	Zero mean Gaussian random variable
Y_i	Random variable (which corresponds to the number of nodes that choose the slot i to start the transmission of the data packet)
α_{add}	Time duration between consecutive Frames
α_{exp}	Path loss exponent
α_{turn}	Turnaround time
$\Delta\sigma$	CW$_1$ time interval
$\Delta\sigma_{max}$	Maximum CW$_1$ time interval
Δt_{SC}	Time interval in which the parent node switches to each channel
η	Bandwidth efficiency
ι_{ret}	Maximum number of retries

λ	Traffic generation rate
λ_{ch}	Mapped slot channel
λ_{DATA}	data packet rate
ω	Clock offset
Π_{max}	Highest interval order
Π_{irmax}	IR Sensing range threshold
ψ	Transmission delay
$\psi_{wake-up}$	Next sender node wake-up time
σ	Standard deviation
$\sigma_{current}$	Current time of the node
τ	Probability that a node chooses a given slot i in CW_2
$\theta_{PAN\_max\_wait}$	Maximum time that PAN waits to initiate slot channel hopping routine
θ_{switch}	Time needed by the sensor node to switch to the chosen channel
φ_{ch}	Random channel chosen by the sender node
$\varphi_{slot}^{CW_1}$	Time slot choice in CW_1
$\varphi_{slot}^{CW_2}$	Time slot choice in CW_2
φ_{slot}^{SYNC}	Slot choice in SYNC CW
ϑ_{fi}	Throughput fairness index
ϑ_s	Achieved throughput

1

Introduction

1.1 Motivation

Wireless Sensor Networks (WSNs) are one of the most contemporary break-through technological innovations in the field of massive Internet of things (IoT) and machine-type communications of our time. Wireless sensor actuator networks (WSANs) are responsible for interconnecting several wireless sensor nodes by providing global ad-hoc communication and computational capabilities. This type of networks consists of a large number of sensing devices called "sensor nodes" and is capable of linking the Physical (PHY) layer with the digital world by sensing, processing, and transmitting the real-world phenomena, and by converting these signals into a form that can be processed, stored and acted upon.

Since this type of networks are battery operated, energy harvesting can be a solution to make WSNs autonomous while enabling a widespread use of these systems in many applications where the energy storage system can consist of a supercapacitor and/or a rechargeable battery. However, one of the components with the largest power consumption in the sensor nodes is the radio transceiver, as it has a decisive influence in network lifetime.MAC protocols will enable to achieve such energy-efficiency while determining the operation mode of the radio.

Recent advances in the field of microelectronic circuits caused an increase on the interest in the development of WSNs. WSNs and WSANs can be deployed in many scenarios such as factory monitoring, healthcare, environment monitoring, logistics, location of persons in commercial buildings, monitoring of building structures and precision agriculture [VDMC08], [RWR04], [ASSC02b].

A set of emergent applications focus on aspects related with the management of the largest parking lots. For years, automotive traffic got all the attention: cars in motion were exciting; slow moving cars were boring.

1

Hardly anyone talked about parking. However, recently, parking has become a paramount concern among city planners and company heads. Space is limited, and yet the demand for parking places keeps increasing. Hence WSNs can be used to monitor the vehicle entrance and exit events that occur in parking structures, wirelessly transmitting the data back to a base station to process the information, and send it to a database that can be access remotely. Smart buildings are also an "hot topic" because about 70 % of the utility bill for the average household could be affected by the lighting and temperature-related WSANs applications. As utility rates and Value added tax (VAT) tends to increase over the years, by using WSANs it is possible to diminish the impact of the increase of the price of electricity by using an energy management system that is responsible for controlling the cooling, heating, and lighting zones, based on real time humidity and temperature measurements.

The European Union is aware of the importance of WSANs and the benefits of these networks with low-complexity and low-cost devices and has funded several ongoing projects such as WISEBED [WIS12], COOLNESS [COO12] and SA-WSN [SAW12], Inter-IoT, BIG IoT, AGILE, symbIoTe, TagItSmart!, VICINITY, bIoTope and NGIoT.

1.2 Challenges and Approach

Several research issues still need to be addressed in the WSN domain, as the current solutions are not enough optimised or are too much constrained. Although in the literature most of the proposed communication protocols improve energy efficiency to a certain extent by exploiting the collaborative nature of nodes within WSNs and its correlation characteristics, the main commonality of these protocols is the compliance to the traditional layered protocol stack, allied to the fact they are mainly implemented in commercially available platforms that employs the IEEE 802.15.4 standard.

In the field of WSNs there is a huge number of proposals for energy-efficient Medium access control (MAC) protocols [HXS+13]. However, these MAC protocols have not enoughly succeeded to be set as real commercial applications due to the lack of standardization. To design an optimized Wireless sensor network (WSN) MAC protocol, the following aspects must be encompassed [DEA+06]:

- Energy efficiency, since there are strong limitations to power supply WSN tiny devices and one of the main goals is to prolong the network lifetime;

- Scalability and adaptability to changes in network size, node density and topology should be handled rapidly and effectively for a successful adaptation. Some causes of these network property changes are node lifetime, addition of new nodes to the network and varying interference which may alter the connectivity, hence change the network topology;
- Quality-of-service (QoS) attributes, such as latency, throughput and bandwidth efficiency are also considered, as they are very common for multimedia and real time applications.

Allied to the aforementioned design issues, IEEE 802.15.4 has been widely accepted as the *de facto* standard for wireless sensor networks. Zigbee Alliance has estimated that over half a billion of IEEE 802.15.4 chips where sold until August 2018 and 4.5 billion will be sold until 2023. In the context of this work, we propose new innovative MAC sub-layer protocols whose relevance justifies that they can also contribute to the non-beacon-enabled mode of the standard itself.

Packet concatenation facilitates the aggregation of several consecutive packets by means of channel reservation and different types of acknowledgment and/or Network Allocation Vector (NAV) procedures. In this context, the Request-to-send/Clear-to-send (RTS/CTS) mechanism enables to reserve the channel and avoids to repeat the backoff phase for every consecutive transmitted frame, probably reducing overhead. In the presence of RTS/CTS signaling, two solution are considered, one with DATA/ACK handshake and another without frame ACKs, simply relying in the establishment of the NAV.

In particular, the Sensor Block Acknowledgment MAC (SBACK-MAC) protocol allows the aggregation of several acknowledgment responses into one special packet *BACK Response* being compliant with the non-beacon mode of the IEEE 802.15.4 standard. Two different solutions are addressed. The first one considers the SBACK-MAC protocol in the presence of *BACK Request* (concatenation mechanism), while the second one considers SBACK-MAC in the absence of *BACK Request* (the so-called piggyback mechanism).

The following objectives have been identified as key points for research:

- Evaluate the IEEE 802.15.4 MAC sub-layer protocol performance by using the RTS/CTS combined with frame concatenation;
- Mathematical derive the maximum average throughput and the minimum average delay for the proposed mechanisms, either under ideal conditions (a channel environment with no transmission errors) or non ideal conditions (a channel environment with transmission errors), by

varying the data payload and the number of transmitted packets. A comparison is made with IEEE 802.15.4 in the basic access mode;

- Address retransmission scenarios for both IEEE 802.15.4 and SBACK-MAC, where two nodes simultaneously, identify the idle channel during CCA and start transmitting, causing data collisions;

The research also involves the proposal of the Multi-channel Scheduled Channel Polling MAC (MC-SCP-MAC) protocol. This new MAC sub-layer scheduled protocol explores the advantages of multi-channel features in addition to the capture effect present in the recent radio transceivers, the Influential Range (IR) concept (to mitigate the energy losses due to overhearing) and cognitive-based capabilities, such as the channel degradation sensing jointly with opportunistic channel selection. A simulation approach is considered, where different aspects are addressed, namely:

- Energy consumption evaluation and comparison (between single and multi-hop topologies) while different traffic generators are considered;
- Evaluation of the expected benefits in terms of packet collision ratio from a multi-channel based protocol with a two-phase contention window mechanism;
- Analysis of the impact of contention windows sizes while varying the number of channels for different traffic profiles (periodic and Poisson);
- Analysis of the impact of the number of channels considered in the energy consumption, delay and aggregate throughput;
- Analysis of the impact in the energy consumption, delivery rate and aggregate throughput while varying the sensor nodes density (dense and sparse scenarios);
- Analysis of the impact of enabling the IR concept in the MC-SCP-MAC protocol in energy consumption, delay and delivery rate;
- Comparison of the performance results for the considered influential range thresholds (for the definition of optimal values);
- Energy efficiency evaluation for the scenarios that consider a cluster-based topology and evaluation and analysis of the short- and long-term throughput fairness index;
- Comparison and analysis of the energy consumption, delivery rate and aggregate throughput for different network sizes while enabling the frame capture capabilities for different values of the reliability;
- Evaluation of the benefits of enabling cognitive-based capabilities.

The proposed MAC protocols have been conceived envisaging to be compliant with the non-beacon mode of the IEEE 802.15.4 standard, since it is envisaged that the proposed solutions can be integrated in the IEEE 802.15.4 standard or serve as the basis for the Wireless Next Generation Networks.

1.3 Structure of the Book

This book is organized into eight Chapters, including this one, and several Appendices.

Chapter 2 discusses different aspects of Medium Access Control sub-layer and Physical Layer in WSNs. Chapter 3 discusses the IEEE 802.15.4 standard PHY and MAC layers in detail. The main focus of the approach followed in the description of the MAC sub-layer is the non-beacon mode.

Chapter 4 describes the Scheduled channel polling MAC (SCP-MAC) protocol [YSH06a] in detail. The motivation to use the SCP-MAC protocol is addressed and the various SCP procedures, such as the two-phase contention window mechanism, and synchronization phase are described. A state transition diagram is proposed, followed by the description of the parameters and definitions, involved in the implementation of the SCP-MAC protocol in the simulation framework. The protocol employs a collision avoidance mechanism with two contention windows, one window for a short wake-up tone and the second one for data.

Chapter 5 explores the SCP-MAC performance evaluation. The chapter starts by showing the simulation results for the performance metrics (e.g., energy consumption, throughput) for single-hop and multi-hop topologies. Heavy and periodic traffic patterns are considered. A lifetime analysis (enabling piggyback synchronization) for the SCP-MAC protocol is proposed by considering a simple battery discharge model, given by Peukert's equation [JH09], followed by the proposal of a stochastic model for the collision probability of the SCP protocol in the saturated and unsaturated regimes, which is applied to derive how and when collisions in the communication between nodes occur, while obtaining the average MAC service time for a successful transmission, and determine the achieved throughput in the same collision avoidance scheme.

Chapter 6 proposes the use of the RTS/CTS mechanism combine with packet concatenation to enhance channel efficiency of the IEEE 802-15-4 MAC sub-layer by decreasing the deferral time before transmitting a data frame. In addition, the SBACK-MAC protocol has been proposed, which allows the aggregation of several acknowledgment responses in one special *BACK Response* frame. The first one considers the SBACK-MAC protocol in the presence of *BACK Request* (concatenation) while the second one considers the SBACK-MAC in the absence of *BACK Request* (piggyback).

Chapter 7 proposes the Multi-Channel-Scheduled Channel Polling (MC-SCP-MAC) protocol, a new protocol based on SCP protocol but envisaging multi-channel features, for different node topologies, while studying the impact of the influential range concept in its energy efficiency.

Finally, Chapter 8 presents the conclusions of the research carried out throughout the book, as well as some final considerations and suggestions for further research.

Detailed information on the IEEE 802.15.4 PHY and MAC sub-layer, and modulations. While Appendix A addresses the PHY layer of IEEE 802.15.4, Appendix B presents additional details of the MAC sub-layer, including the proposal of a taxonomy for underlying protocols. Appendix C addresses the O-QPSK modulation employed by IEEE 802.15.4.

2

Medium Access Control and Physical Layers in WSNs

This chapter is organized into nine sections. Section 2.1 presents the protocol stack for WSNs as well as the different layers that compose it. Section 2.2 briefly discusses the protocol stack commonly assumed for WSNs while presenting a comparison with other protocol stack models. Section 2.3 describes some basic wireless radio system families, as well as the evolution of the IEEE 802 standards until the appearing of the IEEE 802.15.4 standard. Section 2.4 briefly discusses the relationship between the IEEE 802.15.4 standard and ZigBee Alliance. Section 2.5 presents an overview of the IEEE 802.15.4 standard compliant PHY layer, followed by the Section 2.6, which is dedicated to the MAC sub-layer and its different tasks. Section 2.7 presents the state-of-the-art on MAC protocols for WSNs and proposes a taxonomy for MAC protocols (extra details are given in Section B.3 from Appendix B). MAC protocols are classified into different categories, which depend on the technique employed to access the channel, whilst considering that one MAC protocol may utilize multiple techniques. In Section 2.8, a summary of the MAC characteristics is presented to classify and create a taxonomy the state of the art on MAC protocols for WSNs. Finally, conclusions are drawn in Section 2.9 . Complete details on the IEEE 802.15.4 standard PHY layer are addressed in Appendix A. The comparison tables for the MAC sub-layer protocols of the proposed taxonomy are presented in Appendix B.

2.1 Protocol Stack for WSNs

A protocol stack comprises the aggregation of the network protocols that are divided into several distinct layers, depending on their functions. For each layer, a set of interfaces are defined, leading to a more flexible adaptation when there are changes in the software and hardware. Each layer has certain services to offer to the highest layers, while protecting the details (and primitives) of how the services of the layer are implemented. Hence, each layer hides a considerable complexity and can be faced as a "middleware" that offers services to the layer above it. When the layer interacts with the layer below it, then a functionality could be implemented. The model widely used for the protocol stack is the Open Systems Interconnection (OSI) model [PTR], which embraces the following seven layers: i) Physical, ii) Data Link, iii) Network, iv) Transport, v) Session, vi) Presentation and vii) Application Layers. As the WSNs application requirements are quite different from the normal desktop applications usually, the WSNs only make use of the following five layers: Application, Transport, Network, Data Link and Physical Layers. In this simplified representation, an additional Middleware Layer is introduced to implement Applications programming interface (API) for WSN applications. It is presented with a different colour because it can be seen as part of the application layer. Figure 2.1 facilitates the comprehension of the concepts by presenting a

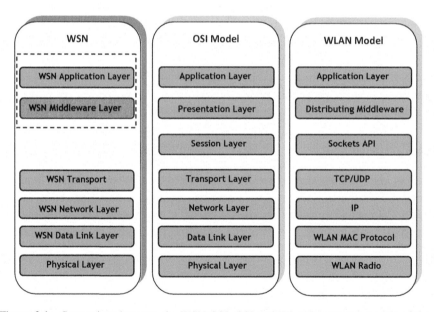

Figure 2.1 Comparison between the WSN OSI, OSI and WLAN computer models of the networking protocol stack.

comparison between the OSI model for WSNs and the standard Open systems interconnection (OSI) model [PTR] and WLAN computer protocol stack [KHH05].

The detailed description of the five layers of the OSI model for WSNs, from top to down, is as follows:

- **Application Layer**: The top layer of the OSI model for WSN is the application layer. It offers network services and the functionalities for the node by using the network applications. This layer contains several processes that could be executed in parallel, e.g., sensing applications for different sensors, actuator, and node diagnostics. The application layer configuration could be changed, depending on the application requirements, e.g., in an office building environment control application, a node can be configured to have any of the four functionalities: control node, sensor node, actuator node and interface node. Besides, this layer defines the format of exchanged messages and the order of message exchanges between different processes. This means that the application layer makes the hardware and software of the lower layers transparent to the end-user.

- **Transport Layer**: The transport layer lays below the application layer. It performs the flow control through the network. The flow control occurs when the receiver of a data stream is temporarily unable to process incoming packets, due to the possible lack of memory or processor power. However, flow control has not been a research issue in WSNs. The transport layer is responsible for upper layer error control to detect and repair losses of packets that were not detected by Data link layer (DLL), by using appropriate mechanisms. Since, in WSNs, the nodes operate at low transmission power levels, in order to save energy, link reliability is worse than in conventional wired and wireless networks. As such, the flow and error control should be done separately for each hop and not end-to-end, as in conventional networks [KW05b]. Another task assured by the transport layer is congestion control. Congestion occurs when more packets are created than the ones the network can carry out. Consequently, the network starts to drop packets, which implies waste of energy and decrease of the reliability (or information accuracy). The transport layer tries to avoid this situation or to solve it in a fair manner. One important way to avoid congestion is to control the rate at which packets are generated by the sensor nodes (flow control). On the one hand, the transport layer performs fragmentation to divide the upper layer application data into small segments, appropriate for DLL. On the other hand, the transport layer reorders and join the received data segments into data packets suitable for the application layer. Another feature of the transport layer is network abstraction. It offers a programming interface to applications, by hiddening diverse complexities of the data transport layer.

- **Network Layer**: The network layer is located below the transport layer and is responsible for network self-configuration and data routing. In order to configure the network topology, the network layer selects an appropriate node operation mode and determines the most suitable neighbours with which to associate and establish communication links. The exchange of routing information process takes place either periodically, set by the network administrator or designer, or is triggered based on events, i.e., when the routing information changes (in order to ensure the network connectivity while optimizing network lifetime by balancing energy consumption among other nodes in the network). The routing protocol is responsible by the decision of the suitable next-hop node in which to forward each data frame, to eventually reach its desired destination. In the overall of the WSN protocol stack a routing protocol executed in the network layer performs end-to-end data routing. In fact, the main function of the network layer is to route packets from the source node to the destination, through the wireless multi-hop network. However, this is a challenging task in WSNs because the computational and energy constraints of the nodes and underlying network-wide functions that they need to accomplish. The routing algorithm is responsible of finding the best possible path from source to destination. It utilises routing information collected by nodes. The best possible path is chosen according to some predefined optimization criteria related to the network or application needs. For example, the simplest routing algorithm consists of finding the shortest path, as it is the path that utilises the least amount of network resources and should provide the shortest delay. Once the routing protocol and algorithms are run, each node should have a routing table that tells the node which is the most appropriate neighbour where to forward the packet to in order to reach the destination through the best possible path. It is also worthwhile to mention that the energy costs of lower protocol layers caused by association and neighbour discovery should also be considered [ZZJ09, KSK+08].

• **Data Link Layer (DLL)**: It is located below the network layer the DLL. It makes the interface between the physical and network layers. It is formed by a MAC sub-layer and a Logical Link Control (LLC) module. The MAC sub-layer provides a fair mechanism to share access to the medium among other nodes and determines how and when to utilise PHY functions. The MAC sub-layer plays a key role in the maximization of a node's energy efficiency. The LLC operates above MAC and is responsible for encapsulating message segments into frames and adding appropriate header information, with destination and source addresses and control and sequencing information and CRC calculation. With the information contained in the frame, it allows for the desired destination node to receive a frame, ensure frame integrity and maintain proper sequencing of frames [KSK+08]. Therefore, the Data Link Layer plays an important role on discovering new nodes. This may be the case in two situations: i) applications with mobile nodes, and ii) in re-deployments, because new nodes are re-deployed to bring the network back to its original operational level. In any case, the topology control thin layer needs to know about the existence of new neighbours before a new topology construction phase is triggered [ZZJ09].

• **Physical Layer**: The lowest layer in the WSN protocol stack is the PHY layer, which implements a network communications hardware that transmits and receives messages, one bit or symbol at a time. In a real sensor node the PHY receives analogue symbols from the medium and converts them to digital bits for further processing in the higher layers of the protocol stack. In most transceivers the PHY functions available are the selection of a frequency channel and a transmit power, the modulation transmitted and demodulation of received data, symbol synchronisation and clock generation for received data. The PHY transceiver may also include extra functions, which could lead to a reducing of processing requirements of the Micro-Controller Unit (MCU). One example is the IEEE 802.15.4 compliant PHY that includes:

- Data frame synchronisation for perceiving the start of an incoming frame;
- Clear Channel Assessment (CCA) for detecting ongoing traffic in a frequency channel;
- Received Signal Strength Indicator (RSSI) and Link Quality Indication (LQI) for measuring signal strength and estimating link quality to neighbouring nodes;

– Cyclic Redundancy Check (CRC) calculation for checking bit errors on received frames;
– Data encryption/decryption for improving network security;
– Automatic acknowledge transmissions after received frames.

All these features mentioned above are implemented most efficiently in physical layer and can contribute to the overall network energy efficiency. But there is a trade-off when including more features at physical level because the increase of complexity can result in an increase of hardware costs. The physical layer includes theoretical aspects like propagation models, energy consumptions models and sensing and error models in WSN [KSK+08].

2.2 Other Protocol Stacks for WSNs

Besides the OSI model for WSN presented here, there are other authors that propose other type of protocol stack. However, although network stacks models are proposed by some authors, [Mah06], uniqueness of addresses may not be feasible or even required. But this protocol stack, is similar to the one used in MANETs, and is also the same stack used in TCP/IP networks, except for the addition of power, mobility and task management planes that operate across all the layers, as shown in Figure 2.2 While this layered approach has been accepted and mostly untouched in MANET networks for quite a long time, most researchers find serious difficulties to apply it in WSNs. This difference exists because WSNs are very application-specific and resource-constrained. Besides, a layered architecture may not be the best way to approach the wide range of applications and optimize the limited resources. Due to this, a cross-layer view of WSN is becoming more and more accepted in the research community.

The proposed WSN protocol stack presents a crossing of all the layers in [ASSC02a] the power, mobility and task management planes, characterized as follows:

• The power plane emphasises the power-awareness that should be included in each layer and across all layers in WSNs. For example, a sensor may keep its radio on after sensing some activity in the channel, or it may turn it off if it is not generating any data, or if it does not belong to any active route. A sensor that is running low in energy may turn off its radio and save its energy for sensing activities only.
• The mobility plane is responsible for maintaining the full operation of the sensor network even in the presence of sensor mobility. Although

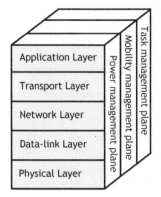

Figure 2.2 Proposed WSN protocol stack.

most sensing applications we can think of are fixed, it cannot be discarded that sooner or later mobile sensing applications will emerge. This could be the case of sensors mounted on mobile platforms such as robots, persons, animals, drones or cars. Routes used to carry information across the network have a limited lifetime and need to be periodically re-established because of node mobility. Even without mobility, routes may change due to the fact that nodes run out of power or follow an awake/sleep duty cycle; hence, a route that is valid at some instant in time may no longer be valid a little bit later [Nic05]. In both cases, the routing layer is mainly responsible for route maintenance.

- The task management plane should be capable of coordinating all nodes toward a common objective in a power-aware mode. For example, some sensors, in each region, may be temporarily turned off if there is enough sensing redundancy from other sensors in that region.

The following description simplifies the understanding of the main role of each of the three management planes:

- **Power management plane**: Manages how a sensor node how uses its energy; Each sensor node manages its own energy;
- **Mobility management plane**: Detects and registers the movement of sensor nodes, know its neighbours and balance their power and task usage; Knows others around;
- **Task management plane**: Balances and schedules the sensing tasks given to a specific region.

Another WSN protocol stack, proposed by some authors, is named as the data-centric architecture, as shown in Figure 2.3, where the WSN is mainly

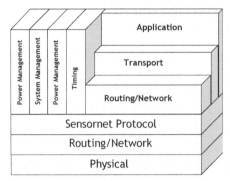

Figure 2.3   Sensor network data-centric stack models (SensorNet example).

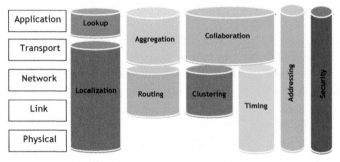

Figure 2.4   Functionality blocks over a WSN stack model.

focused to the data. The main difference between the OSI protocol stack for WSN [GEWD05] is the fact that the blocks needed to build the sensor network usually span themselves over multiple layers while depending on each-other, as WSNs have to provide functionalities that are not present in traditional networks.

Figure 2.4 presents a mapping of the main blocks onto the traditional OSI protocol layers. The service-centric model mentioned by the authors [GEWD05] consists of different layers, namely mission, network, region, sensor and capability layers, as shown in Figure 2.5.

Each layer has associated semantics that use lower level components as syntactic units (except for the capability layer). Within each layer there are four planes or functionality sets normally the communication, the management, the application, and the generational learning ones. The combination of layers and planes enables a service-based visualization paradigm that can provide better understanding of the WSN. These planes do not constitute

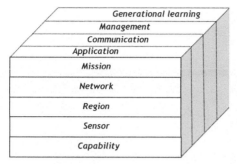

Figure 2.5 Layered model.

vertical segments in the model. Instead, they establish a group with similar functions. This grouping can be used as a reference point for mappings between layers.

The key benefit of considering a service centric-model is a clear separation between the high-level purpose of the WSN (mission/services) and the low-level hardware specific capabilities of an individual sensor node (capability services). A mapping between mission and capability layers, created as a composition of mappings between intermediate layers, provides formalism for a service-centric description and evaluation of the WSN.

Hence, the layered model corresponds to a given level of detail that is used for visualization as follows:

- **Mission:** Presents a general status of the mission using the outline of the geographical location (terrain) where the WSN is located. The displayed information shows the status of a currently processed mission service(s);
- **Network:** Displayed as a collection of regions. Network traffic among regions indicates the overall level of activities in the WSN;
- **Region:** Individual nodes and traffic among nodes in a region are visible;
- **Sensor:** Individual sensor state and characteristics are identified;
- **Capability:** Individual sensor capabilities are identified.

2.3 Evolution of the IEEE 802 Standards

Nowadays, there are various types of wireless networks such as WLANs, Wireless Personal Area Networks (WPANs), Wireless Metropolitan Area Networks (WMANs), Wireless Wide Area Networks (WWANs), MANETs, WSNs and mesh network, is shown in Figure 2.6.

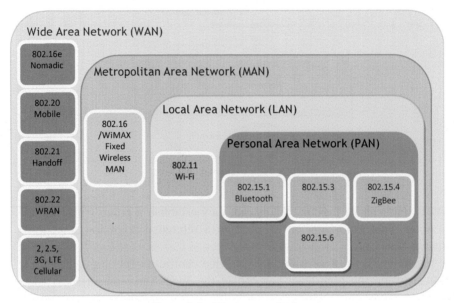

Figure 2.6   Wireless standards.

The evolution starts by IEEE 802.11 WLAN, which was created as the wireless extension of the IEEE 802 wired local area network. The IEEE 802.11b technology supported data rate varies from 2 to 11 Mbps and it has a coverage range up to 100 meters [IEE03c]. Then the development follows two directions from IEEE 802.11, and recently different sub-6 GHz and above frequency bands are considered. One of the directions leads to a larger networking range, higher data throughput and it has QoS characteristics. The main applications purposes are the Internet, e-mail, data file transfer and even Internet Protocol Television (IPTV) in WMAN. In this category the WiMAX IEEE 802.16 standard is also included [IEE04].

Another direction of development led to a smaller networking range (and simple networks). The main targets applications are the WPANs, which are used to transmit information over relatively short distances among the participant devices. The IEEE 802.15.4 family of standards is defined in this category, differentiating from other ones by the data rate supported, the battery drain and aspects of QoS. IEEE 802.15.3 is suitable for multimedia applications that require very high QoS [IEE03a], while IEEE 802.15.1 and Bluetooth are used for cable replacements in electronic devices from the consumer for voice applications [IEE02]. The IEEE 802.15.4 standard is used

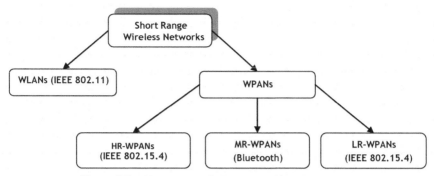

Figure 2.7 Short-range wireless networking classes.

in applications that are not covered by the other 802.15 technologies. WPANS are divided into three classes, as shown in Figure 2.7:

- High-Rate (HR) WPANs;
- Medium-Rate (MR) WPANs;
- Low-Rate (LR) WPANs [Far08].

An example of an HR-WPAN is IEEE 802.15.3, with a data rate of 11 to 55 Mbps. High data rates are helpful in applications such as real-time wireless video transmission from a camera to a nearby TV. Bluetooth, with a data rate of 1 to 3 Mbps, is an example of an MR-WLAN and can be used in high quality voice transmission in wireless headsets.

If the objective is the long term usability, the power consumed by individual nodes in each of these networks needs to be managed efficiently. A subset

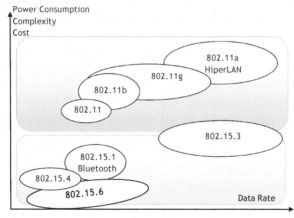

Figure 2.8 Power consumption in IEEE 802 based networks.

of WPANs known as WSNs and are specifically designed for very low power operation and thus deserve special attention.

Figure 2.8 shows how these different types of networks are compared in terms of data rate and power consumption. The IEEE 802.15.4 standard shown in the figure is the one most widely used by WSNs, and will be presented in detail here. In this Figure, IEEE 802.15.6 (a standard for Wireless Body Area Networks) is also considered. It is a low-power and short range wireless standard for communications in on, in or around the human body (but not limited to human) to serve a variety of applications, including medical, consumer electronics and personal communications.

2.4 IEEE 802.15.4 and ZigBee

The first release of the IEEE 802.15.4 standard was issued in October 2003 by the Institute of Electrical and Electronic Engineers (IEEE) [IEE03b]. The IEEE 802.15.4 specified a standard where that enables a sensor rich environment. Then, the IEEE 802.15.4b Group focused on refining the IEEE 802.15.4, where the main objective was to remove and addresse issues raised during early implementation efforts of IEEE 802.15.4 devices, resolving ambiguities, reducing unnecessary complexity, increasing flexibility in security key usage, considerations for newly available frequency allocations, and others. The revisions made by the group ended in 2006 and resulted in the release of an updated version from the original standard. In August 2007, another revision was made to the IEEE 802.15.4 of 2006 where it was added new alternate PHY layers, such as UWB PHY layer at frequencies of 3 GHz to 5 GHz, 6 GHz to 10 GHz and less than 1 GHz and a Chirp Spread Spectrum (CSS) PHY at 2450 MHz. The revision states that UWB PHY supports an over-the-air data rate of 110 kbps, 6.81 Mbps and 27.24 Mbps. The CSS PHY supports an over-the-air data rate of 1000 kbps with option of 250 kbps. The PHY layer depends on the region local regulations, type of application [IEE07].

The IEEE 802.15.4 describes the PHY layer and the MAC layer for WPANs. Usually there is a confusion between IEEE 802.15.4 and ZigBee (ZigBee Alliance 2005), which corresponds to the specification developed by the ZigBee Alliance [Far08]. ZigBee® (ZigBee Alliance 2005) specifies the protocol layers above IEEE 802.15.4 to provide a full protocol stack for low-cost, low-power, low data rate wireless communications. The ZigBee® specification defines the networking, application, and the security layers of the protocol and adopts IEEE 802.15.4 PHY and MAC layers as part of the

ZigBee® networking protocol and any ZigBee-compliant device conforms to IEEE 802.15.4 as well. Therefore, it is possible to create short-range wireless networks using only the IEEE 802.15.4 without implement the ZigBee® layers.

The member of the ZigBee® Alliance adopted themselves the IEEE 802.15.4 standard for WPANs [Far08]. The Stack Architecture of the Zig-Bee® is represented in Figure 2.9 using the IEEE 802.15.4 specification and only defines the network, security and application layers.

The Application (APL) layer is the highest protocol layer in the ZigBee wireless network and hosts the application objects. Manufacturers develop the application objects to customize a device for various applications. Application objects control and manage the protocol layers in a ZigBee device. There can be up to 240 application objects in a single device. The ZigBee standard also offers the option to use application profiles in developing an application. An application profile is a set of agreements on application-specific message formats and processing actions. The use of an application profile allows further interoperability between the products developed by different vendors for a specific application.

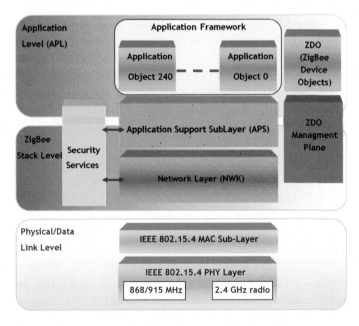

Figure 2.9 ZigBee® functional layer architecture and protocol stack.

The ZigBee Device Objects (ZDO) represents the ZigBee node type of device (End Device, Coordinator or Router), and has a number of initialization and communication roles, while the ZDO Management Plane is responsible by the spanning of the Application Sub Layer (APS) and Network (NWK) layers. It allows the ZDO to communicate with these layers when performing its internals tasks. It also allows the ZDO to deal with requests from applications for network access and security functions using ZigBee Device Profile (ZDP) messages.

The APS will be responsible to communicate with the application. For example to blink a LED the APS relays this instruction to the application using the endpoint information in the message. It is also responsible of maintaining binding tables and sending messages between bound nodes.

The NWK layer of a ZigBee coordinator is responsible for establishing a new network and selecting the network topology (tree, star, or mesh). The ZigBee coordinator also assigns the NWK addresses to the devices in its network. The NWK layer interfaces between the MAC and the APL and is responsible for the management of the network formation and routing. The routing process is used to select the path through which the message will be relayed to its destination device. The ZigBee coordinator and the routers are responsible for discovering and maintaining the routes in the network, the ZigBee end device cannot perform route discovery and the ZigBee coordinator, or a router will perform route discovery on behalf of the end node.

The ZigBee technology incorporates the security functionality, which is managed by the security layer. Different levels of security are available. The security functionality makes use of keys at different levels, as well as challenge-authentication procedures. The keys can be pre-configured within devices to increase security.

The IEEE 802.15.4 standard supports the use of Advanced Encryption Standard (AES) [oST01] to encrypt their outgoing messages.

Since IEEE 802.15.4 was developed independently of the ZigBee standard it is possible to build short-range wireless networking based solely on IEEE 802.15.4 without implementing ZigBee-specific layers. In this case, the users develop their own networking/application layer protocol on top of IEEE 802.15.4 PHY and MAC, as shown in Figure 2.10. These custom networking/application layers are normally simpler than the ZigBee protocol layers and are targeted for specific applications [Far08].

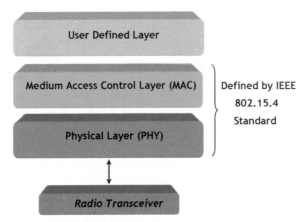

Figure 2.10 Network Protocol based on IEEE 802.15.4 and not conform to the ZigBee Standard.

2.5 IEEE 802.15.4 Physical Layer

The PHY layer main purpose is to establish an interface with the physical medium where the communications are done. The PHY layer is the lowest layer in the protocol stack for WSNs. It is responsible by the control (enable and disable) of the radio transceiver, energy detection, link quality, CCA, channel selection and the transmission and reception of message packets that are exchanged in the physical medium [GCB03].

The IEEE 802.15.4 standard proposes three operational frequency bands: 868 MHz, 915 MHz and 2.4 GHz, as presented in Figure 2.11.

There is room for only a single channel in the 868 MHz band (20 kbit/s). The 915 MHz band has 10 channels (excluding the optional channels) with a bit rate of 40 kbit/s. The total number of channels in the 2.4 GHz band is 16 (250 kbit/s). In IEEE 802.15.4 standard there are three modulation types: BPSK, Amplitude Shift Keying (ASK), and offset quadrature phase shift keying (O-QPSK). In BPSK and O-QPSK, the digital data is in the phase of the signal. While in ASK, the digital data is in the amplitude of the signal. All wireless communication methods in IEEE 802.15.4 utilize either DSSS or Parallel Sequence Spread Spectrum (PSSS) techniques. DSSS and PSSS help to improve performance of receivers in a multipath environment [SW04]. Details about the IEEE 802.15.4 data rates and frequencies of operation along with the supported modulations are given in Appendix A.

The way how channelization is done takes into account that in the 2.4 GHz band there are 16 channels, with each channel requiring 5 MHz of bandwidth,

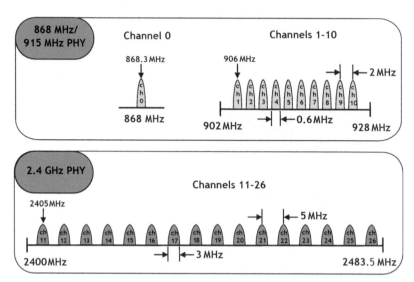

Figure 2.11   Channelization at the 868/915 MHz and 2.4 GHz bands.

as shown in Figure 2.11. And for 915 MHz band there are 10 channels, with each channel requiring 2 MHz of bandwidth. The centre frequency for each channel, for each band, can be calculated, in MHz, as follows:

$$f_c = 868.3, ch = 0 \tag{2.1}$$

$$f_c = (906 + 2(ch - 1)), ch \in \{1, 2, ..., 10\} \tag{2.2}$$

$$f_c = (2405 + 5(ch - 11)), ch \in \{11, 12, ..., 26\} \tag{2.3}$$

where f_c is the centre frequency and ch is the channel number.

2.5.1 IEEE 802.15.4 Device Types and Roles

In IEEE 802.15.4 there are two types of devices that could appear in a wireless network:

- **Full Function Devices (FFDs):** These devices are capable of performing all the tasks described in the IEEE 802.15.4 standard and can play any role in the network;
- **Reduced Function Devices (RFDs):** These devices have limited capabilities in a WSN, since these devices can only communicate with an FFD and they are only meant to do simple tasks such as turning on or off a switch. The memory size is smaller than in an FFD device [Far08];

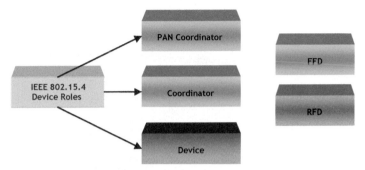

Figure 2.12   Device roles in the IEEE 802.15.4.

In an IEEE 802.15.4 wireless network, an FFD device can support three different roles, as shown in Figure 2.12:

- **Personal Area Network (PAN) coordinator:** is the primary controller of PAN, which initiates the network and often operates as a gateway to other networks. Each PAN must have exactly one PAN coordinator;
- **Coordinator:** Is an FFD device capable of relaying messages using data routing and network self-organization operations to achieve it;
- **Devices:** Devices do not have data routing capability and can communicate only with coordinators [KSK+08].

2.5.2 IEEE 802.15.4 Network Topologies

To overcome the limited transmission range of the wireless devices from a WSN, multi-hop self-organizing network topologies are necessary. The IEEE 802.15.4 standard supports three types of network topologies:

- **Star topology:** In the star topology, Figure 2.13, all devices establish a communication link with a single central controller, called the PAN coordinator.
- **Peer-to-peer topology:** In the peer-to-peer topology, Figure 2.14, there is also one PAN coordinator and in contrast to the star topology each device can communicate directly with any other device if the devices are placed close enough together to establish a successful communication link.
- **Cluster tree topology:** The cluster tree topology, Figure 2.14, is a special case of a peer-to-peer network in which most devices are FFDs and an RFD may connect to a cluster-tree network as a leave node at the end of a branch.

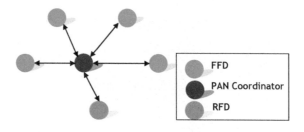

Figure 2.13   IEEE 802.15.4 star topology.

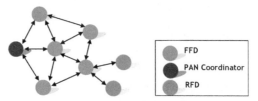

Figure 2.14   IEEE 802.15.4 peer-to-peer topology.

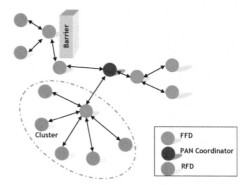

Figure 2.15   IEEE 802.15.4 cluster tree topology.

Further details concerning the IEEE 802.15.4 network topologies are discussed in Appendix A.

2.5.3 IEEE 802.15.4 PHY Specifications

The IEEE 802.15.4 standard not only specifies the PHY protocol functions and interactions with the MAC sub-layer, but also defines the minimum hardware-level requirements. Hence, the physical layer of the IEEE 802.15.4 is responsible by the following tasks:

- **Receiver energy detection (EnD):** When a device plans to transmit a message, it first switches into the receive mode to detect and estimate the signal energy level in the desired channel, known as energy detection (the receiver EnD measurement is used by a network layer as part of the channel selection algorithm).
- **Link Quality Indication (LQI):** The LQI main function is the indication of the quality of the data packets received by the receiver. To obtain the quality of data packets received the device uses the Received Signal Strength (RSS) as a measure of the signal quality. The NWK layer can use for example the LQI information to decide which path to use to route a message.
- **Carrier Sense (CS):** The CS technique is quite similar to the EnD technique and it is used to perform verification if whether a frequency channel is available to use. While in EnD the signal detected in the channel is not decoded, in the CS technique the signals is demodulated.
- **Clear Channel Assessment (CCA):** The CCA mechanism is the first step of the Carrier Sense Multiple Access with Collision Avoidance (CSMA-CA) channel access mechanism when MAC requests the PHY to perform a CCA to the channel, in order to detect that the channel is not in use by any other device. There are three CCA modes defined in the IEEE 802.15.4 standard: i) CCA mode 1: the EnD result is the only one that is taken into account when performing CCA; ii) CCA mode 2: the CS result is the only one that is taken into account when performing CCA; iii) CCA mode 3: This mode is the result of a logical combination (AND/OR) of mode 1 and mode 2 in which the channel is considered busy if the detected energy level is above the threshold and/or a compliant carrier is sensed.
- **Channel Selection:** The IEEE 802.15.4 standard initial release considers a total of 27 channels and the implementation of multiple operating frequencies bands could not be supported. Hence, the physical layer should be able to tune the radio transceiver in a specific channel when requested by a higher layer.

- **Activation and deactivation of the PHY radio transceiver:** The radio transceiver may operate in three different states: transmitting, receiving or sleeping. Upon a request of the MAC sub-layer, the transceiver switches in to one of the three main states: transceiver disabled, transmitter enabled, and receiver enabled. The standard recommends that the turnaround time from transmitting to receiving states, and vice versa, should last, at least, 12 symbol periods.

Details on the different tasks of the IEEE 802.15.4 PHY layer are given in Appendix A.

2.5.4 IEEE 802.15.4 PHY Packet Structure

The PHY Protocol Data Unit (PPDU), Figure 2.16, is the packet data structure at the PHY protocol level that modulates the wireless transmitter.

The structure of the PPDU encapsulates all data structures from the higher levels of the protocol stack. In IEEE 802.15.4, the first bit that will be transmitted is the Least Significant Bit (LSB) of the SHR. The Most Significant Bit (MSB) of the last byte of the PHY payload is transmitted last.

It consists of the following three components:

- **Synchronization Header (SHR):** Consists of two fields, a preamble and a Start-of-Frame Delimiter (SFD). The lengths and durations of the preambles in all PHY layer is addressed in Appendix A;
- **PHY Header (PHR):** It is a single 8 bit field with the MSB reserved and the remaining low order bits used to specify the total number of octets in the PHY payload.
- **PHY payload:** The PHY payload is composed of only one field called the Physical Layer Service Data Unit (PSDU). The PHY payload length can be any value from 0 to 127 bytes.

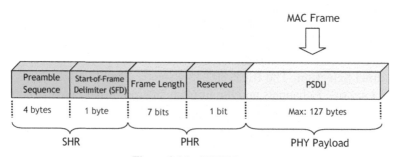

Figure 2.16   PPDU format.

In IEEE 802.15.4, the first bit that will be transmitted is the LSB of the SHR while the MSB of the last byte of the PHY payload is transmitted last.

2.6 IEEE 802.15.4 MAC Sub-layer

The MAC sub-layer provides access control to a shared channel and reliable data delivery between the PHY and the next higher layer above the MAC. It is also responsible by:

- The generation of acknowledgement frames;
- Support of PAN association;
- Disassociation and the security control [Far08].

The IEEE 802.15.4 standard uses a CSMA-CA algorithm, which requires listening to the channel before transmitting in order to avoid collisions with other transmissions that are being performed using the channel and coordinators are required to constantly receive for possible incoming data [KSK+08]. The IEEE 802.15.4 MAC only supplies application support for the creation of the two wireless topologies already mentioned before: **star topology** and **peer-to-peer topology**. However, the management of these topologies is done by the network layer and therefore is beyond the scope of IEEE 802.15.4 standard. The MAC sub-layer only performs functions that are required by network layer. Applications that benefit from a star topology include home automation, Personal Computer (PC) peripherals, toys and games, and personal health care. The peer-to-peer topology is used in industrial and commercial application and then leading to the appearing of a more complex type of self-organizing wireless network (multi-hop/mesh networks). The MAC protocol transfers data as frames with a maximum size of 128 (bytes), which enables a maximum MAC payload of 104 (bytes), turning the IEEE 802.15.4 standard ideal for low data rate systems. The MAC protocol can operate on the following two operational modes:

- **The beacon-enabled:** The beacons are periodically sent by the PAN coordinator and when a device wishes to transfer the data to a coordinator it first listens to the medium for a network beacon frame. The beacon frame is responsible by the limits establishment for the beginning of a superframe and by the establishment of the time interval to exchange packets between different the nodes. The medium access is therefore performed using a slotted CSMA-CA. Since this mode is used in applications that require a certain amount of bandwidth and low latency, the PAN coordinator enables the allocation of some time

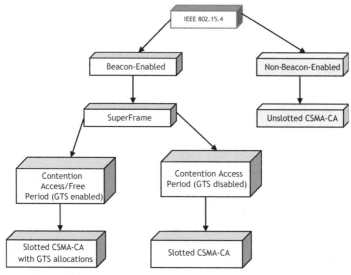

Figure 2.17   IEEE 802.15.4 operational modes.

slots in the superframe. These portions are called Guaranteed Time Slots (GTSs) and they are used when the node needs to have guaranteed services [Far08].

- **The non-beacon-enabled:** the PAN coordinator does not transmit regular beacons and therefore it transmits a data frame using a *unslotted* CSMA-CA and perform only a single CCA operation without synchronization to backoff boundaries. Devices perform only a single CCA operation. If this indicates an idle channel, the device infers success. According to [KW05b], coordinators must be switched on constantly, but devices can follow their own sleep schedule.

The operational modes in IEEE 802.15.4 are shown in Figure 2.17.

2.6.1 IEEE 802.15.4 Beacon-Enabled - Star topology

When considering a beacon-enabled star network topology, the network device that wishes to send data to the PAN coordinator needs first to listen for a beacon. If a GTS is not assigned to the device, it transmits its data frame during the contention access period, after using CSMA-CA. However, if the GTS is assigned to the device, then it waits for the right time within the superframe structure so that it can transmit its data frame. The PAN coordinator after receiving the data frame will sends back an acknowledgment to

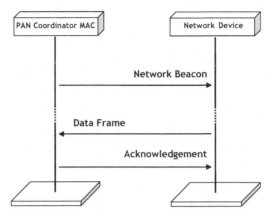

Figure 2.18   Star network data transfer from network device in a beacon-enabled network.

the network device when the data transfer finishes. These message exchanges are shown in Figure 2.18.

When the PAN coordinator has data waiting to be transmitted to a network device, it sets a special flag in its beacon. Then, when the destination network device detects that the PAN coordinator has data pending for him, it sends a data request message back to the PAN coordinator. The PAN coordinator responds with an acknowledgment followed by the data frame. In the end of the transmission the exchanging is finished with an acknowledgment frame sent from the network device. This process is described in Figure 2.19.

The IEEE 802.15.4 design group decided not to implement a combined acknowledgment/data frame to simplify implementation.

2.6.2 IEEE 802.15.4 Non-Beacon-Enabled - Star topology

In non-beacon-enabled star network, the network device that wishes to transfer data sends a data frame to the PAN coordinator using a CSMA-CA mechanism. Then the PAN coordinator responds to the network device, sending an acknowledgment message, as shown in Figure 2.20.

Following the same idea, when the PAN coordinator needs to transfer a data to a network device, it will keep the data until the network device sends a data request message. The PAN coordinator sends an acknowledgment message containing information indicating to the network device if there are data pending to be sent. Then, data will immediately be sent after the acknowledgment message. The transaction is finished when the network device acknowledges the reception of the data frame, as shown in Figure 2.21.

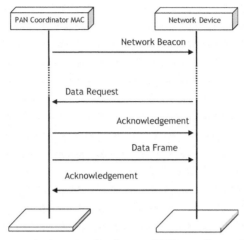

Figure 2.19   Star network data transfer from a PAN coordinator in a beacon-enabled network.

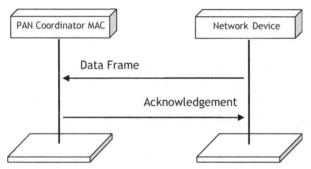

Figure 2.20   Star network data transfer from a network device in a non-beacon-enabled network.

In peer-to-peer topologies, data transfer strategy depends on the network layer that is managing the protocol stack of the WSN. A network device may stay in reception mode while scanning the radio frequency channel for any communications present in channel, or it can send "SYNC" messages, in order to achieve synchronization with other potential listening devices.

2.6.3 SuperFrame Structure

The IEEE 802.15.4 standard includes an optional superframe structure. The superframe is managed by the PAN coordinator after regular intervals.

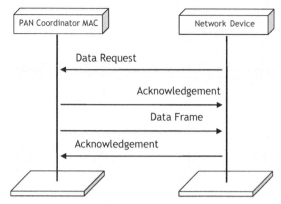

Figure 2.21 Star network data transfer from a PAN coordinator in a non-beacon-enabled network.

The superframe is delimited by two beacon frames. Each beacon contains information that will help network devices synchronize to the network and includes the network identifier, the beacon periodicity and the superframe structure. In a superframe, the following three types of periods can appear: the Contention Access Period (CAP), the Contention-Free Period (CFP) and the inactive period. During the CAP, if a device wants to transmit a frame, then it will use a CSMA-CA mechanism to gain access to a frequency channel.

The probability of a frequency channel be available is equal for all the devices in the same network. When a device starts using am available channel will continue using it until it finishes the transmission of the packet. If the device detects a busy channel, then it will back off a random time period and tries again.

The MAC command frames must be transmitted during CAP. During CAP there is no guarantee for any device that it will use the frequency channel exactly when needed. Nevertheless, during the CFP it has to guarantee a time slot for a specific device and this device will not need to use a CSMA-CA mechanism for channel access.

The use of guaranteed slots in CFP is a good option when considering low latency applications and when a device cannot wait for a random period until channel is available. The use of a CSMA-CA mechanism is not allowed during CFP. The junction of the CAP and CFP is known as the active period and is divided into 16 contiguous equal time slots. The superframe with GTS is presented in Figure 2.22. The IEEE 802.15.4 allows up to seven GTSs in CFP and each GTS can occupy one or more slots. Each GTS is formed by

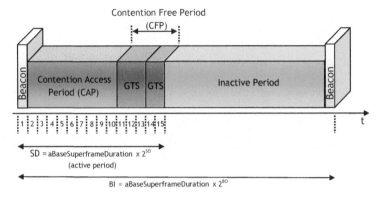

Figure 2.22   Superframe structure with GTS.

an integer multiple of time slots and each time slot is equal to 1/16 of the time between the start of two successive beacons. A superframe may have an inactive period that allows a device to enter power-saving mode. During the execution of power-saving mode, the coordinator can turn off the radio transceiver in order to conserve battery energy [Far08].

The superframe is configured by the Network (NWK) layer using *MLME-START*.request primitive. The Beacon Interval (BI) is defined as the duration between two consecutive beacons and is obtained by using the values of the *macBeaconorder* (BO) attribute and the *aBaseSuperframeDuration* constant. The value (in symbols) is obtained from the equation:

$$BI = aBaseSuperframeDuration \cdot 2^{BO} \qquad (2.4)$$

The *macBeaconorder* can assume any value from 0 to 14 in a beacon enabled network. When the value of *macBeaconorder* is set to 15, the network is assumed to be non-beacon-enabled and the superframe description will not be applied.

The same way as the BI value, the length (in symbols) of the active period of the superframe, Superframe Duration (SD), is obtained from the equation:

$$SD = aBaseSuperframeDuration \cdot 2^{SO} \qquad (2.5)$$

where SO is the value of the *macSuperframeOrder*. The superframe duration cannot exceed the beacon interval, and is given by the following relation:

$$SO \leq BO \qquad (2.6)$$

When considering a non-beacon-enabled network (*macBeaconOrder* is equal to 15), the PAN coordinator will only transmit beacons if receives a beacon request command from a device in its network. A device uses the beacon request command in order to locate the PAN coordinator. The PAN coordinator in a non-beacon-enabled network sets the value of *macSuperframeOrder* to 15. Now, when a beacon-enabled network is assumed, any coordinator besides the PAN coordinator can transmit beacons and create its own superframe.

Figure 2.23 presents the time evolution when both PAN coordinator and another coordinator, located in the same network, are transmitting beacons. The coordinator starts the transmission of its beacon only during the inactive period of the PAN coordinator superframe. While the beacon sent by the PAN coordinator is named as the received beacon, the beacon of the other coordinator is the *transmitted beacon*. Note that the active period for both superframes must have the same length [Far08].

If a device from the network does not use its GTS for an extended period of time, the GTS will reach a timeout and the coordinator can them assign that GTS to a different device.

The inactive period present in Figure 2.23 is an integer multiple of twice the superframe length, while the value of this integer multiple (n^*) depends on the *macBeaconOrder* and is given by the following equation:

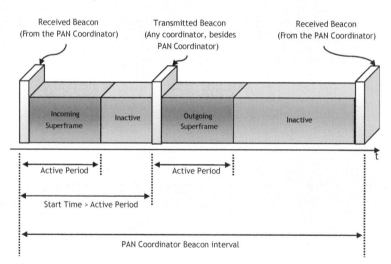

Figure 2.23 The incoming and outgoing Superframes temporal evolution.

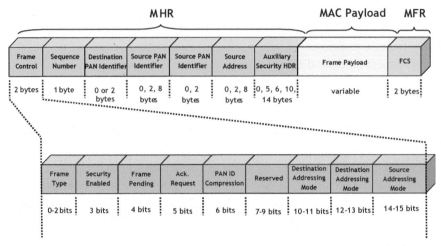

Figure 2.24 General MAC frame format and details of the frame control field.

$$n^* = \left\{ \begin{array}{ll} 2^{8-macBeaconOrder} & ,0 \leq macBeaconOrder \leq 8 \\ 1 & ,8 \leq macBeaconOrder \leq 14 \end{array} \right\} \quad (2.7)$$

The timing parameters in the beacon enabled operating mode are presented in detail in Appendix B. In addition, the InterFrame Spacing (IFS) feature is also discussed in detail in Appendix B.

2.6.4 IEEE 802.15.4 MAC frames and CSMA/CA mechanism

The IEEE 802.15.4 standard [IEE07, Far08] defines four MAC frames structures:

- **Beacon frame:** It is used by the coordinator in order to transmit beacons that are used to synchronize the clock of all the devices located in the same network;
- **Data frame:** It is used to transmit data;
- **Acknowledge frame:** It is used to transmit an acknowledgement when a successful reception of a frame happens;
- **MAC command frame:** It is transmitted using a MAC command frame.

The four different MAC frames structures considered in the IEEE 802.15.4 standard are described in detail in Appendix B.

The IEEE 802.15.4 standard [IEE07] utilizes to access the channel the CSMA-CA mechanism. Nodes in clusters that operate in beacon-enabled mode must utilize the slotted CSMA-CA mechanism, with few exceptions.

When a device is using the CSMA-CA mechanism, whenever it wants to transmit, it performs a CCA first to ensure that the channel is not occupied by any other device. Then after being sure that no other device is using the channel it starts transmitting its own signal.

There are two situations on which a device accesses the channel without using the CSMA-CA algorithm:

- The access to the channel during the CFP;
- Transmit immediately after acknowledging a data request command. This can happen when a device requests data from a coordinator, the coordinator transmits the acknowledgment followed immediately by the data without performing CSMA-CA between these two transmissions, even during the CAP.

The following two types of CSMA-CA can be used:

- **Slotted CSMA-CA:** The slotted CSMA-CA mechanism is defined as a CSMA-CA mechanism that uses a superframe structure for the frames exchanged between the devices. The superframe divides the active period into 16 equal and contiguous time slots and the backoff period has to be aligned to specific time slots;
- **Unslotted CSMA-CA:** The unslotted CSMA-CA mechanism is defined as a CSMA-CA mechanism where there is no superframe structure evolved during the exchanges between the devices and therefore there is no need for backoff slot alignment [Far08, GCB03].

The flowchart presented in Figure 2.25 describes the slotted and unslotted CSMA-CA algorithm which is initiated when a packet is ready to be transmitted.

The CSMA-CA mechanism [IEE07, Far08, GCB03] keeps the following three global variables updated:

- **Backoff Exponent (BE):** It is a variable that determines the allowed range for the random period the CSMA-CA algorithm will wait every time the algorithm faces a busy channel. The initial value of BE is equal to *macMinBE* in an unslotted CSMA-CA channel access. When considering a slotted CSMA-CA, the choice of the Battery Life Extension (BLE) option affects the value of BE. Therefore, if the BLE option is enable, the coordinator turns off its receiver after a period equal

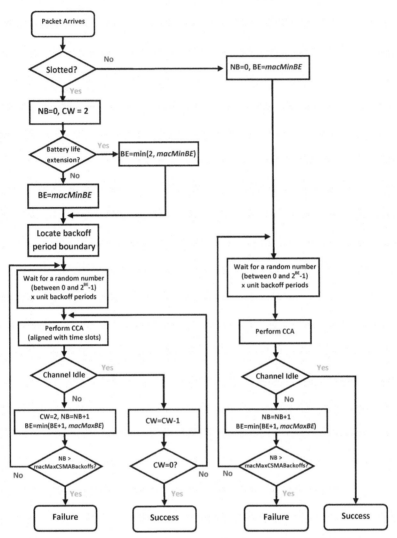

Figure 2.25 Slotted and unslotted CSMA-CA mechanism for MAC Layer.

to *macBattLifeExtPeriods* and transmits a beacon frame right after in order to conserve energy. The range for the backoff period is limited to be the minimum value of 2 or the value of *macMinBE* (BE = min(2, macMinBE)). If the BLE option is disabled, the coordinator is active during the CAP and the value of BE is equal to *macMinBE* and every time the CCA is performed and the channel is busy it is incremented by one unit, while not exceeding the variable *macMaxBE*;

- **Number of Backoffs (NB):** It is a counter that keeps updated the number of times the device backs off and retries to access the channel by means of the CSMA-CA mechanism. This counter at the beginning is equal to zero and each time the device backoff due to the detection of a busy channel, BE is incremented by one unit. When the NB reaches the *macMaxCSMABackoffs* value and still did not got access to the channel then quits the CSMA-CA mechanism and reports failure to access the channel to the NWK layer;
- **Contention Window (CW):** It is used to define the number of backoff periods that the channel must be available before starting to transmit. The CW is only used in the slotted CSMA-CA mechanism. Each backoff period channel sensing, is performed during the 8 first symbols of the BP.

The random backoff time, rdm_{nb}, given by Equation 2.8, results from the random choice of any integer number between 0 to (2^{BE}-1) multiplied by the unit back-off period.

$$Backoff = rdm_{nb} \cdot aUnitBackoffPeriod \qquad (2.8)$$

A detailed description of the slotted and unslotted CSMA-CA algorithm is presented in the Appendix B.

2.7 Taxonomy for Medium Access Control Protocols

When using battery-powered nodes in WSNs, energy efficiency is of primary importance since it is directly related to the lifetime of the network. Major sources of energy waste are idle listening, packet retransmissions (due to packets collisions), unnecessary high transmission powers, overhearing and control overhead [BDWL10, YHE04a]. From all the electronic components of a sensor node, the radio transceiver is the most power consuming component. At the link layer, where the MAC protocol manages the use of the radio transceiver, higher efficiency can be achieved, depending on what performance metrics the MAC protocol intends to improve. MAC protocols from typical ad-hoc wireless networks differ and share some points from the MAC protocols for WSNs. WSNs share some of the objectives of the ad-hoc wireless networks, e.g., error prone channels and limited bandwidth. In WSNs, the MAC protocols must have a built-in power conservation mechanism, mobility module to handle the issues that may arise from the movement of the nodes, and mechanisms that allow the node to recover from failures (e.g., packet transmission failures, out-of-range, ...). All these mechanisms are deeply related with energy-efficiency and network lifetime, which are performance metrics of paramount importance in a WSN. Moreover, important metrics include latency, accuracy, scalability and throughput.

In WSNs research, medium access control protocols, has been a hot topic in wireless networks during the last seven years. Many MAC protocols (with different purposes) have been proposed for WSNs in the literature. From the entire range of WSN MAC protocols, one common feature is the energy efficiency, since the hardware is battery-powered in the majority of the applications.

Furthermore, WSN MAC protocols present more constraints that should be taken into account, e.g., duty cycle mechanisms, necessity of multi-hop operations to forward packets, collision avoidance mechanisms, overhearing, idle listening, overhead, limited memory and processing capabilities.

When a new MAC protocol is proposed to solve one of the sources of energy waste mentioned in the beginning, it is very hard to accomplish the proposed objective. If the source of energy waste is solved, for sure there is other one that is increased. Therefore, instead of solving the chosen source of energy waste, we should try to mitigate at maximum that source of energy waste, while achieving a high energy performance and efficient trade-offs between the considered performance metrics of the respective MAC protocol.

The authors from [YBO09] performed a study on existing WSN MAC protocols where they first outline and discuss the specific requirements and design trade-offs of a WSN MAC protocol, while describing the properties of WSNs that affect the design of MAC layer protocols. Then, a typical collection of WSN MAC protocols presented in the literature are surveyed, classified, and described.

Authors from [BDWL10] presented a state-of-the-art study in which they thoroughly expose the prime focus of WSN MAC protocols, design guidelines that lead to these protocols, as well as disadvantages and liabilities of the solutions. Moreover, in contrast to previous surveys from the literature that classify MAC protocols by the technique being used, this taxonomy is based on the problems being dealt with.

As we have seen before, this research can be categorized into many different ways. For the purpose of this thesis, the classification of the MAC protocols into different categories is performed by the technique to access the channel being used in the protocol, but also taking into account that one MAC protocol may utilize multiple techniques. However, due to the growing number of WSN MAC protocols, many of those apply more than one technique in the MAC layer. Hence, this taxonomy for the classification of MAC protocols extends the classification of WSN MAC protocols for the new ones, which employ multiple techniques to achieve the purposed objective. Moreover, a comparison table is presented for the considered MAC protocols, where the MAC characteristics (e.g., slot assignment, frequencies allocation) are categorized for each one of the MAC protocols.

As presented in Figure 2.26, three main categories of MAC protocols stand out as the most prominent ones: Unscheduled, Scheduled and Hybrid MAC protocols. The former category addresses the MAC protocols that allows for sensor nodes to operate independently, while it conserves energy due to the low complexity of the MAC protocol. The second category is related to the MAC protocols that organize the communication between sensor nodes in an ordered way, while reducing collisions and retransmissions by using synchronization mechanisms. The latter ones use different techniques to conserve energy among nodes, namely by adjusting their behaviour between techniques used in scheduled and unscheduled MAC protocols.

Besides these three main categories, there is the QoS and Cross-layer categories. The QoS category includes the WSN MAC protocols that employ quality of service policies in the communication between sensor nodes, while maintaining the energy efficiency. The last category is dedicated to MAC protocols that do not follow the traditional layered protocol architecture. In

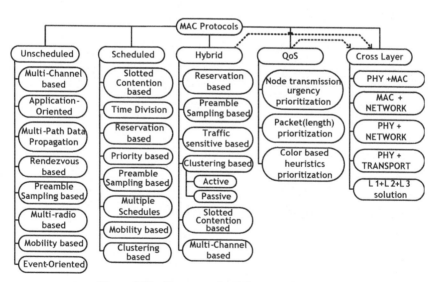

Figure 2.26   Taxonomy for WSN MAC protocols.

contrast with traditional MAC protocols, these ones achieve higher performance metrics due to the joint optimization and design of networking layers (i.e., cross-layer design).

Unscheduled Protocols

As presented in Figure 2.27, the Unscheduled protocols can further be divided into roughly eight categories: Multi-Channel, Application-oriented, Multi-path Data propagation, Rendezvous based, Preamble Sampling based, Multi-radio based, Mobility-based and Event-oriented. The Multi-channel sub-category accounts for the MAC protocols whose main characteristic is the use of multiple radio channels by the sensor nodes, allowing for the node to simultaneously communicate on separate channels. For the Application-oriented sub-category, the application characteristics may be considered to enable the MAC protocol to conserve energy. The Multi-path Data propagation sub-category is dedicated to the MAC protocols where information is broadcasted to more than one node in each single-hop without, however, flooding the network. The receiving nodes continue to broadcast until the message reaches its final destination.

Rendezvous based MAC protocols utilize a method called rendezvous, which allows for nodes to communicate if both of them are powered simultaneously. The study of Preamble sampling-based MAC protocols accounts for the protocols that do not use common active/sleep schedules. Instead, each node chooses its active schedule independently of the other nearby nodes. The Multi-radio based MAC protocols utilize more than one radio transceiver in the communications between the sensor nodes. One is used to communicate the data between nodes while other one is used to wake-up the nodes. The Mobility-based sub-category accounts for the MAC protocols that have a mobility feature embedded in the sensor node architecture. The Event-oriented regards the MAC protocols that are triggered by an event that is sensed by the sensor nodes.

Scheduled Protocols

In Figure 2.28 Scheduled protocols are additionally divided into eight categories: Slotted contention based, Time division based, Reservation based, Priority based, Preamble sampling based, Multiple schedules, Mobility based and Clustering based. Some of these sub-categories have already been previously described. Therefore, only the new ones are going to be further described.

The Slotted contention based sub-category is the one that includes the majority of the MAC protocols. Slotted protocols divide time into frames. Each frame is subdivided into a certain number of slots.

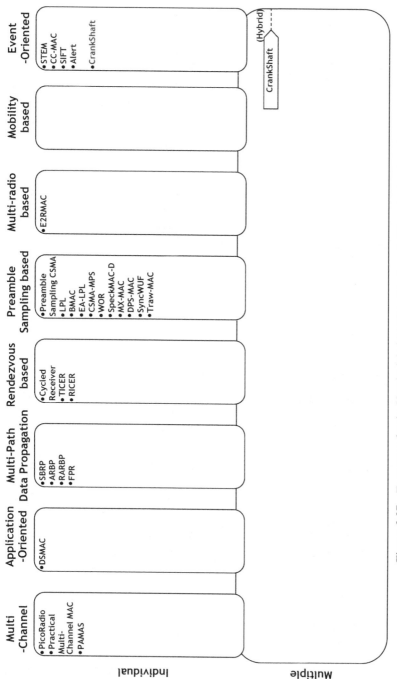

Figure 2.27 Taxonomy for the Unscheduled MAC protocols and different sub-categories.

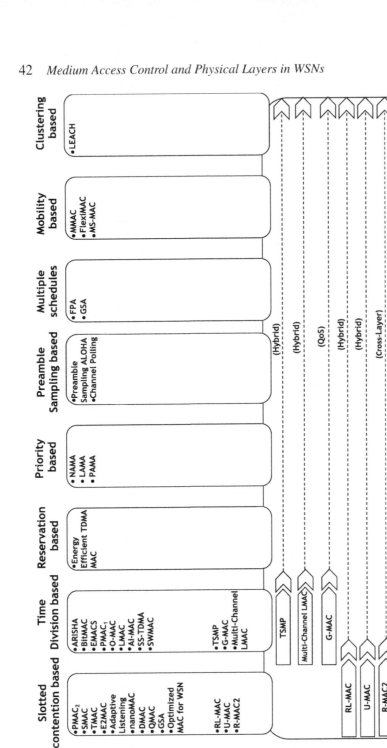

Figure 2.28   Taxonomy for the Scheduled MAC protocols and different sub-categories.

The Time division based MAC protocols divide the channel access time into repeated frames. Each frame is subdivided into N_n time slots, which allocates one slot to only one node. The Reservation based MAC protocols employ a reservation request scheme based on Time Division Multiple Access (TDMA) techniques. The Priority based sub-category aggregates the MAC protocols that utilize a TDMA scheme jointly with a priority based scheme, to decide when packets are transmitted. The study of Preamble sampling-based MAC protocols accounts for the protocols that do not use common active/sleep schedules.

The Multiple schedules sub-category includes the MAC protocols that enables the nodes to have multiple schedules. However, it employs different schemes that allows for the nodes to communicate between them. The Mobility-based sub-category accounts for the MAC protocols that have a mobility feature embedded in the sensor node architecture. The Clustering based sub-category accounts for the MAC protocols that employ an adaptive clustering hierarchy scheme that allows for the nodes to send data to the cluster heads, based on a TDMA approach.

Hybrid Protocols

The Hybrid category results from the combination of different characteristics from the Scheduled and Unscheduled categories, as presented in Figure 2.29. It can be further sub-divided into seven categories: Reservation based, Preamble sampling based, Traffic sensitive based, Clustering based (active and passive), Multi-frequency based, Slotted contention based and Multi-channel based. The Reservation based and Preamble sampling based have been described above. The Traffic sensitive based sub-category corresponds to the MAC protocols that shape their behaviour accordingly with the traffic patterns that the sensor node detects. The Clustering based protocols can be either active, if the protocol utilizes control packets to collect topological information, or passive if no control packets are used to gather this information. The Multi-frequency based MAC protocols correspond to the ones that choose different frequencies which allows for the sensor node to communicate control and data packets in separate frequency bands.

QoS Protocols

The QoS category, presented in Figure 2.30, can be sub-divided into three categories: the Node transmission urgency prioritization, packet (length) prioritization and colour based heuristics prioritization. All the QoS sub-categories account for quality of service policies that may be interpreted as prioritization based schemes.

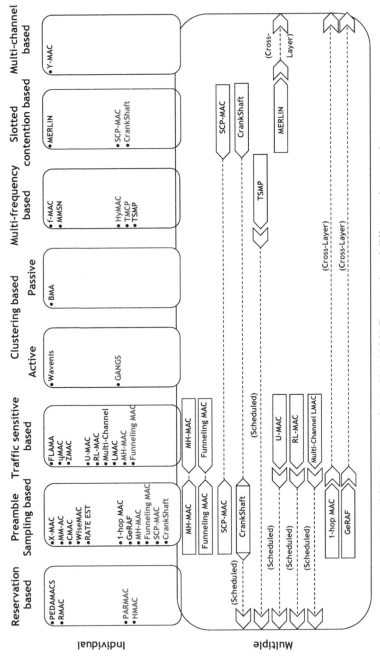

Figure 2.29 Taxonomy for the Hybrid MAC protocols and different sub-categories.

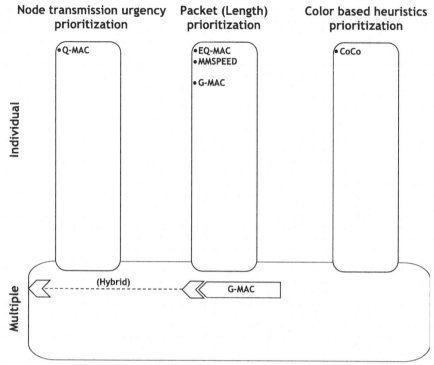

Figure 2.30 Taxonomy for the QoS MAC protocols and different sub-categories.

Cross-layer Protocols

Figure 2.31 presents the cross layer category, which can be divided into five categories: PHY+MAC, MAC+NETWORK, PHY+NETWORK, PHY+TRANSPORT and L1+L2+L3 solutions. Each one of the first four aforementioned sub-categories handles the combination between two layers, while the last one suggests a solution based on the Physical (L1), Link (L2) and Routing (L3) layers. These solutions are becoming the most promising alternative to inefficient traditional layered protocols. In these solutions there is a joint optimization and design of networking layers (i.e., cross-layer schemes).

Extended Taxonomy

Figure 2.32 presents an extended taxonomy of the WSN MAC protocols, from [YBO09], where the protocols are categorized, according to their medium access and control technique and distinguishes the MAC protocols that use more than one medium access technique. The MAC protocols that employ only one medium access technique are represented in the individual

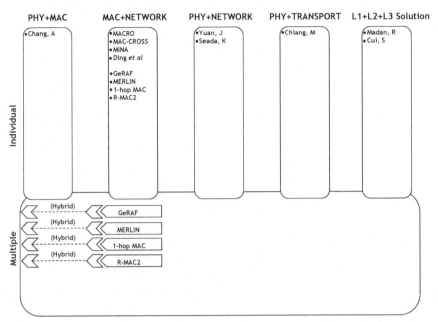

Figure 2.31　Taxonomy for the Cross-layer MAC protocol and different sub-categories.

zone. The ones that utilize more than one medium access technique are represented in the multiple zone of Figure 2.32, where the overlapping regions of two different sub-categories allows for understanding the techniques that are shared between categories and sub-categories, for a specific MAC protocol. By observing Figure 2.32, it is easy to notice that there are only a few QoS and Cross-layer MAC protocols. In contrast to this small number of MAC protocols, the scheduled and unscheduled categories are the ones that present more MAC protocols. Nevertheless, the Hybrid protocols category also presents a notable number of MAC protocols.

Considering the multiple zone of Figure 2.32, the majority of the MAC protocols are shared between the Scheduled, Hybrid and Cross-layer categories. This means that the most recent proposed MAC protocols combine different techniques and methods from different categories in order to accomplish a better energy performance, while maintaining acceptable values for the throughput and latency.

Apart from the taxonomy of WSN MAC protocols presented in Figure 2.32, which provides an idea of the vast number of MAC protocols available nowadays, a few MAC protocols stand out as being the boosters for the field of WSN research. From all the presented MAC protocols there are some ones that have been more referenced

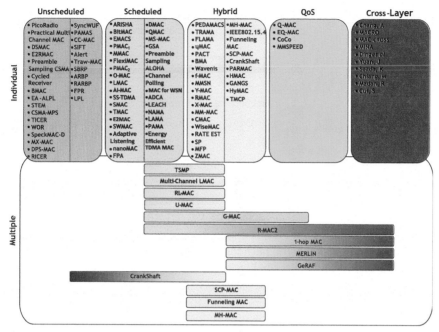

Figure 2.32 MAC protocols distribution among different categories.

and led to the appearing of incremental or derivative MAC proto-
cols all over the last years. These protocols are the following ones:
ALOHA [EH02], S-MAC [YHE04a], T-MAC [DL03], X-MAC [BYAH06],
IEEE 802.15.4 [IEE07], LEACH [HHT02], WiseMAC [EHD04a],
CSMA/CA [TK85], LPL [PHC04] and SIFT [TJB04].

Some of the aforementioned MAC protocols, are mentioned and referenced as related work. However, there is one MAC protocol that the author
of this thesis considers as noteworthy and needs further attention: the SCP-
MAC protocol. Since the SCP-MAC protocol employs some of the most
effective techniques to access the medium, special attention must be given
to it, in order to better exploit upgrades that may be added, to achieve
higher energy performance. The following sections present some of the most
important MAC protocols. The remaining MAC protocols used to build up
the taxonomy for the MAC protocols are presented in Appendix B.

2.7.1 Survey on Unscheduled MAC protocols

Berkeley MAC

Berkeley MAC (BMAC) [PHC04] protocol employs several techniques,
namely CCA and packets backoff for channel arbitration, and Low Power

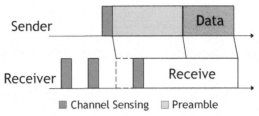

Figure 2.33 BMAC packet exchange.

Listening (LPL) for low power communication. Nodes follow a sleeping schedule independently, by following a duty cycle of the sensor network. BMAC utilizes long beacons/preambles when a message transmission is needed, as presented in Figure 2.33. The CCA technique is one the innovations which evaluates the received signal and the noise floor in particular way. It starts by searching for outliers in the received signal such that the channel energy is significantly below the noise floor. If the node detects an outlier during the sampling, the channel is declared as "clear", because a valid signal will never have outliers significantly below the noise floor. After the nodes sample five times the channel without finding an outlier, then the node declares the channel as busy. The sender node transmits a beacon with a length enough to cover the time when the receiver node wakes up and senses activity. If the node detects activity it remains awake in order to receive the packet that is sent after the beacon or return to sleep if no activity is detected. BMAC allows for controlling several parameters in the protocol. However, the long preambles can lead to higher latencies.

SIFT

SIFT [TJB04] is designed for low latency. Like previously presented protocols, SIFT uses a contention window with a fixed size. The novelty is that, instead of picking a slot randomly and uniformly, nodes use a skewed distribution which gives preference to the slots at the end of the contention window. It is therefore more likely that only one node will have chosen the lowest order slot; SIFT eventually uses that node to transmit data. When the number of contending nodes is low, if no node starts to transmit in the lowest order slot of the Contention Window (CW), each node increases its transmission probability exponentially for the next slot. However, SIFT does not solve the hidden terminal problem and may use a CSMA/CA mechanism.

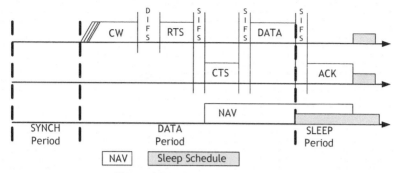

Figure 2.34 S-MAC frame format.

2.7.2 Survey on Scheduled MAC protocols

Sensor-MAC

The Sensor-MAC (S-MAC) [YHE04a] is based on periodic sleep-listen schedules and local synchronizations between nodes. Schedule exchanges are accomplished by periodical SYNC packet broadcasts to the direct neighbours, as presented in Figure 2.34. Collision avoidance is achieved by carrier sense. RTS/CTS packets exchanges can be used for unicast type data packets. A new concept employed in S-MAC is the message fragmentation where long messages are divided into frames and sent in a burst. Periodic sleep may lead to high latency when considering multi-hop routing algorithms. Adaptive listening technique is proposed to improve the sleep delay and consequently the overall latency.

By employing sleeping schedules the idle listening time is reduced and consequently reduces the energy consumption. However, the adaptive listening improves the sleeping delay but causes overhearing or idle listening increase if the packet is not destined to the listening node.

Preamble sampling

Preamble sampling protocols [EH02] do not use common active/sleep schedules. The node spends more time in the sleep mode and it wakes-up only for a short duration, to check if there is an on-going transmission, as depicted in Figure 2.35. To avoid deafness, each data frame is anticipated by a long enough preamble, to make sure that all potential receivers will detect the preamble and wait for the data message that will be sent. The node turns on its radio to sample the channel, according to its duty-cycle periodically. If the node senses an idle channel, it goes back to sleep immediately. However,

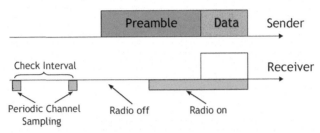

Figure 2.35   Preamble sampling mechanism.

if a busy channel is detected, the node waits for the data message. After the reception of the data packet, the node goes to sleep mode.

2.7.3 Survey on Hybrid MAC protocols

IEEE 802.15.4 MAC

The IEEE 802.15.4 standard specifies the physical and MAC layer for low-rate wireless personal area networks (LR-WPANs). The MAC proposed by the IEEE 802.15.4 standard supports two types of network nodes. On the one hand, the Full-Function Device (FFD), can serve as the coordinator of the PAN and is able to form any type of topology. On the other hand the Reduced Function Device (RFD) can only form star topologies and they can only communicate with FFDs. These last ones can never act as coordinators. The IEEE 802.15.4 standard only allows for the networks to be built as either peer-to-peer or star networks. In every network, it needs at least one FFD to work as a coordinator of the network. In the peer-to-peer topology the multi-hop feature is not supported, but it can be added. Also cluster trees can be formed and can be extended as a generic mesh network, whose nodes are cluster tree networks, with a local coordinator in each cluster, in addition to the global coordinator. In the star topology, the coordinator is the central node. Since the IEEE 802.15.4 standard intends to offer a low-cost and low-speed ubiquitous communications between devices, the proposed MAC protocol must be flexible enough to guarantee the devices requirements. To achieve these objectives the MAC allows for two basic communication modes: 1) beacon-enabled; and 2) non-beacon modes. The latter one is simply the CSMA/CA MAC, and is very useful when the main purpose of the WSN is a simple sensing. The former one is more complex, since it involves the beacon transmission, whilst maintaining a slot structure. This one is similar to the scheduled-based MAC protocols.

As the majority of the IEEE 802.15.4 features are used in the beacon-enabled mode, it will be briefly discussed below. As presented in Figure 2.36,

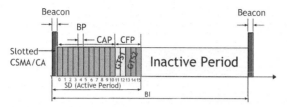

Figure 2.36 IEEE 802.15.4 beacon-enabled superframe structure.

the beacon-enabled mode employs a superframe structure, where the MAC frame is divided into two periods: the active and the inactive periods. Before each active period a beacon frame is transmitted by the coordinator at an interval *BI*, adjustable from 15 ms to 245 s. The length of the active period is determined by *SD*, while the duty cycle is calculated by $2^{SD}/2^{BI}$. The active period is further divided into 16 time slots of duration BP. The active period is composed by the beacon, a CAP and a CFP. The latter one is only available if GTSs are allocated by the coordinator, but only 7 GTSs are allowed in the CFP. First a node listens to the beacon to understand if a GTS has been reserved by the coordinator or not. If so, then it remains powered of until its GTS is scheduled to initiate the transmission of data. If no GTS is reserved, then it uses the CSMA/CA during CAP, where typical back-off procedures are applied.

Scheduled Channel Polling-MAC

The Scheduled Channel Polling-MAC (SCP-MAC) protocol [YSH06a] is an example of a duty-cycle based MAC protocol that specifies awake and asleep time periods within a frame. This duty-cycle based MAC protocol tries to reduce the energy consumption by introducing periodic listen and sleep cycles, while performing contention during the listening period. The SCP-MAC combines scheduling with channel polling in an optimal way, like in the LPL [PHC04] protocol. The SCP-MAC main contribution is the use of a double contention window. The SCP-MAC tries to eliminate the majority of the sources of energy waste: collisions, overhearing, control overhead and mainly, idle listening. It tries to solve packet collisions by using the two phase contention window scheme (when the sender node wants to transmit a data packet).

As depicted in Figure 2.37, if a node that wants to send a packet, it will choose a random slot within the first contention window (following a uniform distribution). The potential node that wants to send a packet switches on its radio, and checks the channel. If it senses the channel free it sends a preamble,

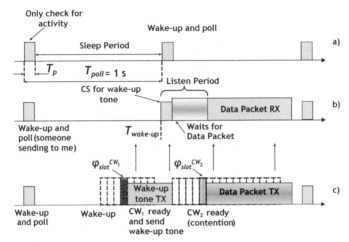

Figure 2.37   SCP-MAC two-phase contention window mechanism.

which acts like a busy tone, and continues to send it until the end of the first contention window, announcing to other potential senders that a transmission is ongoing. After it sends the preamble, all nodes except the one that "won" the channel, will wake-up and poll the channel to sense the medium for a preamble. If the node detects a preamble it will stay awake to receive the packet. If no preamble is detected the node goes immediately to the sleep mode. The node that has gained the control of the channel will perform another random slot choice into the second contention window and sense the medium. If it is free then the packet will be sent.

One of the applied techniques is the channel polling synchronization of the neighbouring nodes, using SYNC packets and optimal intervals for scheduling synchronization. This process of sending SYNC packets is repeated after a synchronization period, T_{sync}, but may be suppressed by piggybacking the schedules in the data packets when the packets exchanged are of the broadcast type.

Two contention windows are used because the collision probability is lower than the one obtained by using a single contention window whose size is the sum of the sizes from CW_1^{max} and CW_2^{max}. The two-phase contention window reduces the effective collision probability compared to a single contention window of equal duration, since only nodes that succeed in CW_1 will enter into the CW_2. After the first phase of the contention, only the successful nodes from CW_2 contend in the CW_2. With fewer contending nodes, the effective collision probability is greatly reduced. With periodic traffic, the energy cost can be minimized by using a scheduled listening algorithm, and the synchronization of the neighbour's channel polling time is maintained like in Sensor-MAC (S-MAC) [YHE04a], while adjusting the duty cycle to variable traffic conditions. When the data rates are lower than synchronization periods, explicit SYNC packets are used. An adaptive channel polling mechanism can be used when it detects bursty traffic. This mechanism adds additional wake-ups in the interval between two regular wake-ups. However, this optional mechanism may increase idle listening in case the information transmitted can fit in a single packet, due to a node that receives a packet during its normal channel polling that systematically wakes-up in the subsequent adaptive slots.

Crankshaft

The main characteristic of Crankshaft [HL07] is the time being decomposed into frames, which are composed by slots. Some slots in the beginning of the frame are assigned for unicast traffic, while in the end of the frame some slots are assigned to broadcast traffic. The way how the slots are of type receiver-based, where each node has its own slot during which it wakes-up to receive data. When considering broadcast slots, all nodes wake-up. Crankshaft employs a collision avoidance mechanism that is employed prior to the transmission and each transmission is ended with as acknowledgement packet. This collision avoidance mechanism is similar to the one used in SCP-MAC protocol, where a sender transmits at randomly chosen times a preamble. The sender node that starts before the others to transmit wins the contention, because a carrier sense is performed before transmitting any packet to sense for the presence of a signal in the shared medium. If the channel is sensed as busy the contention is stopped. The slot' choice used to perform the carrier sense, gives preference to the higher order time slots (uses a skewed distribution). This leads to a reduction of the length of the preamble and therefore it reduces the energy consumption. Although Crankshaft reduces idle listening, some overhead is added when sampling all the broadcast slots, leading to an increase of the energy waste. Moreover, the use of a TDMA approach it decreases the flexibility of the protocol in the presence of sporadic traffic.

2.7.4 Survey on QoS MAC protocols: EQ-MAC

EQ-MAC [YBO08] is another MAC protocol that implements quality of service policies. It intends to reduce energy consumption, while providing QoS guarantees by using a service differentiation mechanism. EQ-MAC consists of two sub-protocols: the Classifier MAC (C-MAC), and the Channel Access MAC (CA-MAC). The first one is responsible by the classification of the data collected among the sensor nodes based on the importance of the data, while storing it in the appropriate queue of the node's queuing system. The second one is a hybrid collision avoidance system of both scheduled and unscheduled schemes that allows for the node to save energy, and thus extending the network's lifetime. The energy savings comes from the different procedures that are applied, depending if the the exchanged messages are long or small. Long messages are assigned to scheduled slots with no contention, where time slots are only assigned to nodes that have data to send (leading to an energy reduction when using TDMA slots), while for small messages (periodic control messages) are assigned random access slots.

2.7.5 Survey on Cross-Layer MAC protocols: MERLIN

MERLIN [ROJ08] integrates MAC and routing features into a single architecture. Compared with other sensor network protocols, it employs a multicast upstream and multicast downstream approach that allows relaying packets to and from the gateway. Simultaneous reception and transmission errors are notified by using asynchronous burst ACK and negative burst ACK. The network is divided into time zones, together with an appropriate scheduling policy, enables the routing of packets to the closest gateway. MERLIN employs a TDMA/CSMA based approach, in which it divides the time into slots. A scheduling table allows allocating timeslots to the nodes in the network to assign periods of node activity and inactivity. The scheduling allows synchronizing neighbouring nodes for transmission and reception in the same time zones.

All wireless nodes must be well synchronized between themselves and with the gateways. As a drawback, MERLIN suffers from the hidden terminal problem and collisions occur often in the network. Moreover, it needs more than one base station in the network in order to manage the input of user data into the network and presents high delays. Figure 2.38 presents the mechanism employed in MERLIN to avoid collisions when nodes try to access the medium to transmit data packets.

2.7.6 Survey on Multiple based MAC protocols: 1-hop MAC

1-hop MAC [WBD+06] protocol is based on communication architecture grouping MAC and routing layers which avoids 1-hop neighbourhood knowledge. It combines L2 (link) and L3 (network/routing) protocols. The basic

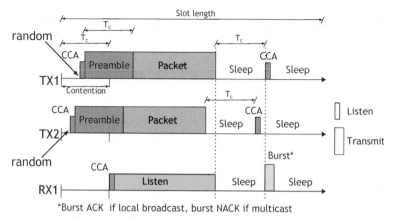

Figure 2.38 Transmission mechanism for collision avoidance of MERLIN.

procedure of 1-hop MAC protocol initiates with a node that sends out a request message and is received by its 1-hop neighbourhood. Each neighbour will answer this request after a time proportional to its metric. The requesting node can then elect the most suitable node based on when it has received the answers. Three 1-hop protocol variants are proposed, apart from the basic one. The first one tries to reduce the sender node listening period, the second one tries to avoid multiple ACK messages while the last one allows for sending data packets during the "do not answer anymore" message period of the protocol. It can dynamically swap between the 1-hop variant 1 and 1-hop variant 3, by following specific rules that allows for achieving energy efficiency. By using this approach, it can be viewed as a multi-mode protocol where the decision on which mode to use is distributed and no signalling messages are necessary.

2.8 Classification of MAC Protocols Characteristics

This section is dedicated to the description of some of the most prominent MAC characteristics that have been identified. These characteristics are related with several features that characterize a MAC protocol and allows for a user to distinguish between the MAC protocols. Tables for the MAC protocols taxonomy are presented in detail in Appendix B.

The characteristics of the MAC protocols are as follows.

1. **Action against energy waste**: This characteristic is related to the main purpose of the MAC protocol or on what sources of energy waste it tries to mitigate. Possible values are the following:

- Reducing collisions (Co);
- Reducing Overhead (Ov);
- Reducing Overhearing (Oh);
- Reducing Idle Listening (Il);
- Narrow Band Interference (NaI).

2. **Type of scheduling**: This feature concerns the type of scheduling protocol that the MAC protocol employs. Possible values are the following:

- Global Schedule (Glob);
- Centralized Scheduling (Cent);
- Distributed Scheduling (Dist);
- Localized Collision free based (Local);
- Staggered based (Stag).

3. **Roles for the nodes**: This characteristic helps to identify if the sensor
nodes have always the same functionality along the WSN. Possible
values are the following:

- Rotative (Rt);
- Fixed (Fi);
- Adaptive (Adapt);
- Static (St).

4. **Node traffic change behaviour**: It establishes if the MAC protocol
adapts to the traffic patterns variation. Possible values are the following:

- Adaptive (Adapt);
- Fixed (Fi).

5. **Node traffic exchange**: This characteristic defines if the traffic is
variable or uniform. Possible values are the following:

- Uniform (Uni);
- Variable (Var).

6. **Slot assignment**: If the MAC protocol is slot-based type, this feature
defines how the slot assignment is performed. Possible values are the
following ones:

- Receiver-based (Rc);
- Neighbour-based (Nghb);
- On-demand based (On).

7. **Frequency allocation**: It defines if the MAC protocol utilizes one or
more frequency bands:
- Single-frequency (Single);

- Multi-frequency (Multi);

8. **Frame subinterval time reduction**: It identifies what part(s) of the MAC frame protocol is reduced. Possible values are the following ones:

   - Active Period (Act);
   - Synchronization Period (Synch);
   - Sleep Delay (Sleep);
   - Schedules Period (Sched);
   - Sampling Duration (Samp).

9. **Duty cycle tuning**: This feature is related to the way how the duty cycle is adjusted in MAC protocol. Possible values are the following ones:

   - Statistical Function (Stat);
   - Utilization Function (Util);
   - Pre-defined sets of duty cycles (Pre);
   - Mobility adjust (Mob);
   - Range (Rang);
   - Application Demand (App);
   - Traffic Load (Tf).

10. **Preamble length reduction**: It defines the technique employed by the MAC protocol (if present) to reduce the preamble length. Possible values are the following ones:

    - Packetization (Pack);
    - Piggybacking Synchronization Information (Pigg);
    - Data frame preamble (DpR);
    - Dynamic size (Dis).

11. **Channel allocation**: It defines how the channel allocation is performed in the MAC protocol. Possible values are the following ones:

    - Single-Channel (Single);
    - Dual-Channel (Dual);
    - Multi-Channel (Multi).

12. **Preamble reception filtering**: It defines a special feature that may be employed by the MAC protocol in the preamble reception when Preamble Reception Filtering is present. The only possible value is the following one: Micro-Frame (Micro).

13. **Communication modes**: This characteristic defines if the MAC protocol employs or not beacon packets. Possible values are the following ones:

- Beacon mode (Be);
- Non-Beacon mode (n-Be).

14. **Multiplex (channel access) technique**: It defines the method by which multiple analog message signals or digital data streams share the medium and are possibly combined into one signal over the shared medium. Possible values are the following ones:

    - CSMA;
    - TDMA;
    - CDMA;
    - FDMA.

15. **Wake-up techniques**: It defines what type of technique is employed in the MAC protocol that needs to be awakened. Possible values are the following ones:

    - Wake-up radio (Radio);
    - Wake-up tone (Tone);
    - Wake-up schedule (Schd).

16. **Contention window sizes**: In contention-based MAC protocols it defines how many contention windows and if the size of the contention window is variable or not. Possible values are the following ones:

    - Single CW & Fixed size (Sg & Fi);
    - Single CW & Variable size (Sg & Var);
    - Double CW & Fixed size (Db & Fi);
    - Double CW & Variable size (Db & Var).

17. **Type of traffic**: This characteristic defines the type of traffic which is handled by the MAC protocol. Possible values are the following ones:

    - Converge cast (Conv);
    - Sensor (Sensor);
    - Multimedia (Media).

18. **Traffic load**: This feature defines the traffic load supported by the MAC protocol. Possible values are the following ones:

    - High Traffic (Hi);
    - Medium traffic (Me);
    - Low traffic (Lo).

19. **Cluster**: It defines if the MAC protocol allows for the sensor nodes to form clusters. Possible values are the following ones:

- Yes;
- No.

20. **Channel assignment**: This feature defines how the channel assignment is performed by the sensor nodes in multi-channel based MAC protocols. Possible values are the following ones:

    - Receiver-based (Rc);
    - Neighbour-based (Nghb);
    - On-demand based (On).

21. **Synchronization**: It defines if the sensor nodes perform schedules synchronization. Possible values are the following ones:

    - Yes;
    - No.

22. **Channel technique**: It defines how the channel technique is employed by the MAC protocol to control the access to shared medium. Possible values are the following ones:

    - Clear Channel Assessment (CCA);
    - Preamble Sampling (PrS);
    - Data Packet Sampling (DtS);
    - Busy Tone (Busy);
    - Carrier Sense (CS);
    - Data Prediction (DaP);
    - Framelets Division (FrD);
    - Low Power Listening (LPL);
    - Channel Assessment (CA);
    - Schedule Channel Polling (ScP);
    - Channel Detection (CD);
    - Frequency hopping (FrH);
    - Cross-Layer (CL);
    - Node Activation based (NoB);
    - Link Activation based (LiB).

2.9 Summary and Conclusions

A large amount of MAC sub-layer protocols have been proposed in the literature to mitigate or try to solve several waste energy problems, whilst optimizing some performance parameters. Key MAC protocols have been classified and a taxonomy has been proposed.

MAC protocols can be categorized into many different ways. However for the purpose of this book, the classification of the MAC protocols into different categories is performed by the technique to access the channel being used in the protocol, along with the possibility of the MAC protocol to use multiple techniques. Due to the large amount of work that has been produced over the past two decades on WSNs, many of these categories apply more than one technique in the MAC sub-layer. Hence, this taxonomy classification extends the classification of WSN MAC protocols for the new ones that employ multiple techniques to achieve the purposed objective.

Three main categories of MAC protocols stand out from the ones presented throughout this chapter: Unscheduled, Scheduled and Hybrid MAC protocols. The former category addresses the MAC protocols that allows for sensor nodes to operate independently, while it conserves energy due to the low complexity of the MAC protocol. The second category is related to the MAC protocols that organize the communication between sensor nodes in an ordered way, while reducing collisions and retransmissions by using synchronization mechanisms. The latter one considers different techniques to conserve energy among nodes, namely by adjusting their behaviour between techniques used in scheduled and unscheduled MAC protocols. Moreover, comparison tables for the considered MAC protocols are presented from Table B.3 up to Table B.14 (in Appendix B), where the MAC characteristics (e.g., slot assignment, frequency bands allocation) are categorized for each one of the MAC protocols.

From this intensive analysis of the MAC techniques and protocols one of the most promising ones is SCP-MAC. Therefore, it is worthwhile to assume SCP-MAC as a solid basis for the future development of a novel MAC protocol. These initial assumptions are going to be further demonstrated in the following chapter by a simulation tool based on the OMNeT++ event driven simulator developed throughout the research, where different studies have been conducted to evaluate potential improvements for our new MAC protocol. There are three main reasons to choose the SCP-MAC protocol as a starting point to the development of a new MAC protocol:

- Employs two-contention windows to contend for the medium;
- Synchronized channel polling is combined with reservation mechanism;
- A low power listening protocol is considered.

3

Further Insights into the IEEE 802.15.4 Standard

This Chapter presents the PHY and MAC layer specifications for the IEEE 802.15.4 standard and is organized into two parts: Section 3.1 addresses the PHY aspects whilst Section 3.2 discusses the MAC sub-layer aspects.

3.1 Physical Layer

The IEEE 802.15.4 PHY layer is responsible for providing the medium interface where the actual communication occurs. The PHY layer is the lowest component in the WSN protocol stack and is in responsible for providing control (activation and deactivation) of the radio transceiver, energy detection, link quality indication, CCA, channel selection, and data transmission/reception [GCB04].

3.1.1 Channel Assignment

IEEE 802.15.4 networks can operate in several different frequency channels that are defined through a combination of channel numbers and pages allowing for identifying the frequency bands as well as the modulation employed. The channel pages were introduced in the IEEE 802.15.4-2006 standard allowing for distinguish between the different supported PHY layers as well as modulation schemes. In the previous release (i.e., the IEEE 802.15.4-2003 standard) there were no optional PHY layers with multiple frequency bands. One of the problems of the IEEE 802.15.4-2003 standard is that the modulation schemes were different for the 2.4 GHz, 868 MHz and 915 MHz frequency bands. Therefore, there was a need to change the local oscillator for each frequency as well as the modulator/demodulator block responsible for processing the signals. This in turn, imposes the use of separate processing blocks in order to support the different frequency bands. A larger portion of the radio was unused all the time. Since the popularity of the IEEE 802.15.4-2003 standard started to increase dominating the product offerings for low-rate wireless personal area networks (LR-WPANs), these PHY layers were not enough to respond to the markets demands. Moreover, there was also a need of better coexistence between IEEE 802.15.4 compliant devices and other Radio frequency (RF) technologies, such as IEEE 802.11 (i.e., WiFi) and Bluetooth.

The IEEE 802.15.4-2006 specification enables more flexibility to the radio transceivers, where different frequency bands can use the same modulation scheme (e.g., O-QPSK). The new modulation schemes were introduced for the radios operating in the 868 MHz and 915 MHz frequency bands, most likely because nobody was using them. This has the benefit of allowing the radio transceiver to span all three frequency bands.

In the IEEE 802.15.4-2006 standard, the PHY layer offers 20 kb/s of bit rate by using a single channel in the frequency range from 868 MHz to 868.6 MHz. This range of frequencies is used in Europe for applications, such as short-range wireless networking. The 915 MHz and 2.4 GHz are part of the Industrial, scientific and medical (ISM) frequency bands. The 2.4 GHz band is used worldwide, and the 915 MHz band is used mainly in North America. The IEEE 802.15.4 standard requires simultaneous and joint support for the 868 MHz and 915 MHz frequency bands.

The 868/915 MHz Band binary phase-shift keyin (BPSK) PHY, originally specified in 2003, offers a trade-off between complexity and data rate. The optional PHYs offers a much higher data rate than the one given by the 868/915 MHz BPSK PHY, which provides a data rate of 20 kb/s in the 868 MHz band and 40 kb/s in the 915 MHz band. The Amplitude shift keying (ASK) PHY offers data rates of 250 kb/s in both the 868 MHz and 915 MHz bands, the same data rate of the 2.4 GHz band PHY. In the 915 MHz band the O-QPSK PHY offers a signalling scheme identical to the one of the 2.4 GHz band PHY and a data rate equal to the one of the 2.4 GHz PHY band. In terms of data rate, the O-QPSK PHY in the 868 MHz band supports a data rate of 100 kb/s.

Table 3.1 presents an overview of the channel pages and numbers for the IEEE 802.15.4-2006 standard. For sake of simplicity from now on we will use the term IEEE 802.15.4 when referring to the IEEE 802.15.4-2006 standard.

The frequencies for the IEEE 802.15.4 standard are divided by three different frequency bands with a total of 27 channels. The central frequencies are given by equations (2.1), (2.2) and (2.3).

Table 3.1 Channel assignments for the IEEE 802.15.4 standard.

Channel page	Channel number	Frequency band	Modulation	Data rate (kb/s)	Symbol rate (ksymbols/s)
0	0	868 MHz	BPSK	20	20
	1-10	915 MHz	BPSK	40	40
	11-26	2.4 GHz	O-QPSK	250	62.5
1	0	868 MHz	ASK	250	12.5
	1-10	915 MHz	ASK	250	50
	11-26	2.4 GHz	O-QPSK	250	62.5
2	0	868 MHz band	O-QPSK	100	25
	1-10	915 MHz band	O-QPSK	250	62.5
	11-26	Reserved	-	-	-
3	Reserved	Reserved	-	-	-

The 27 half-duplex channelization specified by the IEEE 802.15.4 standard shown in Figure 3.1 is organized as follows:

- The 868 MHz band, ranging from 868.0 MHz to 868.6 MHz is used in Europe. It adopts a binary phase shift keying (BPSK) modulation format, with a DSSS at a chip-rate of 300 kchip/s. Only a single channel with data rate of 20 kb/s is available and devices shall be capable of achieving a sensitivity of -92 dBm or better. A pseudo-random sequence of 15 chips is transmitted within 50 μs symbol period.

- The 915 MHz band, ranging from 902 MHz to 928 MHz is used in the North American and Pacific area. It adopts a BPSK modulation format, with DSSS at a chip-rate of 600 kchip/s. Ten channels with data rate of 40 kb/s are available and the devices shall be capable of achieving a sensitivity of -92dBm or better. A pseudo-random sequence of 15 chips is transmitted in a 25 μs symbol period.

- The 2.4 GHz ISM band, ranging from 2400 MHz to 2483.5 MHz, adopts an O-QPSK modulation format, with a DSSS at 2 Mchip/s. Sixteen channels with data rate 250 kb/s are available and devices shall be capable of achieving a sensitivity of -85 dBm or better. A pseudo-random sequence of 15 chips is transmitted in a 16 μs symbol period.

The IEEE 802.15.4a-2007 [WPA07b] was released, in order to support high-precision ranging capability (1m accuracy and better), high aggregate throughput, and ultra low power consumption (mainly due to low transmit power levels, typically under -10 dBm). It supports two additional PHY layers: 1) an Ultra-wideband (UWB) that operates in three different bands and 2) a Chirp spread spectrum (CSS) PHY in the 2.4 GHz ISM band. The choice

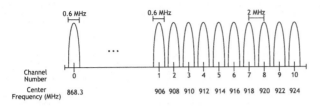

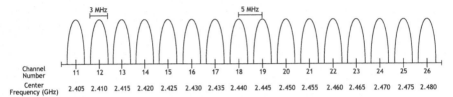

Figure 3.1   IEEE 802.15.4 channelization at the 868/915 MHz and 2.4 GHz bands.

of adopting UWB has many advantages when compared with narrowband systems as follows:

- **Improved channel capacity**: UWB technology enables the increase of channel capacity, since it offers a bandwidth of at least 500 MHz for high data rate communications. However, coverage range is limited up 10 m due to the low power levels mandated by Federal communications commission (FCC) and the Radio spectrum committee (RSC) of the European commission (EC).

- **High-precision and ranging capabilities**: The very short pulses used in UWB enables precision ranging measurements whith high accuracy. Therefore, UWB can be applied in geolocation systems with an achievable ranging accuracy from 15 cm to 50 cm [A Non-Coherent 802.15.4a UWB Impulse Radio].

- **Low power consumption**: Since UWB radios have a simpe architecture, being the information transmitted by using low energy pulses with a duration of approximately 2 ns, this allows for energy and cost savings in addition to a reduction in the implementation complexity.

- **Resistance to multipath fading**: The UWB signals are very robust in terms of multipath fading due to its large bandwidth. The narrow pulses prevent multiple reflections, avoiding that multiple signals reach the receiving antenna by multiple paths, which can cause interference and phase shifting of the signals.

UWB signals can operate in the following frequency bands:

Table 3.2 UWB PHY band allocation.

Band group	Channel number	Center frequency (MHz)	Bandwidth (MHz)	Mandatory/ optional
0	0	499.2	499.2	Mandatory bellow 1 GHz
1	1	3494.4	499.2	Optional
	2	3993.6	499.2	Optional
	3	4492.8	499.2	Mandatory in low band
	4	3993.6	1331.2	Optional
2	5	6489.6	499.2	Optional
	6	6988.8	499.2	Optional
	7	6489.6	1081.6	Optional
	8	7488.0	499.2	Optional
	9	7987.2	499.2	Mandatory in high band
	10	8486.4	499.2	Optional
	11	7987.2	1331.2	Optional
	12	8985.6	499.2	Optional
	13	9484.8	499.2	Optional
	14	9984.0	499.2	Optional
	15	9484.8	1354.97	Optional

- A sub-gigahertz band from 249.6 MHz to 749.6 MHz;
- A low band from 3.1 GHz to 4.8 GHz;
- A high band from 6.0 GHz to 10.6 GHz.

The UWB spectrum allowed by the Federal Communication Commission (FCC) is divided into 16 channels as shown in Table 3.2. A compliant device shall support at least one of the mandatory channels (i.e., channels 0, 3 and 9). All the other channels are optional, where channels 4, 7, 11, and 15 are differentiated by having a larger bandwidth (>500 MHz).

3.1.2 Carrier Sense

The Carrier sense (CS) mechanism is responsible for verifying the frequency channel availability, and report the sensing to the radio transceiver. Therefore, every time a node has a packet to transmit, it enters into the RX mode for detecting for any possible signal that might be present in the desired frequency channel. In contrast with the energy detection mechanism, where there is no attempt to identify or decode signals, during CS the signal is demodulated for verifying if the signal modulation and spreading characteristics belong to the PHY layer being currently utilized by the device. If the occupying signal is compliant with the IEEE 802.15.4 PHY layer, the device might choose to consider the channel busy regardless the signal energy level.

3.1.3 Received Signal Strength Indication

The quantized signal energy of each received packet allows for the creation of the RSSI. This signal strength indicator does not care about the "quality" or "correctness" of the signal. The RSSI availability indicates that a location-based system can be implemented without the need of any additional hardware for the individual nodes in the network. As presented in Table 3.3, there are four parameters associated with the RSSI as follows [Far11]:

- **Dynamic range**: The dynamic range specified in dB is responsible for indicate the minimum and maximum received signal energy. If the RSSI has a range of 100 dB (i.e., from -100 dBm to 0 dBm), the minimum signal energy that the receiver can measure is -100 dBm, whilst the maximum signal energy reported as RSSI is 0 dBm;
- **Accuracy**: The RSSI accuracy indicates the average error associated with the received signal strength. The CC2420 transceiver is capable of providing an accuracy of ± 6 dB;
- **Linearity**: The RSSI linearity indicates the relationship between maximum deviation of the RSSI versus the actual received signal power within the limits of ± 3 dB;
- **Average period**: The averaging period is required by IEEE 802.15.4 compliant transceivers to generate the Link quality indicator (LQI) being obtained by measuring the received signal strength over the first 8 symbol periods (128 μs).

By considering a FIFO queue, like the one presented in Figure 3.2, the RSSI value is obtained as follows:

- The time at which the packet was received (i.e., timestamp) and the RSSI are passed to the MAC and upper layers for analysis;
- When the packet is read from the FIFO queue the second last byte contains the RSSI value (i.e., one byte value as a signed 2's complement value) that was measured after receiving 8 symbols of the actual packet;

Table 3.3   RSSI parameters [Ins07].

Parameter	Typical value	Unit
Dynamic range	100	dB
Accuracy	± 6	dB
Linearity	± 3	dB
Average period	128	μs

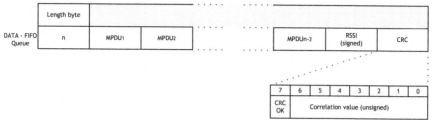

Figure 3.2 Data buffering in a FIFO queue.

- If the RSSI value was captured at the same time as the data packet being received, the RSSI value will reflect the intensity of received signal strength at that time, not necessarily the signal power belonging to the received data;
- When multiple nodes access to the medium at the same time, the RSSI value being captured could be erroneous (i.e., the value represents the overlay of several different signals).

As described above the RSSI is represented as signed 2's complement value. However, the actual value can not be read and interpreted as the received signal strength. To convert the actual value to the received signal strength an offset must be added. The CC2420 radio transceiver offset taken by the data sheet is approximately -45 dB [Ins07]. However, this value is dependent on the actual antenna configuration.

WSN MAC protocols use RSSI measurements for localization, sensing, transmission power control and packet reception ratio modelling. To obtain the RSSI values two can methods can be explored as stated in [VVL+11].The first involves using the physical (theoretical) relationship between the RSSI and the distance. Therefore, free space and two ray ground models are used to obtain the theoretical RSSI. The second (experimental) method considers an empirical RSSI database filled with measurement records during an extensive calibration phase, and the location is estimated by fitting the measured RSSI to the database. However, since the RSSI provided by the redio transceivers has limited accuracy, which directly impacts on the estimated distance between the nodes, the path-loss determined experimentally can be a major source of errors [Far11].

Empirical studies [NH93], have shown the log-normal shadowing model [R+02] provides more accurate multi-path channel models than Nakagami and Rayleigh for indoor environments. Therefore a good approximation for obtaining the RSSI, in dBm, is by considering the log-normal shadowing model as follows:

$$RSSI(d) = P_t - P_L(d) \qquad (3.1)$$

where, P_t is the transmit power and $P_L(d)$ is the path loss at a given distance, d.

The $P_L(d)$ is given as follows:

$$P_L(d) = P_L(d_0) + 10\alpha log10\left(\frac{d}{d_0}\right) + X_\sigma \qquad (3.2)$$

where, $P_L(d_0)$ is known as the $d_0 = 1$ m loss, or insertion loss that arises due to free-space path loss and antenna inefficiencies (i.e., reference value of path loss), α is the signal propagation constant (also named path loss exponent) and d is the distance between the transmitter and the receiver. The shadow effect is described by a zero mean Gaussian random variable, X_σ, with standard deviation, σ, that models the random variation of the RSSI value.

The IEEE 802.15.4 standard suggests that a good approximation for obtaining the $P_L(d)$ at 2.4 GHz band is given by:

$$P_L(d) = \begin{cases} 40.2 + 20log_{10}d + X_\sigma & d \leq 8m \\ 58.5 + 33log_{10}d + X_\sigma & d > 8m \end{cases} \qquad (3.3)$$

By analysing Equation (3.3) we conclude that the path loss model described in IEEE 802.15.4 has a two-segment function with a path loss exponent of 2.0 for a distance less or equal than 8 m and a path loss exponent of 3.3 for a distance longer than 8 m. Figure 3.3 presents the RSSI as a function of the distance (we assume $P_t = 0$ dBm and $X_\sigma = 0$). By increase the distance we decrease the RSSI value.

The LQI is not a signal strength indicator like RSSI, instead is a metric of the current quality of the received signal. The LQI gives an estimate of how easily a received signal can be demodulated by accumulating the error magnitude between ideal constellations and the received signal over the 64 symbols immediately following the sync word. LQI is commonly used as a relative measurement of the link quality since the value is dependent on the modulation scheme. In real deployment scenarios there are five "extreme situations" that can be used to describe the behaviour of the RSSI and LQI as follows:

• A weak signal with noise may give low RSSI and high LQI;

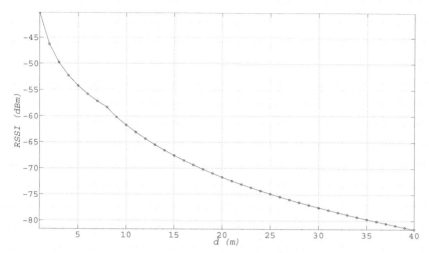

Figure 3.3   RSSI as a function of the distance.

- A weak signal without noise may give low RSSI and low LQI;
- As strong noise coming from an interferer may give high RSSI and high LQI;
- A strong signal with low noise may give high RSSI and low LQI;
- A very strong signal that causes saturation at the receiver may give high RSSI and high LQI.

3.1.4 Clear Channel Assessment

When performing CSMA-CA, the MAC sub-layer request the PHY layer to perform CCA in order to detect if the medium is found to be busy or idle. The CCA period has a duration that is equal to $T_{CCA} = 8 \times T_{symbol}$ (i.e., 128 μs), where T_{symbol}=16 μs is the symbol time. The CCA operation has three operational modes as follows:

- **Energy Detection mode** - In this mode the CCA shall reports a busy medium if the received energy is above a given threshold, referred as energy detection threshold;
- **Carrier Sense mode** - The CCA reports a busy medium only if it detects a signal with the modulation and the spreading characteristics of IEEE 802.15.4 standard and which the signal may be higher or lower than the energy detection threshold;
- **Carrier Sense with Energy Detection mode** - The IEEE 802.15.4 defines 27 different wireless channels. A network can choose to operate

within a given channel set. Hence, the PHY layer should be able to tune its transceiver into a specific channel upon the reception of a request from a higher layer.

3.2 Medium Access Control Sub-layer

The IEEE 802.15.4 MAC sub-layer is designed to support a large number of applications, such as industrial, medical and home applications for control and/or monitoring. The main function performed by the MAC layer is the access control to the physical radio channel. It is responsible for the generation of ACK frames, support of PAN association and disassociation and the security control. The method used for random medium access is based on the IEEE 802.15.4 non-beacon-enabled mode. It uses unslotted CSMA-CA to transmit frames. The core idea is the following: when one node needs to send data, it will compete for the wireless channel. If collisions occur nodes will backoff for a random period of time before attempting to access the channel. Depending on the application requirements, the IEEE 802.15.4 standard considers two topologies (i.e., star and peer-to-peer), as presented in Figure 3.4.

The star topology is formed around a Full function device (FFD), so the communication is established between nodes and a single central controller, called the PAN coordinator which is the only node allowed to form links

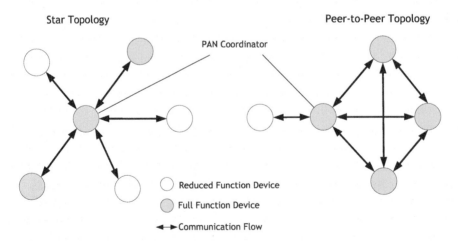

Figure 3.4   Examples of star and peer-to-peer topologies.

with multiple devices. The function of the PAN coordinator includes not only running a specific application but also be used to initiate, terminate or route the communications in the network. So the PAN coordinator acts like the primary controller of the personal area network. All devices operating on a network of any type of topology mentioned shall have a unique 64-bit address. This address may be used for direct communication within the PAN, or a short address may be allocated by the PAN coordinator when the device associates and used instead. The PAN coordinator often will be powered by a continuous power supply, while the devices will most likely be battery powered. Applications that benefit from a star topology include: home automation, Personal Computer (PC) peripherals, toys and games, and personal health care [WPA07a]. The IEEE 802.15.4 MAC protocol supports two operational modes: the non-beacon-enabled mode and the beacon-enabled mode which can be selected by the PAN coordinator. In the non-beacon-enabled mode the PAN coordinator do not transmit regular beacons, and so it transmits a data frame using a non-slotted CSMA-CA. In the beacon-enabled mode the beacons are periodically sent by the PAN coordinator, so when a device wishes to transfer the data to a coordinator it first listen the medium for a network beacon frame. The beacon frame is responsible by the boundaries establishment for the beginning of a superframe while defining a time interval to exchange packets between different nodes. The medium access is basically a slotted CSMA-CA. This mode is used in applications that require a certain amount of bandwidth and low latency so that the PAN coordinator enables the allocation of some time slots in the superframe. These portions are called Guaranteed Time Slots (GTSs) and they are used in the situation when the node needs to have guaranteed services.

As described before the IEEE 802.15.4 has different types of topologies, so the data transfer model is always related with the network topology. In the peer-to-peer mode a device will communicate with other devices in its vicinity while in the star networks the communication exchange will occur between a PAN coordinator and a network device. In the presence of a star network two types of data transfer mechanism could exist, depending on whether the PAN coordinator is beacon-enabled or non-beacon-enabled. Details on the IEEE 802.15.4 topologies were given in Chapter 2. MAC data frame and PHY packet were also introduced in Chapter 2.

Beacon-enabled - star topology

In the presence of a beacon-enabled star topology, a network device that wants to send data to the pan coordinator needs to listen for a beacon. In

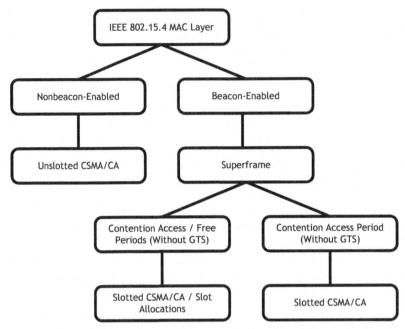

Figure 3.5 IEEE 802.15.4 operational modes.

this case if a device does not have any Guaranteed time slot (GTS) assigned, it transmits the data frame in the contention access period in accordance with the CSMA-CA procedure. If the device has already a GTS assigned, it needs to wait for the appropriate time within the superframe structure in order to transmit its data frame. After receiving the data frame, the PAN coordinator sends an acknowledgement reply to the network device, as presented in Figure 3.6. in the cases when the PAN coordinator has data to send to a network device, it will set a special flag in its beacon. when the network device detects that the PAN coordinator has pending data for him, it will sends back a "data request" message and the PAN coordinator responds with an acknowledgement followed by the data frame, and finally an acknowledgement is sent from the network device in order to finish the transmission, as presented in Figure 3.7.

Non-beacon-enabled - star topology
in the case with a non-beacon-enabled star network, a network device that wants to transfer data sends a data frame to the pan coordinator by using the CSMA-CA procedure. after correctly receive the data frame the PAN

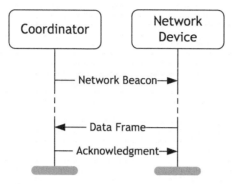

Figure 3.6   Star topology - communication to a coordinator in a beacon-enabled PAN.

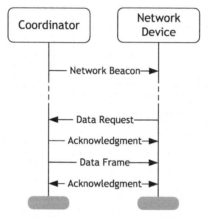

Figure 3.7   Star topology - communication from a coordinator in a beacon-enabled PAN.

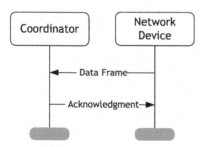

Figure 3.8   Star network - communication to a coordinator in a non-beacon-enabled PAN.

coordinator will reply to the network device, sending back an acknowledgement message, as presented in Figure 3.8.

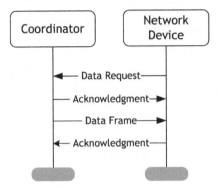

Figure 3.9   Star network - communication from a coordinator in a non-beacon-enabled PAN.

In the cases the PAN coordinator requires a data transfer to a network device, it will keep the data until the network device sends back a data request message. Then, the PAN coordinator sends an acknowledgement followed by the data frame. Finally, the network device acknowledges the reception of the data frame, as presented in Figure 3.9.

Peer-to-peer data transfer
As presented before, there are different types of data transfer transactions. In the first type, the data is transferred to a coordinator and the network device transmits the data. In the second one the exchanging of the data is transmitted from a coordinator in which the device receives the data. Finally, in the third one, the exchanging of the data is between two peer devices. When a star topology is used, only two first types of these transactions are used because data may be exchanged only between the coordinator and a device. In a peer-to-peer topology, data may be exchanged between any two devices on the network and consequently all three transactions may be used in this topology. Hence, in the peer-to-peer topology, the strategy is ruled by the specific network layer that is responsible for managing the WSN. A given network device may stay in reception mode, while scanning the radio channel for ongoing communications or it can send periodic "SYNCH" messages with other potential listening devices in order to achieve synchronisation.

3.2.1 MAC frames

The IEEE 802.15.4 standard defines four MAC frame structures:

- Beacon frame;

- Data frame;
- Acknowledgement frame;
- MAC command frame.

The beacon frame is used by the coordinator in order to transmit beacons. This type of frame is used to identify the network and its structure, wake-up devices from the sleep mode to the listening mode and synchronise devices in the network, assuming an important role in the mesh and cluster-tree networks topology, especially because it can reduce energy consumption and extend battery lifetime due to synchronization. The entire MAC frame is used as a payload in a PHY packet. The MAC beacon frame structure is described in Figure 3.10. The active part of the beacon frame is constituted by three parts: the MAC header (MHR), the MAC payload and the MAC footer (MFR). The MHR contains information about the MAC frame control field, Beacon sequence number (BSN), addressing fields, and optionally the auxiliary security header. The MFR contains a field of a 16-bit Frame check sequence (FCS).

The data frame of the IEEE 802.15.4 standard is presented in Figure 3.11. The data payload passed from the network to the MAC sub-layer is referred as MAC service data unit (MSDU). The MHR contains the frame control field, Data sequence number (DSN), addressing fields, and optionally the auxiliary security header. The MHR, MAC payload, and MFR together form the MAC data frame, (i.e., MAC protocol data unit, MPDU). The MPDU is passed to

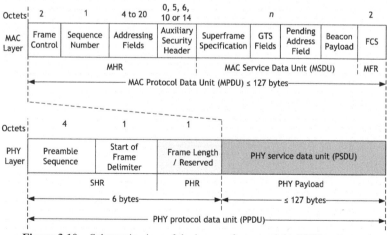

Figure 3.10   Schematic view of the beacon frame and the PHY packet.

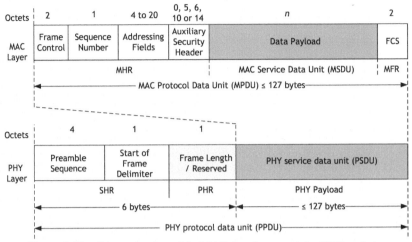

Figure 3.11   Schematic view of the MAC data frame and the PHY packet.

the PHY as the PHY service data unit (PSDU), which becomes the PHY payload [WPA07a].

The maximum allowed PHY protocol data unit (PPDU) from the IEEE 802.15.4 standard is 133 Bytes, Figure 3.11. the PHY header is 6 Bytes. Therefore, the MAC protocol data unit (MPDU) is 127 Bytes. The minimum and the maximum MHR and MFR without security are 9 and 25 Bytes respectively. Consequently, the maximum payload could range between 127-25=102 and 127-9=118 Bytes. In order to ensure compliance with the IEEE 802.15.4-2003 release, the maximum data payload is assumed to be 102 Bytes.

The acknowledgement frame, presented in Figure 3.12, has a MHR and a MFR field, and does not have a MAC payload. The MHR contains the MAC frame control and DSN fields, and the MFR is composed by a 16-bit FCS. This type of frame has the simplest MAC frame format in the IEEE 802.15.4 standard. The acknowledgement frame is sent whenever the destination device successfully receives a packet that was previously sent to it.

The MAC command frame presented in Figure 3.13 is used to request association or disassociation. The MAC payload contains the command type field that is responsible to determinate if the type of the command is an association request or a data request, while the command payload field contains the command itself.

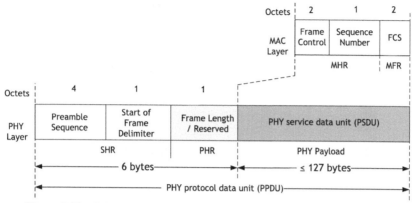

Figure 3.12   Schematic view of the acknowledgement and the PHY packet.

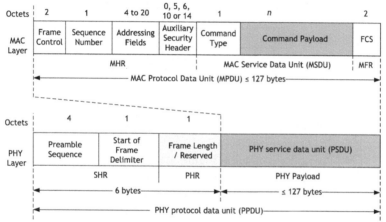

Figure 3.13   Schematic view of the MAC command frame and the PHY packet.

All the command frames types defined by the MAC layer that a reduced function device may send or receive are described in Table 3.4. The FFD device could transmit and receive all the command frame types shown in Table 3.4, with the exception of the GTS request command.

In the IEEE 802.15.4 MAC layer between two successive frames transmitted an IFS period must be inserted. The IFS depends on whether the transmission transaction is acknowledged or unacknowledged. When the acknowledgement is received the IFS follows the acknowledgement frame and when the frame length do not exceeds the *aMaxSIFSFrameSize*, the acknowledgement must be followed by a short IFS (SIFS) period and the

Table 3.4   Command frame types.

Command frame identifier	Command name	Reduced function device (RFD)	
		TX	RX
0x01	Association Request	X	
0x02	Association Response		
0x03	Disassociation Notification	X	X
0x04	Data Request	X	X
0x05	PAN ID Conflict Notification	X	
0x06	Orphan Notification	X	
0x07	Beacon Request		
0x08	Coordinator Realignment		X
0x09	GTS Request		
0x0a - 0xff	Reserved		

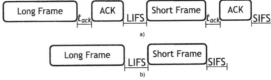

Figure 3.14   Interframe Spacing (IFS) in a) Acknowledged Transmission and b) Unacknowledged Transmission.

duration should be at least *aMinSIFSPeriod*. In the cases that the frame length exceeds *aMaxSIFSFrameSize*, the acknowledgement must be followed by a long IFS (LIFS) period as shown in Figure 3.14.

The minimum LIFS and SIFS periods for the three different PHY layers specified by the IEEE 802.15.4 standard are presented in Table 3.5.

3.2.2 Carrier Sense Multiple Access with Collision Avoidance

The IEEE 802.15.4 standard allows multiple devices to share the same frequency channel for data transmission whilst enabling reliable data delivery. The considered medium access mechanism, is the Carrier Sense Multiple Access with Collision Avoidance (CSMA-CA). The CSMA-CA, requires

Table 3.5   Minimum LIFS and SIFS period.

PHY	*macMinLIFSPeriod*	*macMinSIFSPeriod*	Units
868-868.5 MHz BPSK	40	12	Symbols
902-928 MHz BPSK	40	12	Symbols
2400-2483.5 MHz O-QPSK	40	12	Symbols

PHY	*macMinLIFSPeriod*	*macMinSIFSPeriod*	Units
868-868.6 MHz BPSK	40	12	Symbols
902-928 MHz BPSK	40	12	Symbols
2400-2483.5 MHz O-QPSK	40	12	Symbols

Figure 3.15 Minimum LIFS and SIFS period.

listening to the channel before transmitting in order to reduce the collision probability, being used before the transmission of data or MAC command frames within the Contention access period (CAP), unless the frame can be quickly transmitted due to an ACK of a data request command. The CSMA-CA algorithm shall not be used for the transmission of beacon frames in a beacon-enabled PAN, acknowledgement frames or data frames transmitted in the Contention-Free Period (CFP) [WPA07a]. Two versions of the CSMA-CA mechanism were created: (i) slotted CSMA-CA algorithm used in the beacon-enabled mode and (ii) non-slotted CSMA-CA algorithm used in the non-beacon-enabled mode. Both approaches use a basic time unit, called Backoff period (BP), which is equal to *aUnitBackoffPeriod* = 20 symbols (i.e., 0.32 ms).

When using slotted CSMA-CA, each operation (channel access, backoff counter and CCA) can only occur at the boundary of a BP. Additionally, the BP boundaries must be aligned with the slot boundaries of the superframe time [WPA07a]. In non-slotted CSMA-CA the backoff periods of one node are completely independent of the backoff periods of any other node in a PAN. The CSMA-CA algorithm, represented by the flowchart presented in Figure 3.16, is invoked when a packet is ready to be transmitted. This algorithm, maintains three variables for each packet:

- **Number of Backoffs (NB):** number of times the CSMA-CA algorithm was required to experience backoff due to unavailability. It is initialised to zero before each new transmission attempt.
- **Backoff Exponent (BE):** enables the computation of the backoff delay, and represents the number of backoff periods that need to be clear of channel activity before a transmission can occur. The backoff delay is a random variable between $[0, 2^{BE} - 1]$.
- **Contention Window (*CW*):** the number of backoff periods during which the channel must be sensed idle before accessing the channel. This

type of variable is only used with slotted CSMA-CA. This value shall be initialized with the value $CW = 2$ before each transmission attempt and reset to two each time the channel is assessed to be busy (corresponding to 2 backoff periods). Each backoff period channel sensing, is performed during the 8 first symbols of the BP.

The unslotted CSMA-CA can be summarised in five steps, as follows:

- In the first step the CW variable is not used, since the non-slotted has no need to iterate the CCA procedure after detecting an idle channel. Hence, in Step 3, if the channel is found to be idle, the MAC protocol immediately starts the transmission of the current frame. Second, the non-slotted CSMA-CA does not support the battery life extension mode and BE is always initialised with the *macMinBE* value [WPA07a].
- Steps 2, 3, and 4 are very similar to the slotted CSMA-CA algorithm. The only difference is that the CCA starts immediately after the expiration of the random backoff delay generated in Step 2.
- In step 5 the MAC sub-layer starts immediately transmitting its current frame after the channel is found to be idle during the CCA procedure.

For slotted CSMA-CA used in the beacon-enabled mode, nodes must perform two backoff periods (i.e., $CW = 2$) before trying transmitting the packet. This value shall be initialized to two before each transmission attempt and must be reset to two each time the channel is found to be busy. Then, if the channel is found to be idle, CW is decremented by 1 and compared with 0 [step (5)]. If CW is equal to 0 the CSMA-CA algorithm will finish by entering in the "Success" state, otherwise the algorithm will return to step (2) and the sensing phase (i.e., CCA) is repeated. Moreover in the beacon-enable mode the backoff period boundaries of a WPAN must be aligned with the superframe slot boundaries of the coordinator. Therefore, the beginning of the first backoff period of each node is aligned with the beginning of the beacon transmission, and the transmissions will start on the boundary of a backoff period [CBF11].

3.2.3 Non-beacon-enabled operation

A Non-beacon-enabled network allows every node participating in the network to transmit at any time the channel is found to be idle. Therefore, for such kind of networks, the Non-beacon-enabled mode seems to be more adapted to the scalability requirement than the beacon-enabled mode, in the former case all nodes are independent from the PAN coordinator and

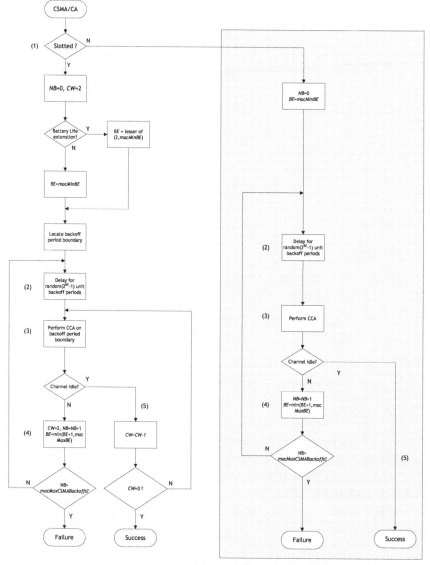

Figure 3.16 CSMA-CA algorithm [WPA07a].

the communication is completely decentralised. In addition, the unslotted CSMA-CA in the Non-beacon-enabled mode will enable a better flexibility for large-scale IEEE 802.15.4-compliant peer-to-peer networks [Mah07]. Before, each transmission the MAC sub-layer exchange messages with the

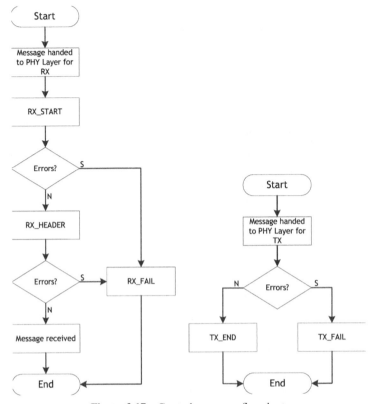

Figure 3.17   Control messages flowchart.

PHY layer for packet transmission/reception. Figure 3.17 presents the algorithms flowchart showing the interaction between the different packet types (e.g., DATA and ACK) and the control messages, being invoked when a packet is transmitted/received.

The PHY layer and MAC sub-layer exchange control messages every time an event occurs, as follows:

PHY − > MAC

- **RX_START:** Start of message indicator;
- **RX_FAIL:** Failed to receive message after RX_START. The message can fail because Cyclic redundancy check (CRC) or collision;
- **TX_END:** Message being transmitted has completed;
- **TX_FAIL:** End of transmission (i.e., like TX_END) but the message has failed to be transmitted. For most radio transceivers, this should never happen (but there are valid cases for packet-based radios, e.g., CC2420).

After starting carrier sense, one of the following messages must be sent to the MAC layer as follows:

- **CHANNEL_IDLE:** If the specified "packet length" has been processed, and the carrier sense returns channel not "busy";
- **CHANNEL_BUSY:** If the carrier sense returns channel "busy".

MAC − > PHY

- **SET_TRANSMIT:** Switch the PHY layer to the transmit mode;
- **SET_LISTEN:** Switch the PHY layer to the listen mode;
- **SET_SLEEP:** Switch the PHY layer to the sleep mode;
- **START_CARRIER_SENSE:** Start carrier sense.

3.2.4 Beacon-enabled operation

In the beacon-enabled mode the PAN coordinator periodically sends a beacon control frame to identify the PAN inside a given cluster and to synchronise all nodes that are associated to. The time period between two successive beacon frames is known as the superframe, also known as the Beacon interval (BI) and it includes an active period and an inactive period as shown in Figure 3.18.a).

Within a give cluster all the communications occurs during the active portion of the superframe. Nodes can send/received data to the coordinator, being these two communication flows referred as uplink and downlink, respectively.

The active part , which represents the Superframe duration (SD) is divided in 16 contiguous time slots. If the inactive period exists, nodes can enter in the sleep mode and save energy while they are inactive. The SD and the BI are calculated by using two parameters: the Superframe order (SO) and the Beacon order (BO). The Beacon Interval is given by:

$$BI = aBaseSuperframeDuration \times 2^{BO}, \ 0 \leq BO \leq 14 \quad (3.4)$$

The Superframe Duration, which corresponds to the active period is given by:

$$SD = aBaseSuperframeDuration \times 2^{SO}, \ 0 \leq SO \leq BO \leq 14 \ (3.5)$$

The *aBaseSuperframeDuration* can be defined as follows:

$$aBaseSuperframeDuration = aBaseSlotDuration \times 16 = 960 \ symbols$$
$$(3.6)$$

Where *aBaseSlotDuration* is the number of symbols forming a superframe slot when the superframe order is equal to 0 and has a value of 60 symbols.

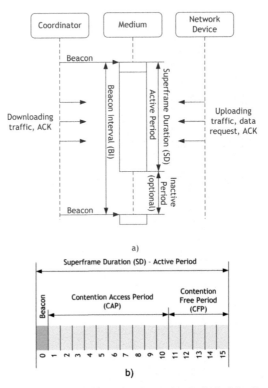

Figure 3.18   a) Superframe Structure b) Active Period of the Superframe

As shown before the *aBaseSuperframeDuration* is equal to 960 radio symbols (where a symbol has a time duration of 16μs), corresponding to 15.36 ms (assuming a bit rate of 250 kb/s in the 2.4GHz frequency band) and each time slot as a duration of 15.36/16=0.96 ms.

In Figure 3.18.b) the active part of a superframe is divided into 16 contiguous time slot s that are divided in to three parts: the beacon, the CAP and optionally the Contention free period (CFP). The beacons are used to synchronise the attached devices, to identify the PAN and to describe the structure of the superframes. The CAP is the period of time immediately following a beacon frame during which devices wish to transmit will compete for channel access using a slotted CSMA-CA mechanism, and the CFP is an optional feature in the IEEE 802.15.4 MAC and it follows immediately after the CAP, extending it to the end of the active portion of the superframe as shown in Figure 3.18.b). If we want to allocate any GTS they must be located within the CFP and the primary objective is the use of these time slots in applications that require bandwidth for delay critical applications.

3.2.5 Hidden and Exposed terminal problems

One of the main contributions to attenuate the received power (P_r) is the path loss. The Friis free-space equation relates the power received (P_r), at a distance, d, with the transmitted power (P_t), as follows:

$$P_r = P_t \cdot G_t \cdot G_r \cdot \left(\frac{\lambda}{4\pi d} \right)^2 \tag{3.7}$$

where, G_t and G_r are the antenna gains of the transmitter and receiver, respectively. Based in equation (3.7), we can observe that, the received power depends on the frequency (the higher the frequency is the lower the received power is), and decreases with the square of the distance (i.e., path-loss exponent equal to 2).

This path loss combined with the fact that any transceiver needs a minimum signal strength to demodulate signals successfully, leads to a maximum range that a sensor node can reach with a given transmit power [KW05a]. If the nodes are out of the influential range, there is no way they can listen to each other. As a consequence, the hidden and exposed -terminal problems may occur, Figure 3.19.

The hidden terminal problem is one of the problems associated with the IEEE 802.15.4 CSMA-CA algorithm [WPA07a]. Consider the example presented in Figure 3.19. In this case, both nodes D and A are located too far from each other. Therefore, the power detection mechanism does not detect the presence of another signal and assumes that the wireless medium is

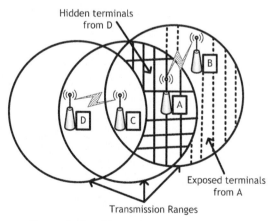

Figure 3.19 Hidden and exposed terminals.

available. However, both nodes are able to communicate with node C. If both node D and A simultaneously transmit packets to node C at the same time, by performing CCA they will both find the wireless medium available and start transmitting the packets concurrently. However, this will create a collision of packets in node C. To overcome the hidden-terminal problem, authors from [Far11] have proposed two solutions. The first solution considers changing the location of the nodes or increasing the output power (of the hidden nodes) to ensure that both nodes D and A are able to hear each other. The second solution considers the use of the RTS/CTS handshake mechanism, which is not supported by IEEE 802.15.4.

The exposed terminal problem occurs when node C transmits a packet to D, and some moment later A wants to transmit a packet to B. Although both nodes D and B would received their packets with no erros, the CSMA-CA performed by node A will prevent node A's transmission, since is in the radio influence range of node C. This suppression could lead to an needless waiting. Therefore, using the solutions presented for the hidden-terminal problem, are also suggested for the exposed-terminal problem.

3.2.6 Coexistence in the 2.4 GHz ISM band

With the appearance of more and more wireless devices operating in the 2.4 GHz ISM band portion of the radio spectrum, there was an increase of the signals from other interfering sources. Moreover, in most of the cases, there is no collaboration between these independent wireless networks, and the operation of one network may adversely affect the others. Therefore, there is a need for allowing a better coexistence between the different wireless devices operating in the same frequency band, whilst ensuring QoS. In order to cope with the design requirements and allow a better coexistence between the existing 2.4 GHz systems like Cordless Phones, Bluetooth, Wi-Fi (i.e., IEEE 802.11b/g/n) and IEEE 802.15.4, MAC management efficient techniques must be addressed allowing for mitigate the interference from other sources under hostile conditions. Figure 3.20 presents the utilized channel bandwidth as well as the expect output power for each wireless system.

Coexistence with Wi-Fi

The IEEE 802.15.4 standard was developed to address low data rate wireless communications. However, due to its low power, IEEE 802.15.4 is potentially vulnerable to interference by other wireless technologies having higher

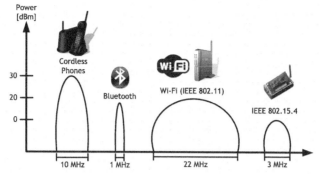

Figure 3.20 Wireless systems operating in the 2.4 GHz ISM band.

transmission powers and working in the same 2.4 GHz ISM band, such as IEEE 802.11b/g/n. The IEEE 802.11b/g/n wireless nodes have a typical output power ranging from 12 to 18 dBm [WPA07a], but can achieve values form the order of magnitude of 30 dBm. This is significantly higher than the typical 0 dBm output power of IEEE 802.15.4 nodes.

There are 14 channels defined for Wi-Fi in the 2.4 GHz ISM band. As shown in as shown in Figure 3.21, not all of the channels are allowed in all countries: 11 are allowed by the FCC in the North American domain, 13 are allowed by the European telecommunications standards institute (ETSI) in Europe and in Japan there is an extra channel (i.e., channel 14) for the IEEE 802.11b standard. The center frequency (f_c) channels are spaced 5 MHz apart (with the exception of a 12 MHz spacing between the last two channels), where each channel as a bandwidth of 22 MHz (after spreading using DSSS) as follows:

$$f_{c[MHz]} = \begin{cases} 2412 + 5 \times (k-1), \ 1 \leqslant k \leqslant 11 \ in \ North \ America \\ 2412 + 5 \times (k-1), \ 1 \leqslant k \leqslant 13 \ in \ Europe \\ 2412 + 5 \times (k-1), \ 1 \leqslant k \leqslant 14 \ in \ Japan \end{cases} \quad (3.8)$$

Table 3.6 presents the fourteen IEEE 802.11 frequency channels available around the globe (fetails are given in Figure 3.21).

The IEEE 802.11 channels 1, 6 and 11 or 2, 7 and 12, or 3, 8 and 13 or 4, 9 and 14 (if allowed) can be used together as set of 3 non-overlapping channels. Since, by default Wi-Fi routers set channel 6 as the default channel, for this particular case the set of 3 non-overlapping channels are 1,6 and 11. As presented in Figure 3.1, the IEEE 802.15.4 standard defines 16 frequency channels working in the 2.4 GHz ISM band. Therefore, in order to avoid

Table 3.6 Wi-Fi channel frequencies for the 2.4 GHz ISM Band.

Channel number	Lower frequency [MHz]	Center frequency [MHz]	Upper frequency [MHz]
1	2401	2412	2423
2	2406	2417	2428
3	2411	2422	2433
4	2416	2427	2438
5	2421	2432	2443
6	2426	2437	2448
7	2431	2442	2453
8	2436	2447	2458
9	2441	2452	2463
10	2446	2457	2468
11	2451	2462	2473
12	2456	2467	2478
13	2461	2472	2483
14	2473	2484	2495

interference there is a need of choosing the non-overlapping ones in the proximity of an IEEE 802.11 network. This coexistence performance, can be accomplished by using a channel alignment scheme. Figure 3.21 presents the IEEE 802.15.4 and IEEE 802.11b/g/n channel overlapping in detail for the 2.4 GHz ISM band.

The IEEE 802.15.4 frequency channels 15, 20, 25, and 26 are the ones that will suffer the least interference from IEEE 802.11b/g in North America compared to other IEEE 802.15.4 frequency channels. In Europe the IEEE 802.15.4 channels that will suffer the least interference from IEEE 802.11 b/g are the channels 15, 16, 21 and 22.

By considering a scenario where both IEEE 802.11b/g and IEEE 802.11n wireless nodes exists simultaneously, as shown in Figure 3.21. In this case, there is the possibility that IEEE 802.11n nodes consider the use of a channel bandwidth of 40 MHz in order to achieve higher data throughput. However, for this particular situation there will be a reduction of the total number of channels available. Thus, we can conclude that is 2 channels (e.g., 3 an 11) of 40 MHz are used by IEEE 802.11n nodes, the 2.4 GHz ISM band will be heavily congested, being almost impossible the coexistence between the different wireless systems. Therefore, a good approach is to use only a single 40 MHz channel (if needed) when using 2.4 GHz ISM band by considering IEEE 802.11n.

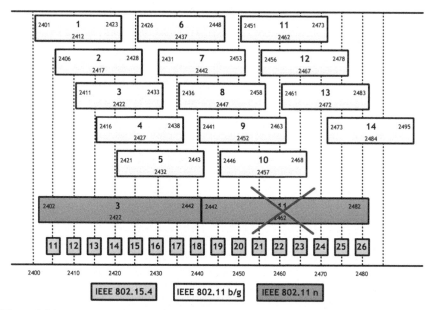

Figure 3.21 IEEE 802.15.4 and IEEE 802.11b/g/n channel overlapping in the 2.4 GHz ISM band.

Coexistence with Bluetooth

The Bluetooth (or IEEE 802.15.1) is a standard for short-range, low-cost and low-rate data communications. It operates in the 2.4 GHz ISM band, and uses the Frequency hopping spread spectrum (FHSS) mechanism for spreading the signals. The Bluetooth devices use a channel bandwidth of 1 MHz, being the frequency channel selected by using a pseudo-random sequence. The maximum number of hops in Bluetooth is 1600 hops per second in the connection state [Far11]. Therefore, the retransmission mechanism employed by IEEE 802.15.4 ensures the delivery of packets corrupted by the Bluetooth interference. Normally, if a Bluetooth device interferes with IEEE 802.15.4 during the first transmission attempt, and a collision occur, by using frequency hoping mechanism it goes to a different part of the radio spectrum for retransmission. There are 79 channels defined for Bluetooth as follows:

$$f_{c[MHz]} == 2402 + (k), \ 0 \leqslant k \leqslant 78 \tag{3.9}$$

where k in an integer that represents the channel number.

As presented in equation (3.9) the Bluetooth channels are numbered from 0 to 78. Each channel has a bandwidth of 1 MHz and the f_c are spaced 1 MHz apart.

Bluetooth devices generate an output power that is generally less than 4 dBm for the most commonly class 2 devices (e.g., wireless headsets and keyboards), and the transmission range is about 10 m. The class 1 devices can achieve and output power up to 20 dBm and typically have a transmission range of 100 m. Although not mandatory for class 2 devices, almost all devices have embedded a power control unit in order to achieve energy efficiency. Therefore, the output power is often less than 4 dBm and can be as low as -30 dBm if the devices are close in proximity. The Bluetooth can achieve data rates of 1 Mb/s and 3 Mbps for versions 1.2 and 2.0 respectively. To improve performance and mitigate the impact of interference caused by IEEE 802.11b/g/n networks an AFH mechanism has been introduced by the Bluetooth Special interest group (SIG).

The Bluetooth products before the creation of the AFH scheme employed a RFH scheme. Therefore, the first generation os Bluetooth devices was designed for using the 79 channels given by equation (3.9). Therefore, when another wireless device tried to transmit this type of hopping results in occasional collisions. Figure 3.22 presents an RFH example by considering Bluethooth, IEEE 802.15.4 and IEEE 802.11b/g/n devices working in the 2.4 GHz ISM band.

As presented before, although RFH helps avoiding collisions, some collisions still take place due to lack of adaptation to its environment. Therefore, the AFH was created in order to better adapt to the channel conditions by identifying the fixed sources of interference and excluding them from the list of available channels. This re-mapping process also involves reducing the

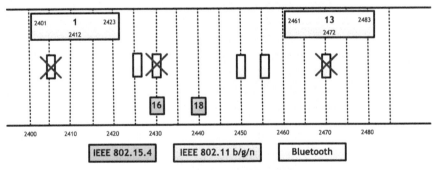

Figure 3.22   Bluetooth RFH collisions.

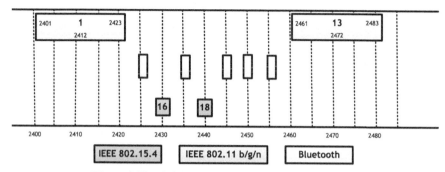

Figure 3.23 Collisions avoided by the AFH mechanism.

number of channels to be used by the Bluetooth devices. Figure 3.23 presents the same channel condition by considering the AFH mechanism.

Since the IEEE 802.15.4 standard employs an DATA/ACK handshake ACK mechanism for data retransmissions in case there are packet collisions with Bluetooth devices. This AFH is particular useful in dealing with frequency hopping interference problems. Since, the Bluetooth device by means of AFH, will hopped to another parth of the spectrum for retransmissions, which also enables IEEE 802.15.4 channel access with no errors. Moreover, the AFH mechanism not only improves the performance of the Bluetooth devices,but it also reduces the effect of the Bluetooth network on other nearby networks that are not Bluetooth compliant [Far11].

Other Sources of Interference

Since the cordless phones operating in the 2.4 GHz ISM band share channels frequencies with IEEE 802.11b/g/n networks, the interference from the 2.4 GHz ISM band cordless phones can completely stop a Wi-Fi network. Therefore, in order to improve the network performance mechanisms such as FHSS and DSSS are employed. For example if a cordless phone uses a DSSS, the channels used by the cordless phone and Wi-Fi network can be configured to avoid interference due to channel overleaping. The phones that use DSSS typically have a "channel" selection button allowing for manually change the frequency channel if need, whilst the phones with FHSS do not have such kind of button, since they are constantly changing from channel to channel. Another way to avoid the interference is to use cordless phones that operate in the 5 GHz band.

4

Scheduled Channel Polling MAC Protocol

The SCP-MAC protocol is noteworthy to be analyzed and in detail, to provide contributions to the design and proposal of a new MAC protocol. This chapter is divided into six sections. Section 4.1 gives the motivation to use the SCP-MAC protocol as a comparison protocol with other ones. Sections 4.2 and 4.3 describe in detail the SCP procedures, namely the two-phase contention window mechanism and the synchronization phase. In the fourth section is proposed a state transition diagram for the SCP protocol (Section 4.4). Section 4.5 describes how the implementation of the SCP in the simulation framework is carried out, in which the SCP simulator parameters and general definitions along with the description of the different SCP simulator layer modes available are presented and discussed. In Section 4.6 the chapter ends with a brief summary and considerations about this simulation implementation.

4.1 Context and Motivation

WSNs answer the need for nomadicity and mobility in the context of *in situ* event monitoring. The main purpose of this class of networks is to provide quick and easy access to the information gathered from a set of sensor nodes scattered in a geographical region.

In WSNs, energy efficiency is of primary importance since it is directly related to the lifetime of the network formed by mostly battery-powered nodes. Major sources of energy waste are idle listening, frame retransmissions (e.g., due to frame collision), unnecessary high transmission powers, overhearing and control overhead [YHE04a, BDWL10]. Frame collision occurs when frames are transmitted by two nodes and collide at a common receiving node. This may occur if transmitting nodes are out of range of one another but within range of a common destination node (known as the hidden terminal problem). In this case, if both transmitters transmit at time intervals close enough for their frames to overlap in time, a collision occurs at the receiving node. Both frames are corrupted and the physical layer is unable to decode them. Since they have not been received correctly, the frames need to be retransmitted, which increases the energy consumption.

In contrast to previous surveys that have focused on classifying MAC protocols according to the employed medium access scheme, authors from [BDWL10] provide a thematic taxonomy in which protocols are classified according to the problems they address. Therefore, the MAC protocols can be classified as scheduled protocols, protocols with common active period, preamble sampling protocols and hybrid protocols. Hybrid protocols include protocols that employ contention-free or contention-based medium access.

Some MAC protocols that rely on contention-based medium access are based on slotted CW. Sensor-MAC (S-MAC) [YHE04a] is one of the first proposed slotted CW protocols. It is based on network-wide periodic sleep and listening periods, which requires nodes to be synchronized. Synchronization is achieved by periodically exchanging synchronization frames between neighbour nodes.

SIFT [JBT06] was designed for low latency. Like S-MAC, SIFT adopts a fixed-size contention window. Instead of randomly picking a slot, nodes use a skewed distribution which gives preference to slots at the end of the contention window. Other MAC protocols, such as CrankShaft [HL07], have adopted a contention method with non-uniform distribution.

Timeout-MAC (T-MAC) [HvDL05] reduces idle listening by continuously adapting the length of the listening time of the frames. In T-MAC, nodes perform a procedure similar to the one from S-MAC. However, instead of adopting a fixed listen and sleep period durations, nodes adapt these depending on the frame length. Since all nodes wake-up simultaneously, if a node receives a frame during the listening time it waits for a certain timeout. If the node does not receive anything, it assumes that there are no more frames to be received, and goes to the sleep mode, reducing the inactive time compared to S-MAC.

In the aforementioned protocols, there only is one contention window (with possibly a variable duration). However, work has been performed for the IEEE 802.11 DCF protocol addressing the use of the two-phase collision avoidance mechanism with some modifications. The authors from [HNS+06b] propose a two-phase collision avoidance scheme to reduce the collision probability and enhance the throughput performance. Contention among stations is resolved in two phases: SuperSlots and SubSlots. A truncated backoff mechanism is used to increase the throughput by reducing the idle time slots.

The work from [YV06a] applies "pipelining" techniques to the design of multiple access control protocols so that channel idle overhead could be (partially) hidden, and the collision overhead reduced. In particular, an implicitly pipelined Dual-Stage Contention Resolution (DSCR) MAC protocol is proposed. The authors also propose partial pipelining where the two channels are used. The contention resolution procedure is split into two phases, where pipelined stage 1 includes only contention resolution phase 1 and is performed on the busy tone channel. Contention resolution phase 2 and frame transmissions are performed on the data channel in pipelined stage 2. With DSCR, only one channel is needed, while the scheme with busy tone is left aside. The fundamental difference between DSCR and conventional two-stage contention resolution algorithms is that, in DSCR, stage 1 proceeds in parallel with stage 2 which includes the contention resolution phase 2 and frame transmissions. In the case where the channel activities are observed, channel resources are only utilized in stage 2. In addition, contention windows have dynamic sizes, which depend on the success or failure of data frame transmission.

The Scheduled Channel Polling (SCP) [YSH06a] protocol goes one step further by combining scheduling with channel polling to minimize energy consumption. Compared to [HNS+06b, YV06a], the SCP's two-phase collision avoidance mechanism employs a wake-up tone on a single channel in between the two contention windows and does not distinguishes the slots into SuperSlots and SubSlots. The wake-up tone works as a busy tone to inform the other listening nodes of an ongoing transmission. Potential senders that overhear this preamble will postpone their own transmissions.

SCP starts by dividing time into time frames. It achieves synchronization by piggybacking the schedules in broadcast data frames. Once neighbour nodes are synchronized, if a node has data in its Transmission (Tx) queue, it applies the two-phase contention window mechanism depicted in Figure 4.1. The data frame is transmitted in the next time instant the receiver node wakes-up, $T_{wake-up}$. By default, nodes wake up and poll the channel for activity. A sending node reduces the duration of the wake-up tone by starting it just before the receiver starts listening. SCP avoids frame collision by using a two-phase contention window scheme. By orchestrating senders to contend for channel access prior to a poll, SCP-MAC can operate very efficiently for low traffic loads. However, the main limitations of SCP-AC is that it does not provides no implicit mechanism to ensure fairness in the channel utilization, as well as mechanism to cope with overhearing problems.

4.2 Two-Phase Scheduled Channel Polling Mechanism

SCP [YSH06a] uses two contention windows, CW_1 and CW_2. CW_1 is divided into CW_1^{max} time slots of equal duration, as shown in Figure 4.1. When a node wants to transmit a data frame, it randomly chooses a time slot $\varphi_{slot}^{CW_1}$ in CW_1, with a uniform distribution, whilst sensing the shared medium for the duration of that time slot. The choice of the slot (from the ones available in the interval) follows a uniform distribution:

$$\varphi_{slot}^{CW_1} \in [1; CW_1^{max}] \tag{4.1}$$

All nodes contend to access the channel to transmit frames. If no channel activity is detected (idle channel), a node sends a wake-up tone. The wake-up tone is sent after the time slot chosen in CW_1 and has a minimum duration of 62 Bytes (2ms @ 250 kbps). This length is extended for the duration of the remaining time slots (T_{subs_slots}) in CW_1, as presented in Figure 4.2. This wake-up tone announces to other potential senders that a node is preparing to send data. The node enters CW_2 immediately after the wake-up tone transmission. The second contention window is divided into CW_2^{max} time slots of equal duration. Only the nodes which have data to be sent utilize the two-phase contention window mechanism, while the remaining ones poll the channel.

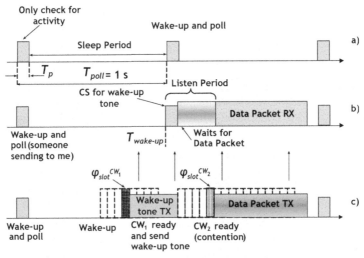

Figure 4.1   SCP time frame period structure.

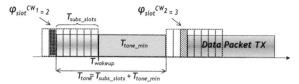

Figure 4.2 Wake-up tone duration as a function of the CW_1 chosen slot.

In both contention windows, a collision occurs if more than one node chooses the same slot (i.e., the wake-up tones collide, or simultaneous transmissions start in the same slot). Only nodes that succeed in CW_1 enter CW_2. When more than one node succeeds in CW_1 (at least two nodes transmit the wake-up tones in the same lowest order slot), "effective" data frame collision occurs in CW_2 if the nodes that have been successful in CW_1 randomly choose the same time slot in CW_2. This simultaneous choice of a time slot corresponds to an effective data frame collision event.

If a node succeeds in sending the wake-up tone during CW_1, it randomly chooses a slot $\varphi_{slot}^{CW_2}$ from CW_2 by using a uniform distribution:

$$\varphi_{slot}^{CW_2} \in [1; CW_2^{max}] \tag{4.2}$$

The node then starts a carrier sense procedure in CW_2 (with the same duration of CW_1), as shown in Figure 4.1. If the channel is sensed idle, it starts transmitting its frame.

After sending/receiving the data frame, the sender/receiver node checks whether there are any new data frames pending to be sent. If so, the node repeats the steps described above before going into sleep mode.

4.3 Synchronization Phase

The original SCP-MAC protocol distributes schedules as it was developed in S-MAC, where each node broadcasts its schedule in a SYNC frame to its neighbours every synchronization period. In our simulation framework, we defined a maximum time interval equal to 50 ms for the synchronization period. This time interval is used by the node to send a SYNC frame to its neighbours or to receive the SYNC frame from a neighbour node. When the SCP-MAC schedule algorithm is enabled and the sensor network simulation environment starts all nodes choose randomly a value between $t_{wake-up}$ and $t_{init_{CS}} = t_{wake-up} + t_{sync} - t_{CS}$:

$$t_{init\_CS} \in [t_{wake-up}; t_{wake-up} + t_{sync} - t_{CS}] \tag{4.3}$$

where t_{CS}= 2 ms and t_{sync}= 50 ms. The value chosen by the sensor node, $t_{init_{CS}}$ corresponds to the simulation time when the sensor node performs the CS of the radio channel for collision avoidance. After performing the CS of the channel (it takes 2 ms) and if the medium is sensed to be free, the sender gets the medium and can start the SYNC frame transmission. This SYNC frame contains fields of the address of the sender and the time of its next wake-up. Initially when the nodes choose randomly a value for $t_{init_{CS}}$ the one that chooses the smallest value is the one that initiates and finishes first the CS gaining the opportunity to transmit the SYNC frame to its neighbours. Immediately after the CS the sensor node transmits the SYNC frame (18 Bytes). Since no nodes have any information about their neighbour's schedules, each node listens to the channel for a certain amount of time. The sensor node wakes-up at time $t_{wake-up}$ and the CC1100 transceiver initiates in RX mode, but it only starts performing the CS at time $t_{init_{CS}}$, as presented in Figure 4.3. The scheme proposed in our simulation framework divides the listen interval into two parts: i) interval for receiving SYNC frames, and ii) interval for CS and send SYNC frame, if an idle channel is detected.

The first part of the interval is time variable because it depends on the value of $t_{init_{CS}}$ chosen randomly by the sensor node and is given by Equation 4.4:

$$t_{RX\_SYNC\_pkt} = t_{init\_CS} - t_{wake-up}|_{t_{init\_CS\_max}=t_{wake-up}+t_{sync}-t_{CS}} \quad (4.4)$$

If the sensor node detects a signal during the first interval it receives the frame and verifies what type of frame was received by the sensor node. If the type of frame received was a SYNC frame the node updates the schedule table, while storing the time of the next wake-up from the synchronizer node and assuming the schedule from the synchronizer node as his schedule until the next synchronization period. This schedule table is a timetable

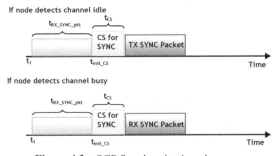

Figure 4.3   SCP Synchronization phase.

for recording neighbours' working schedules. When the timeout timer for synchronization period triggers the nodes will choose again a random value for $t_{init_{CS}}$ and the one that chooses the smallest value is the one that initiates and finishes the CS first, gaining the opportunity to transmit the SYNC frame to its neighbours.

The original SCP protocol for schedule synchronization offers to all sensor nodes the opportunity to be the synchronizer node. The schedule synchronization mechanism implemented in our simulation framework for the distribution of schedules follows all the same procedures described above. However, when the node chooses a random value for $t_{init_{CS}}$ and if it gains the opportunity to transmit the SYNC frame to its neighbours, this node will be always the synchronizer node while the others are the follower's nodes, until the end of the simulation. In our simulator this mode is called as the *sync slave* mode. The follower node sets its synchronization timer and its wake-up schedule according to the information sent by the synchronizer. Then, the node will rebroadcast the schedule to its neighbours. Nodes that receive the rebroadcasted frame with synchronization information while having the same schedule value as the received SYNC frame, discard these frames and will not rebroadcast the SYNC frame (in order to prevent from message flooding).

Some nodes may miss the synchronization with other nodes at beginning of the synchronization period due to signal collisions that could occur on the radio channel. These nodes can be synchronized by using SYNC frames rebroadcasting scheme. However, in our simulation, since we are evaluating a small scale network this feature is disabled. This feature is designated for networks that present a large density of sensor nodes (per unit of area) since the probability of two or more nodes to choose the same $t_{init_{CS}}$ value is higher. By periodically updating each of their schedule tables a long-time clock drift will be prevented.

This process of sending SYNC frames is repeated after a synchronization period, T_{sync}, but can be suppressed by piggybacking the schedules in the data frames when the frames exchanged are of the broadcast type. After the wake-up schedules of all the neighbour nodes from synchronizer node are synchronized, the nodes will wake-up after a certain channel polling period, T_p, and poll the channel for activity sensing. However, maintaining schedule synchronization leads to a penalty in the overall consumption of the node, as the node must resynchronize after some time, in order to have very little accuracy errors between the schedules clock of all the nodes.

This verification for activity lasts 2 ms for the CC1100 or CC2420 transceivers. If the node detects an idle channel and there is no data to be sent, the node will schedule its next wake-up time instant and goes immediately to the sleep mode after verifying the TX MAC queue. If the node has data in the MAC queue it schedules all the process, as described in Figure 4.1, enabling the data frame transmission in the next time instant the node wakes-up, $t_{wake-up}$, where scheduling the next wake-up (before going to the sleep mode again).

Other type of frame that could be added to overcome the control overhead problem is the SYNC one. Nevertheless, all these explicit SYNC frames can be suppressed by employing the piggybacking technique, when the network is exchanging broadcast traffic. This technique is used to minimize the cost of synchronization by avoiding the explicit SYNC frames. SCP-MAC piggybacks the synchronization information in broadcast frames without increasing the size of the frame length. The SCP-MAC header includes the following three fields: frame type, source address and destination address. In this case, the address field is used to piggyback the schedule information. This technique is more efficient.

4.4 State Transition Diagram for SCP

Before we started to implement the SCP [YSH06a] in the Mobility Framework we had proposed a possible Finite State Diagram (FSM). The objective of this FSM is to give a detailed specification of why and how the simulator changes between the different states, while explaining the associated actions and events as shown in Figure 4.4. The corresponding transition events and actions are listed in Table 4.1. This state machine is based on seventeen states, described as follows:

- **START:** The node is switched on;
- **WAIT_SYNC:** The node is waiting for SYNC frames;
- **SYNC:** If the node does not receive a SYNC frame it performs a CS; if detects an idle channel it sends a SYNC frame for the neighbouring nodes;
- **SLEEP:** The node "turns off" the radio and puts CPU in low power;
- **IDLE:** The node is waiting for a task to perform;
- **READING_FRAME:** After receiving a frame, the node has to check if it was successfully received;
- **NAV_SLEEP:** The node goes to the sleep mode until the end of the remaining transmissions (after receiving a wake-up tone not for itself);
- **READING_WAKEUP_TONE:** After receiving the wake-up tone, the node has to check the destination address contained in the wake-up tone frame;

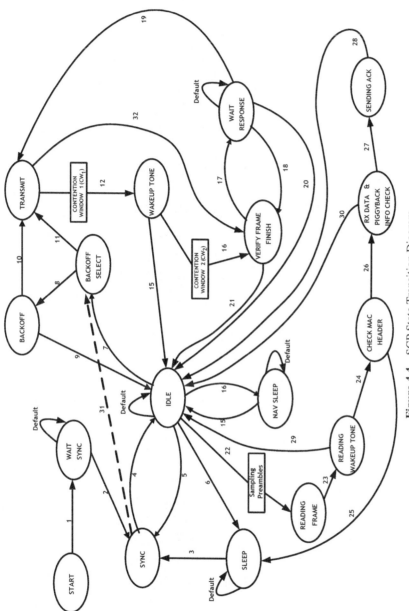

Figure 4.4 SCP State Transition Diagram.

- **CHECK_MAC_HEADER:** To avoid overhearing, the node performs a MAC header checking when it starts receiving the frame;
- **BACKOFF:** The node activates the "backoff time";
- **BACKOFF_SELECT:** The node selects a "backoff time" depending on the contention window or if it has data or not in MAC queue to send;
- **TRANSMIT:** The node transmits the frame;
- **WAKEUP_TONE:** The node transmits the wake-up tone;

Table 4.1 SCP-MAC events and actions.

Events	Actions
1	**State**: WAIT FOR SYNC **Schedule**: WAIT_RX_SYNC_TRANSMISSION; SYNC_TIMEOUT_START; **Save**: Node_ID;

2	**State**: SYNC
	Schedule: TIMER_TRANSITION_TO_IDLE;
	   FINISH_SYNC; SYNC_BACKOFF;
	   Checks if the sync broadcast frame was received with success.

3	**State**: SYNC **Schedule**: SLEEP_TIMEOUT
4	**State**: IDLE **Schedule**: TIMER_TRANSITION_IDLE
5	**State**: SYNC **Schedule**: IDLE_TO_SYNC **Update**: Schedule table
6	**State**: SLEEP **Schedule**: WAKEUP_TIMEOUT
7	**State**: BACKOFF_SELECT **Schedule**:SELECT_BACKOFF **Save**:Type of frame (unicast or broadcast), CW_1 slot, CW_2 slot
8	**State**: BACKOFF **Schedule**: BACKOFF_REQUEST;

9	**State**: IDLE
	Schedule: BACKOFF_TIMEOUT;
	   TRANSMIT_FRAME
	   After backoff timeout, it returns to IDLE state

10	**State**: TRANSMIT **Schedule**: BACKOFF _COMPLETE; TRANSMIT_FRAME
11	**State**: TRANSMIT **Schedule**: REQUEST_TRANSMIT_FRAME

(Continued)

Table 4.1 *Continued*

Events	Actions
12	**State**: WAKEUP_TONE **Schedule**: CONTENTION_WINDOW_1_INITITATE **Save**: The wake-up tone length. **Update**: For each data frame sent the receiver adds three slots time. Performs CS, chooses a slot time for the CW_1.
13	**State**: IDLE **Schedule**: STOP_IDLE
14	**State**: NAV_SLEEP **Schedule**: COMMUNICATION_ONGOING; NAV_TIMEOUT
15	**State**: IDLE **Schedule**: MEDIUM_BUSY Stops the transmission of the frame and goes do IDLE and starts to receive.

16	**State**: VERIFY_FRAME_FINISH
	if (unicast_packet== true && channel_still_idle==true
	&& cw1_length_finish==true && data_piggyback_info_ready==true)
	    Checks if the unicast packets were correctly transmitted
	
	if (broadcast_packet==true && channel_still_idle==true &&
	data_piggyback_info_ready==false)
	    Complete the transmission of the broadcast packets
	Schedule: CONTENTION_WINDOW_2_INITITATE;

17	**State**: WAIT_RESPONSE if (unicast packet==true) wait for the CTS due to RTS and ACK due to DATA This event is optional. Only if RTS/CTS/DATA/ACK mechanism is enabled.
18	**State**: VERIFY_FRAME_FINISH **Schedule**: WAIT_ACK; PERFORM_BACKOFF; TRANSMIT_FRAME The frame transmitted could be SYNC/WAKEUP_TONE/CTS/RTS/ACK/DATA
19	**State**: TRANSMIT **Schedule**: TRANSMIT_FRAME **Save**:NUM_ATTEMPTS
20	**State**: IDLE **Schedule**: IDLE_TO_SYNC; UNICAST_PACKET_COMPLETE **Save**: NUM_ATTEMPTS=0
21	**State**: IDLE **Schedule**: BROADCAST_PACKET_COMPLETE; TRANSMIT_FRAME
22	**State**: READING_FRAME **Schedule**:MEDIUM_BUSY; RECEIVE_FRAME SYNC_TIMEOUT_START; **Save**: Node_ID;

(Continued)

Table 4.1   *Continued*

Events	Actions
23	**State**: READING_WAKEUP_TONE **Schedule**: RECEIVE_FRAME
24	**State**: READING_WAKEUP_TONE if (unicast frame ==true) Check if MAC header is correct **Schedule**: RTS/CTS_OFF
25	**State**: SLEEP **Schedule**:UNICAST_frame_FOR_OTHER; TIMER_TRANSITION_IDLE
26	**State**: RXDATA_&_PIGGYBACK_INFO_CHECK **Schedule**: RX_DATA_&_PIGGYBACK_EXTRACT; SEND_ACK (if enabled) **Save**: Data frame **Update**:Schedule table from the piggyback info (piggyback enabled). Schedule table from the SYNC frame received periodically.
27	**State**: SENDING_ACK if (RTS/CTS mechanism ==true) Send ACK frame **Schedule**:SEND_ACK
28	**State**: IDLE **Schedule**: WAKEUP_TIMEOUT; IDLE_TO_SYNC
29	**State**: IDLE **Schedule**: RTS/CTS_OFF; PIGGYBACK_EXTRACT **Save**: Data frame **Update**: Schedule table (piggybacking info from the data frame)
30	**State**: BACKOFF_SELECT **Schedule**: PERFORM_BACKOFF; TRANSMIT_FRAME
31	**State**: VERIFY_FRAME_FINISH **Schedule**: WAKEUP; CHECK_MAC_QUEUE This event occurs when the frame is not a data frame.

- **VERIFY_FRAME_FINISH:** After sending the frame, the node verifies if the frame was correctly transmitted;
- **WAIT_RESPONSE:** When the RTS/CTS/DATA/ACK mechanism is enabled the node waits for a confirmation of a successful transmission; if the mode is disabled it directly do idle state and goes to sleep;
- **RX_DATA_&_PIGGYBACK_INFO_CHECK:** The node receives the data frame and extracts the schedule information included in the broadcast frame (due to the use of the piggybacking technique), suppressing the sending of explicit SYNC frames;

- **SENDING_ACK:** The node sends an ACK frame to the node that sends the data frame to it, in order to confirm the successful reception; it only sends the ACK if the RTS/CTS/DATA/ACK mechanism is enabled.

4.5 Implementation of the SCP Simulation Framework

4.5.1 SCP Simulator Parameters and General Definitions

SCP was implemented in the OMNeT++ simulator [VH08, Var10], using the Mobility Framework from [REHD08]. This Mobility Framework supports the CC1100 [CC107] and CC2420 [Ins07] radio energy consumption models, as well as several propagation models. We use the free space propagation model, and implemented the two-phase contention window mechanism.

Our simulator considers single and multi-hop network topologies. However, to test the two-phase contention window collision avoidance mechanism, the considered topology was single-hop, where the number of contending nodes is increased until there are 100 % of frame collisions, and all the nodes are simultaneously sink and sources (whilst sending broadcast frames). The nodes may communicate only with reachable neighbour nodes.

Nodes are deployed randomly, and the positions are generated by a Matlab script for a deployment area of 1200×1200 m^2, as shown in Figure 4.5. We vary the number of contending nodes.

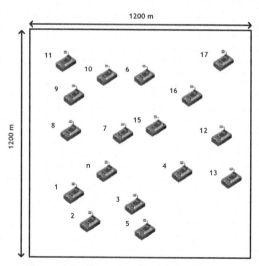

Figure 4.5 Example of node's deployment.

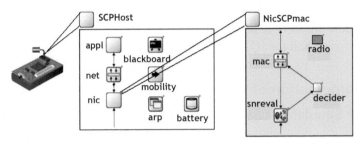

Figure 4.6   OMNeT++ SCP protocol stack and the different layers from a sensor node.

For each deployment, with a certain number of contending nodes, a set of six seeds are chosen for the random generator of SCP. All the seeds were chosen such that they guarantee the full delivery of the data frames. The maximum simulation time depends on the last frame to be received by a node.

Other parameters are also defined in the simulation configuration file, as presented in Tables 4.2 and 4.3. Figure 4.7 presents the sensor node protocol stack employed in the simulation framework. Each module is connected with the higher and lower layers. *APPL, NET* and *NIC* stand for the application, network and PHY/MAC layer modules, respectively.

Besides these modules, a battery module is included in order to simulate the network or node lifetime, while varying the capacity of the batteries. Nodes are assumed static. In our set of SCP simulations, we consider the radio transceivers available in the Mobility Framework [REHD08] and MiXiM [WSKW09], the CC1100 and CC2420, which have a maximum bit rate of 250 kbps. The CC1100 does not necessarily comply with the

Table 4.2   Parameters defined in the configuration file for the different sensor node layers.

Module	Parameter	Value
ChannelControl [REHD08]	carrierFrequency	868 MHz
	pMax	110.11 mW
	Sat	-120 dBm
SnrEvalRadioAccNoise3 [REHD08]	transmitterPower	1.0 mW
	thermalNoise	-110 dBm
	snrThresholdLevel	10 dB
	channelModel	free-space
	propagation_coef	2
DeciderRadioAccNoise3 [REHD08]	berLowerBound	10^{-8}
SCPMacLayer [REHD08]	busyRSSI	-97 dBm
	queueLength	50 frames
	poll_period	Varying

Table 4.3 Symbols used in radio analysis, and typical values for CC1100 radio working at 868 MHz and at full data rate.

Symbol	Meaning	CC1100
t_{cs1}	Average carrier sense time [ms]	7
t_{p1}	Average time to poll channel [ms]	3
T_p	Channel polling period [s]	Varying
T_{data}	Data frame period [s]	Varying
λ_{data}	Data frame rate [$1/T_{data}$]	Varying
L_{data}	Data frame length [Bytes]	50
t_{tone}	Duration of the wake-up tone [s]	Varying
$r_B = \lambda_B$	Bit rate (max.) [kbps]	250
n	Number of nodes	Variable
P_{tx}	Power in transmitting [mW]	50.7
P_{sleep}	Power in sleeping [μW]	60
$P_{rx} = P_{listen}$	Power in receiving/ listening [mW]	49.2
P_{poll}	Power in channel polling [mW]	0.0294
L_{sB}	Piggybacked Bytes length	2
L_{sync}	SYNC frame length [Bytes]	18
λ_{sync}	SYNC frame rate [$1/T_{sync}$]	Varying
T_{sync}	SYNC frame period	Varying

IEEE 802.15.4 standard and has the lower energy consumption; it operates at 868 MHz.

pMax corresponds to the maximum transmission power allowed in the channel; *sat* one is the signal attenuation threshold (in dBm) which controls the transmission distance. *transmitterPower* is the transmission power, while *thermalNoise* is the electronic noise generated by the thermal agitation. *snrThresholdLevel* is used to decide whether a frame is received correctly.

Only when *snr* is less or equal than *snrThresholdLevel* the frame is lost. *channelModel* is the propagation model used in the channel, while *propagation_coef* is the propagation coefficient for the free-space path loss formula. *berLowerBound* is the lower bound of the bit-error-rate, and *busyRSSI* is used to evaluate whether the channel is busy: only if the RSSI is less than *busyRSSI* the channel is clear. *queueLength* is the maximum number of frames waiting to be transmitted allowed in the queue. *poll_period* is the time interval between each time the node wakes up and checks the channel for activity. In our simulation, the CC1100 radio model (from OMNeT++) does not include the poll state [VH08, Var10]: when a node wakes-up and does not have data frames in its MAC queue, it polls the channel during T_{poll}, while the power consumption is the same as in the receiving state. The main differences between the CC1000 and the CC1100 radio transceiver is the maximum data rate achievable by each radio, the CC1100 can be put into sleep mode from which it wakes automatically on receiving any signal, and the CC1100 consumes three times less energy per each bit it receives or sends. For the CC1000 the maximum data rate is 76.8 kbps while for the CC1100 is 500 kbps. The CC1100 radio chip we use in our simulation is the sub-GHz equivalent of the 2.4 GHz CC2420 used in the original SCP.

4.5.2 SCP Simulator Layer Modes

As the SCP was being implemented in OMNeT++, implementing some modes in the simulator was required, in order to optimize the simulator. These modes are employed to conduct diverse types of experiments regarding single and multi-hop topologies, different traffic patterns generated by the application layer (heavy and periodic), and the need of sending acknowledgment frames as a response to successful reception of data frames. These requirements lead to the addition of the following modes in our simulator.

RTS/CTS mode

The overhearing avoidance problem is solved by the MAC frame headers. The receiver node examines the destination address of the frame immediately after receiving its MAC header. If the frame destination is for other node, it immediately stops receiving the frame and goes to sleep.

Since control frames cause protocol overhead, for the control overhead problem, SCP has the option to enable or disable the RTS/CTS/DATA/ACK, or simply the DATA/ACK exchange scheme for unicast traffic. When the RTS/CTS mechanism is enabled, overhearing avoidance is performed in the same way as in S-MAC. In our simulator, this option is named as the *acknowledgement frame* one.

Throughput mode

Another mode that is implemented in the application layer of the SCP simulator is named as the *throughput* one. It is useful to test the network's response capacity as the number of transmitting nodes increases. When this mode is enabled, the application layer generates a data frame, then sends it to the lower layer, until it reaches the MAC sub-layer, where it is stored in the TX MAC queue. Then, the node schedules the next wake-up to send the data frame received from the upper layers as quickly as possible, starting to send the next frame as soon as the prior frame is sent. As soon as the node sends the frame the physical layer informs the MAC sub-layer that the transmission has finished. With the *throughput* mode enabled, the MAC sub-layer sends the message which indicates the end of frame transmission to the application layer. When the application layer receives this message it builds another data frame and sends it to the lower layers until it reaches the MAC sub-layer, repeating all the scheduling process that is needed to send a data frame through the network.

Synchronization mode

In the SCP simulator there are two modes to choose the synchronization type: the *sync slave* mode, which uses explicit SYNC frames, and the *piggyback* mode, which piggybacks the schedule information into the broadcasted data frames.

Multi-hop mode

Another mode available in the SCP simulator is the *multi-hop* one, which enables the nodes to organize themselves as presented in Figure 4.7.

Figure 4.7 presents a linear chain of a WSN, where d is the distance between the sensor nodes, D is the distance between the source and sink, and n is the number of nodes. Only the application layer from node 1 generates data frames to be sent to other nodes, the node n receives data frames and answers with ACK frames when a data frame is received correctly. The remaining nodes are the so called forwarder's nodes, which receive and forward the data frames as well as it sends and receives ACK frames. For this mode, we defined which node was the source and which one was the sink. The remaining ones only had to forward the frames to next sensor node with the identifier i_d+1.

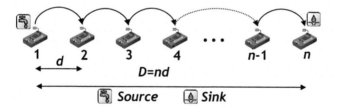

Figure 4.7 Simple linear chain wireless sensor network.

4.6 Summary and Conclusions

This chapter provides a detailed description of the mechanisms employed in SCP, along with the lessons learned during our own experience trying to implement the SCP protocol in the Mobility Framework. Moreover, the possible SCP state transition diagram offers to the user the chance to follow closely how the simulator changes between the different simulations states. In order to conduct diverse types of experiments regarding single and multi-hop topologies, different traffic patterns generated by the application layer (heavy and periodic), and the need of sending acknowledgment frames as a response to successful reception of data frames, different modes of simulation were implemented in the simulator. The use and functioning of these modes was described throughout the chapter.

5

Performance Evaluation of the SCP-MAC Protocol

This chapter presents the implementation of the SCP-MAC protocol in the OMNet++ simulator, considering the Mobility framework, and also using the MiXim framework, original contributions from this research, followed by its performance evaluation. The chapter is divided into six sections. Section 5.1 starts by presenting the simulation results for single-hop topology in terms of power consumption (with and without piggyback), energy consumed and throughput performance, whilst considering heavy and periodic traffic patterns. Section 5.2 addresses the multi-hop (in a linear chain scenario) performance for the node energy consumption and latency metrics. Section 5.3 introduces a lifetime analysis (enabling piggyback synchronization) for the SCP protocol, by considering a simple battery discharge model, given by Peukert's equation. Section 5.4 addresses the proposal of a stochastic model for the collision probability in the collision avoidance mechanism with two contention windows utilized in the SCP protocol, whilst considering the saturated and unsaturated cases. Analytical results are matched against simulation results for both saturated and unsaturated regimes. A discussion of the results is also provided. Section 5.5 introduces another main contribution of this research. The proposed model for the collision probability is applied to theoretically derive the average MAC service time for a successful transmission as well as the achieved throughput. Numerical results for both service time and throughput are verified by simulation and conclusions are extracted at the end of the section. Finally, in Section 5.6, a comparison of the performance metrics of the SCP while considering an IEEE 802.15.4 compliant and IEEE 802.15.4 agnostic physical layer is presented. Section 5.7 presents final remarks for these tests.

5.1 Single-hop Performance Results

The authors from SCP [YSH06a] define the expected energy consumption per node, for both LPL and SCP, as the sum of the expected energy spent in each state, as follows:

$$
\begin{aligned}
\overline{E_{cons}^{SCP}} &= \overline{E_{listen}^{SCP}} + \overline{E_{tx}^{SCP}} + \overline{E_{rx}^{SCP}} + \overline{E_{poll}^{SCP}} + \overline{E_{sleep}^{SCP}} \\
&= P_{listen}t_{CS} + P_{tx}t_{tx} + P_{rx}t_{rx} + P_{poll}T_p + P_{sleep}t_{sleep} \quad (5.1)
\end{aligned}
$$

where E_{cons}^{SCP} is the total energy consumed by the node, E_{listen}^{SCP} is the energy consumption in the node listening process, E_{tx}^{SCP} is the energy consumption in the node transmission process, E_{rx}^{SCP} is the energy consumption in the node reception process, E_{sleep}^{SCP} is the energy consumption in the node sleeping process, P_{listen} is the energy consumption in the node sleeping process, t_{CS} is the expected time a node spends in carrier sense, P_{tx} is the power in transmitting, t_{tx} is the time that last the transmission, P_{rx} is the power in receiving, t_{rx} is the receiving time, P_{poll} is is the power in channel polling, T_p is the channel polling period, P_{sleep} is the power in sleeping and t_{sleep} is the duration of the sleeping period. N.B.: $t_{CS} = t_{CS1}/T_{DATA} = t_{CS1}\lambda_{DATA}$, where t_{CS1} is the average carrier sense time and λ_{DATA} is the data packet rate.

This analytical model is based on three assumptions: i) polling time is set to its optimal value, if the traffic rate is known (an unrealistic hypothesis), ii) channel access failures during contention windows are neglected, and iv) the setup times between operation modes in radio are also neglected.

By considering the best case scenario, all the synchronization information is piggybacked on the data frames. The authors define that all synchronization information can be included in the data frame when $r_{data} \geq r_{sync}$. We define both the lower and higher bound for the energy consumption in scheduled channel polling with piggybacked synchronization, considering the maximum and minimum contending times each node can wait. The lower bound for the energy consumption with piggybacked synchronization is given by:

Lower Bound: $\varphi_{slot}^{CW_1}=8 \wedge \varphi_{slot}^{CW_2}=1$

$$
\begin{aligned}
\overline{E_{cons\_lw}^{SCP}} &= P_{listen}2t_B\lambda_{data} \\
&+ P_{tx}(t_{tone} + L_{data}t_B)\lambda_{data} \\
&+ nP_{rx}(t_{tone} + \varphi_{slot}^{CW_2}t_B + L_{data}t_B)\lambda_{data} \\
&+ P_{poll}t_{p1}/T_p
\end{aligned}
$$

$$+ nP_{sleep}[1 - 2t_B\lambda_{data} + (t_{tone} + L_{data}t_B)\lambda_{data}$$
$$+ n(t_{tone} + \varphi_{slot}^{CW_2}t_B + L_{data}t_B)\lambda_{data} + t_{p1}/T_p)] \tag{5.2}$$

where t_B is the time to transmit/receive a byte, t_{tone} is the duration of the wake-up tone, L_{data} is the data frame length, in Bytes, and t_{p1} is the average time to poll the channel.

The upper bound for the energy consumption with piggybacked synchronization is given by:

Upper Bound: $\varphi_{slot}^{CW_1}=1 \land \varphi_{slot}^{CW_2}=16 \land CW_1^{max}=8$

$$\overline{E_{cons\_hg}^{SCP}} = P_{listen}2t_B\lambda_{data}$$
$$+ P_{tx}(t_{tone} + (CW_1^{max} - \varphi_{slot}^{CW_1})t_B + (\varphi_{slot}^{CW_2} - 1)t_B$$
$$+ L_{data}t_B)\lambda_{data}$$
$$+ nP_{rx}(t_{tone} + (CW_1^{max} - \varphi_{slot}^{CW_1})t_B + \varphi_{slot}^{CW_2}t_B$$
$$+ L_{data}t_B)\lambda_{data} + P_{poll}t_{p1}/T_p$$
$$+ P_{sleep}[1 - (2t_B\lambda_{data} + (t_{tone} + (CW_1^{max} - \varphi_{slot}^{CW_1})t_B$$
$$+ (\varphi_{slot}^{CW_2} - 1)t_B + L_{data}t_B\lambda_{data}$$
$$+ n(t_{tone} + (CW_1^{max} - \varphi_{slot}^{CW_1}))t_B$$
$$+ \varphi_{slot}^{CW_2}t_B + L_{data}t_B)\lambda_{data} + t_{p1}/T_p)] \tag{5.3}$$

In the worst case scenario, if no piggybacking is possible, all the synchronization must be performed with SYNC frames. The authors from SCP [YSH06a] also define the optimal synchronization period time. In this work, we define both the lower and upper bounds for the energy consumption in SCP without piggybacked synchronization, considering the maximum and minimum contention times for each node. On the one hand, the lower bound for the energy consumption without piggybacked synchronization is given by:

Lower Bound: $\varphi_{slot}^{CW_1}=8 \land \varphi_{slot}^{CW_2}=1$

$$\overline{E_{cons\_lw}^{SCP}} = P_{listen}(t_B\lambda_{data} + t_{cs1}\lambda_{sync})$$
$$+ (P_{tx} + nP_{rx})(t_{tone} + L_{data}t_B)\lambda_{data}$$
$$+ (P_{tx} + nP_{rx})(t_{tone} + L_{data}t_B)\lambda_{sync}$$
$$+ P_{poll}t_{p1}/T_p$$

$$
\begin{aligned}
&+ P_{sleep}[1 - ((t_B\lambda_{data} + t_{cs1}\lambda_{sync}) \\
&+ (n+1)(t_{tone} + L_{data}t_B)\lambda_{data} \\
&+ (n+1)(t_{tone} + L_{data}t_B)\lambda_{data} + t_{p1}/T_p)]
\end{aligned}
\tag{5.4}
$$

On the other hand, the upper bound for the energy consumption without piggybacked synchronization is given by:

Upper Bound: $\varphi_{slot}^{CW_1}=1 \wedge \varphi_{slot}^{CW_2}=16 \wedge CW_1^{max}=8$

$$
\begin{aligned}
\overline{\overline{E_{cons\_hg}^{SCP}}} &= P_{listen}(2t_B\lambda_{data} + t_{cs1}\lambda_{sync}) \\
&+ (P_{tx} + nP_{tx})(t_{tone} + (CW_1^{max} - \varphi_{slot}^{CW_1})t_B \\
&\quad + (\varphi_{slot}^{CW_2} - 1)t_B + L_{data}t_B)\lambda_{data} \\
&+ (P_{tx} + nP_{tx})(t_{tone} + (CW_1^{max} - \varphi_{slot}^{CW_1})t_B \\
&\quad + (\varphi_{slot}^{CW_2} - 1)t_B + L_{data}t_B)\lambda_{sync} \\
&+ P_{poll}t_{p1}/T_p \\
&+ P_{sleep}[1 - ((2t_B\lambda_{data} + t_{cs1})\lambda_{sync}) \\
&+ (n+1)(t_{tone} + (CW_1^{max} - \varphi_{slot}^{CW_1})t_B + (\varphi_{slot}^{CW_1} - 1)t_B \\
&+ L_{data}t_B)\lambda_{data} + (n+1)(t_{tone} + (CW_1^{max} - \varphi_{slot}^{CW_1})t_B \\
&+ (\varphi_{slot}^{CW_2} - 1)t_B + L_{data}t_B)\lambda_{data} \\
&+ (n+1)(t_{tone} + (CW_1^{max} - \varphi_{slot}^{CW_1})t_B + (\varphi_{slot}^{CW_2} - 1)t_B \\
&+ L_{data}t_B)\lambda_{data} + (n+1)(t_{tone} + (CW_1^{max} - \varphi_{slot}^{CW_1})t_B \\
&+ (\varphi_{slot}^{CW_2} - 1)t_B + L_{data}t_B)\lambda_{sync} + t_{p1}/T_p)]
\end{aligned}
\tag{5.5}
$$

5.1.1 Power Consumption without Piggyback and Periodic Traffic

It is worthwhile to compare simulation and analytical results for the average power consumption for each node. When no piggybacking is considered after we defined the "adapted" analytical model (lower and upper bounds), we ran the simulator and extracted the values. For these tests the network is composed by five nodes arranged in star topology. In this simulations, the *sync slave* mode is enabled, the synchronization time is defined as $t_{sync}=50$ ms and the inter arrival period is variable. Moreover the simulator also considers, t_{sync}, the synchronization period, which depends on the inter arrival period (already defined by the authors from [YSH06a]), T_{sync}. As the

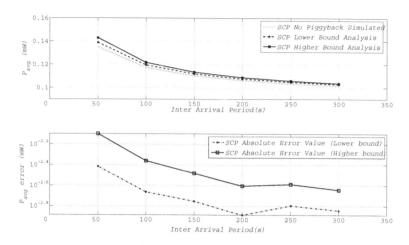

Figure 5.1 Comparison between the lower/upper bound and the simulation results for the power consumption per node set of SCP simulations with no piggybacking.

synchronization period increases the wake-up tone duration also increases. We consider that the maximum contention window sizes is $CW_1^{max} = 8$ and $CW_2^{max} = 16$. For this experiment, all five nodes from the network are placed onto a single-hop configuration, where each node generates and broadcasts 50 data frames (50 Bytes each). Since the inter-arrival period varies between 50 and 300 s, the traffic load is considered to be light. Simulations are run five times for each of the six seeds, generating a total of 30 experiences. For each experiment the average power consumption of each node is obtained, as shown in Figure 5.1. The achieved standard deviations are negligible.

Figure 5.1 presents a comparison between the simulations and analytical results (higher and upper bound), as a function of the inter arrival period. Considering $CW_1^{max} = 8$ and $CW_2^{max} = 16$, it can be observed that the simulation curve is similar to the lower bound curve defined by the analytical model. Besides, the Mean Square Error (MSE) is calculated with respect to the analytical lower bound curve, and a value of 4.8155×10^{-6} mW is achieved. This shows a high similarity between the simulation and the analytical results.

Another comparison between the simulation and the upper/lower bound curves is the absolute error for each analytical bound, as shown in Figure 5.1. As the error decreases while the inter arrival period is increasing, the similarity between the simulation and the analytical lower bound results increases.

To show how the t_{sync} parameter influences the power consumption with no piggybacking performance results, changed its value from 50 to 60 ms, whilst maintaining the remaining parameters. With this modification, the simulation results for the power consumption becomes closer to the lower bound (in comparison to the ones obtained when t_{sync} = 50 ms). We also computed the MSE for t_{sync} = 60 ms, obtaining a value of 1.2808×10^{-6} mW. From these results, one may conclude that the power consumption with the *sync slave* mode enabled presents better results when t_{sync}= 60 ms. However, for values of t_{sync} lower than 50 ms it was not possible to achieve stable results with the SCP simulations. Finally, it is worthwhile to note that for values of t_{sync} longer than 60 ms higher values for the SCP power consumption are achieved.

5.1.2 Power Consumption with Piggyback and Periodic Traffic

The lower and upper bound for the adapted analytical model are also defined for the case when synchronization schedules are piggybacked. The simulator topology and MAC parameters considered for this test are the same as the ones previously considered. However, since all the exchanged traffic is of the broadcast type, all the SYNC frames can be suppressed and the schedule synchronization is performed by using the piggyback technique. Hence, the *piggyback* mode is enabled for this experiment. There is no need to define the values for the parameter t_{sync}, as the schedule information is already piggybacked into the data frames. Therefore, since the time spent during SYNC frames exchange is not wasted in this experiment, there is an increase of the efficiency. This leads to a consequent node energy consumption diminishing.

For this set of simulations, we define the synchronization period as $T_{sync}=T_{DATA}$, where T_{DATA} is the transmission time for data, meaning that the wake-up tone duration is inversely proportional to T_{DATA}, i.e., as T_{DATA} increases, the wake-up tone duration decreases. The dependence of the average power consumption of each node on the inter arrival period is shown in Figure 5.2, where the simulations and analytical results (upper and lower bounds) are also presented for comparison purposes. The values of the standard deviation are negligible.

For CW_1^{max}= 8 and CW_2^{max}= 16, the obtained simulation results are similar to the lower bound curve defined by the analytical model until the inter arrival period achieves 150 s. For a inter arrival period longer than 150 s, the simulation results are similar to the upper bound curve. The MSE value obtained for this experiment with respect to the lower bound curve is equal

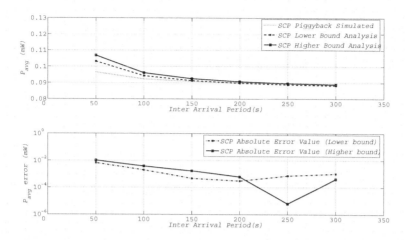

Figure 5.2 Comparison between the lower/upper bound and the simulation results for the power consumption per node set of SCP simulations with piggybacking.

to 8.3763×10^{-6} mW. This value is higher than the one obtained with explicit SYNC frames. This is due to the simulation results similarity with the lower bound, which only occurs until an inter arrival period of 150 s. For inter arrival periods longer than 150 s, the simulation results diverges from the analytical lower bound, leading to an increase of the MSE (with respect to lower the bound).

The absolute error between the simulation values and the upper and lower analytical bound curves is also determined as the error decreases (while the inter arrival period is increasing), the similarity between simulation and analytical lower bound results increases.

By comparing the power consumption with explicit SYNC frame and with piggyback synchronization, we observe that the piggyback technique is truly efficient, since the power consumption with piggybacking achieves lower values than the one with explicit SYNC frames.

To show how the size of the wake up influences the simulation results performance, we define the wake-up tone duration always equal to 2 ms. The only variable parameter is $T_{DATA} \in \{50; 100; 150; 200; 250; 300\}$.

With this modification, the value of the MSE decreases to 7.9825×10^{-6} mW. By comparing the power consumption in the presence and absence of a fixed wake up tone duration, the energy saving obtained with a fixed wake-up tone duration are adequate (approximately 0.005 mW).

5.1.3 Throughput Performance with Heavy Traffic Load

In the previous sections, the simulations were performed by applying periodic traffic load. However, in real world applications, WSNs do not have the capability to predict what type of traffic load is being exchanged with high accuracy. For example, in an healthcare scenario, the sensor nodes deliver the monitoring data from the vital signs of the patients at a lower data rate (in usual situations). However, when a patient presents anomalies in his/her vital signs, the sensor node may deliver data at a higher rate, leading to the need for higher network throughput.

For this set of experiments, we have enabled the *throughput* mode at the application layer. This is useful to test the network's response capability when the number of transmitting nodes increases. By varying the number of transmitting nodes, the contending time of the network will increase, leading to an increase of the collision probability. Each node generates 20 frames (with 100 Bytes each). For these tests the a network is composed by five nodes arranged in star topology. In these experiences the nodes could operate at a duty cycle of 0.2 %, when there is no data to be sent or received, while polling the channel each second. Besides the variation of the number of transmitting nodes, the CW_1 and CW_2 contention window sizes also vary. Figure 5.3 shows the average throughput for three experiments: the first one considers $CW_1^{max} = 8$ and $CW_2^{max} = 16$, second one considers $CW_1^{max} = CW_2^{max} = 78$, while the third one considers $CW_1^{max} = CW_2^{max} = 100$.

By analysing Figure 5.3, the throughput achieves its maximum (for all contending time configurations) when there is only one node transmitting, as expected, since there is no more nodes competing for the channel to send frames. When two nodes are competing for the channel, the throughput drops to 20 - 25 % of the maximum achievable throughput, because there is more than one node trying to transmit the data frame. As the number of transmitting nodes increases, the contending time will also increase, leading to a collision probability increase.

For a $CW_1^{max} = 8$ and $CW_2^{max} = 16$ configuration the throughput curve is lower than the other ones, because for this configuration there is less contention window time slots, increasing the chance of nodes to choose the same time slot and therefore increase the collision probability. For a $CW_1^{max} = CW_2^{max} = 78$ and a $CW_1^{max} = CW_2^{max} = 100$ contention windows configuration the values of the achieved throughput are higher and more stable (80 B/s) than the first one. This is due to the larger maximum contention windows sizes.

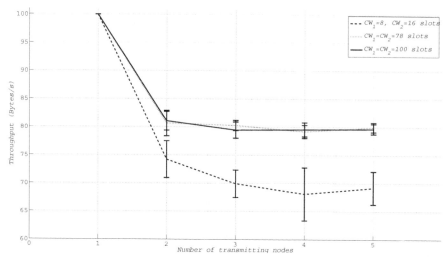

Figure 5.3 SCP-MAC throughput performance considering high traffic load for different contention windows sizes configurations and number of transmitting nodes.

The standard deviations are presented for all the experiments. The curve that presents higher deviations is the one that presents lower values for the throughput (first experiment).

5.2 Multi-hop Energy Efficiency – Linear Chain Scenario

For this set of experiments, we intend to evaluate the SCP energy and end-to-end latency performance for multi-hop networks in a linear topology for the nodes. Besides the energy consumption per node, curves for the multi-hop experiments (obtained from our simulator), an analytical model is proposed for multi-hop communications in the linear chain. It accounts for the MAC protocol contention time.

For multi-hop experiments we enabled the *multi-hop* mode in the SCP protocol stack, where only the application layer from node 1 generates data packets to be sent to the other nodes. The node n receives data packets and answers with ACK packets when a data frame is received correctly. The remaining nodes are the so called forwarder's nodes, which receive and forward the data packets and the answer and receive ACK packets through the linear chain (see Figure 4.7 from Chapter 4). In the evaluation of the analytical model the parameter T_{sync} is equal to zero, meaning that the wake-up tone duration is always equal to 2 ms. The polling interval for this analytical evaluation is set to one second.

In addition to the energy model derived for the single-hop topology in [YSH06a], we can derive the energy model boundaries (higher and lower bound) for a multi-hop topology in a linear chain. In our analytical model for a linear multi-hop chain, we define (in the equation) the X's nodes as source(s), the Y's nodes as forwarders, and Z's nodes as sink(s). In our topology we deploy (*n*-2) Y nodes, one Z node, and one X node. For the multi-hop evaluation the acknowledgement packets were enabled with data frame re-transmissions up to three retries.

For the energy consumption lower bound in a linear multi-hop chain, we consider that all the packets are sent at the first time. The power with piggy-backed synchronization in a linear multi-hop chain lower bound considering maximum timeout to receive de ACK frame, $t_{to_{ack}}$, is given by:

Lower Bound: $\varphi_{slot}^{CW_1}=8 \wedge \varphi_{slot}^{CW_2}=1 \wedge t_{to_{ack}}=10\times10^{-3}$s

$$
\begin{aligned}
\overline{P_{multi\_lw}^{SCP}} = {}& (n-2)(P_{tx}Y_{tx}^{d\_ack} + P_{rx}Y_{rx}^{d\_ack} \\
& + P_{listen}Y_{listen}^{data} + P_{poll}Y_{poll}^{data}) \\
& + (P_{tx}X_{tx}^{data} + P_{listen}X_{listen}^{data} + P_{poll}X_{poll}^{data} + P_{rx}X_{rx}^{ack}) \\
& + (P_{rx}Z_{rx}^{data} + P_{listen}Z_{listen}^{data} + P_{poll}Z_{poll}^{data} + P_{tx}Z_{tx}^{ack}) \\
& + P_{sleep}[1 - ((n-2)(Y_{tx}^{d\_ack} + Y_{rx}^{d\_ack} + Y_{listen}^{data} + Y_{poll}^{data}) \\
& + (X_{tx}^{data} + X_{listen}^{data} + X_{poll}^{data} + X_{rx}^{ack}) \\
& + (Z_{rx}^{data} + Z_{listen}^{data} + Z_{poll}^{data} + Z_{tx}^{ack}))]
\end{aligned}
$$

$$
Y_{tx}^{d\_ack} = \frac{(t_{tone} + (L_{data} + L_{ack})t_B + t_{to_{ack}})}{T_p},
$$

$$
Z_{tx}^{ack} = \frac{L_{ack}t_B + t_{to_{ack}}}{T_p},
$$

$$
X_{tx}^{data} = (t_{tone} + L_{data}t_B)\lambda_{data},
$$

$$
Y_{rx}^{d\_ack} = \frac{(t_{tone} + (L_{data} + L_{ack})t_B + t_{to_{ack}})}{T_p},
$$

$$
X_{rx}^{ack} = (L_{ack}t_B + t_{to_{ack}})\lambda_{data},
$$

$$Z_{rx}^{data} = \frac{t_{tone} + L_{data}t_B}{T_p},$$

$$X_{listen}^{data} = 2t_B\lambda_{data},$$

$$Y_{listen}^{data} = Z_{listen}^{data} = \frac{P_{listen}2t_B}{T_p},$$

$$X_{poll}^{data} = Y_{poll}^{data} = Z_{poll}^{data} = \frac{t_{p1}}{T_p} \tag{5.6}$$

where n is the number of nodes in the topology for SCP-MAC energy efficiency analysis in the linear chain scenario, as shown in Figure 4.7 from Chapter 4. Note that t_B is the time to transmit a Byte and X, y and Z are auxiliary variables.

We consider the worst-case scenario for the energy consumption higher bound in a linear multi-hop chain higher bound, we consider the worst scenario, when all the nodes reach the maximum number of retries, ι_{ret}, meaning that each data frame is send or forward after three retries. The higher bound for the power with piggybacked synchronization in a linear multi-hop chain is given by:

Higher Bound: $\varphi_{slot}^{CW_1}=1 \wedge \varphi_{slot}^{CW_2}=16 \wedge CW_1^{max}=8 \wedge \iota_{ret}=3$

$$\overline{P_{multi\_hg}^{SCP}} = \iota_{ret}[(n-2)(P_{tx}\widehat{Y_{tx}^{d\_ack}} + P_{rx}\widehat{Y_{rx}^{d\_ack}} + P_{listen}\widehat{Y_{listen}^{data}}$$
$$+P_{poll}\widehat{Y_{poll}^{data}})$$

$$\tag{5.7}$$

which depends on the auxiliary variables just defined above.

In multi-hop networks, there is no optimal working point. Hence, network performance depends on what application is supported by the network. The users configure the MAC layer according to the application needs while, depending on the tradeoffs that can be established between the maximum end-to-end latency and the energy consumption, in order to obtain the network or node lifetime (suitable for the application).

Another feature that is still not implemented in our simulator is the adaptive channel polling. It allows for sending a stream of packets from the source to the sink by switching to high duty cycles (it is designed to reduce the latency of the packets). This also enables the transmission of multiple packets in one polling interval.

The topology used in the performance evaluation is the one presented in Figure 4.7 from Chapter 4, with a network composed by a 9-hop linear topology with ten nodes (each one with a separation of 50 m from each other). The source node will generate 20 packets (50 Bytes each), with a variable inter arrival period between 0 and 10 s. Nodes will rebroadcast the data packets along the linear chain until they reach the most distant node from the chain. All the packets are sent as unicast ones and the RTS/CTS mechanism is not enabled. Only the *acknowledgement* mode, with up to three retries, is enabled in our simulations.

By varying the maximum sizes for the contention windows it is possible to increase or diminish the probability of frame collisions. Each node polls each second. For comparison purposes (with the SCP original multi-hop experiment [YSH06a]), one has defined $CW_1^{max}=CW_2^{max} = 208$, so that the maximum contending time of our simulations could be more or less equal to the original SCP maximum contending time. The other contention windows size configurations presented here were the ones that we achieved better results.

By considering the inter arrival time equal to zero means the application layer generates a data frame, and sends it to the lower layers, until it reaches the MAC layer, where it is stored in the TX MAC queue. Then, the node schedules the next wake-up to send the data frame received from the upper layers the fastest way possible, starting to send the next frame as soon as the prior frame is sent. When the node sends the frame, the physical layer informs the MAC layer that the transmission had finished and with the *throughput* mode enabled the MAC layer sends the message that indicates the end of frame transmission to the application layer. If the application layer receives this message it will generate another data frame and will send it to the lower layers until it reaches the MAC sub-layer, repeating all the scheduling process needed to send a data frame throughout the network. This evaluates the performance of the multi-hop network when non periodic traffic is generated (near the real world). In order to test the multi-hop network performance for periodic traffic, the inter arrival periods vary between 1 and 10 s .

By observing Figure 5.4, one observe that the energy curve for CW_1^{max} = CW_2^{max}=208 is the one that presents a higher energy consumption per node, because of the maximum contending time used by the nodes to avoid packet collisions. When comparing with the energy consumption per node for multi-hop experiments in the original paper [YSH06a], energy consumption is lower. The other contention windows configurations present a lower energy consumption per node (less than 0.05 J). For all the configurations the energy

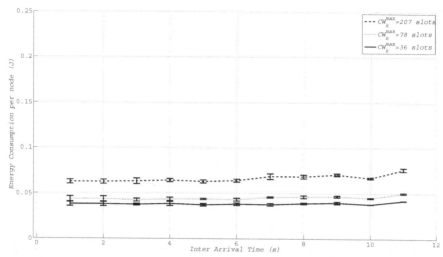

Figure 5.4 Mean energy consumption per node for multi-hop experiments with different contention windows sizes configurations (20 data frames over 9 hops), $k \in \{1, 2\}$.

consumption behaviour is considerably stable when the inter arrival time increases.

5.3 Lifetime Analysis with piggyback (Periodic Traffic)

In this section we describe a model that allows us to calculate and set SCP-MAC's parameters with the purpose of optimize the overall energy consumption. Using the model, we intend to illustrate the effect of different application variables such as the contention window maximum sizes, inter arrival periods, network density, and polling period.

To calculate the lifetime of an application that employs the SCP-MAC protocol with piggybacked synchronization, we consider a periodic sensing application that sends data to a base station. The categories for WSN lifetime are *sensornoderelated* or *networkrelated*. The sensor node related interpretation establishes the WSN lifetime as the time until the first sensor node "dies" and stops transmitting/receiving frames. On the other hand, in the network related interpretation, the WSN lifetime is defined as the time until a certain percentage of sensor nodes "dye" and stop receiving/transmitting frames.

The node's lifetime is given by the overall power, which is defined in millijoules per second, or milliwatts. The total energy consumed by the node can be obtained by multiplying the power ($\overline{P_{cons}^{SCP}}$) by the node lifetime, t_{node}. Since we want to calculate the maximum lifetime a sensor node can achieve, when employing the SCP-MAC protocol with piggybacked synchronization, we also need to add the energy consumed by the sensors, which are attached to the Mica2 mote. By considering the application deployed from [MCP$^+$02] and also mentioned by the authors in [YSH06a], where the node's lifetime with BMAC protocol is estimated, we also assume that each node takes 1100 ms to start, sample, and collect the data from its sensors. For this test we defined that data is sampled every minute ($r_{dat} = 1/(1\times60)$ min^{-1}).

Therefore the total energy consumption is given by:

$$
\begin{aligned}
\overline{E_{cons}^{SCP}} &= \overline{E_{listen}^{SCP}} + \overline{E_{tx}^{SCP}} + \overline{E_{rx}^{SCP}} + \overline{E_{poll}^{SCP}} + \overline{E_{sleep}^{SCP}} + \overline{E_{sensor}^{SCP}} \\
&= P_{listen} t_{CS} + P_{tx} t_{tx} + P_{rx} t_{rx} + P_{poll} t_{poll} + P_{sleep} t_{sleep} \quad (5.8)
\end{aligned}
$$

where E_{sensor}^{SCP} is the energy associated with sampled data.

To calculate the energy consumption of the sensors (and the node's lifetime) we need to specify the amount of current that is drained by the sensors. These values are presented in Table 5.1.

The energy associated with the sampled data, E_{dat}, is given by:

$$
E_{dat} = t_d \times r_{dat} \overline{E_{sensor}^{SCP}} = t_d i_{dat} V_{cc}
$$

The energy consumed by the other components that contribute to the overall energy consumption of the sensor node have also been calculated by considering the higher bound equation of the SCP protocol with piggybacked synchronization. The energy consumed during the node listening process is

Table 5.1 Parameters for a monitoring application that uses Mica2 motes.

Symbol	Parameter	Value
C_{batt}	Capacity of the battery [mAh]	2500
V_{cc}	Voltage [V]	3
i_{dat}	Sample sensors current [mA]	20
i_{tx}	Current to send a byte [mA]	16.9
i_{rx}	Current to receive a byte [mA]	16.4
i_{sleep}	Current in sleeping mode [μA]	20
t_{dat}	Time to sample sensors [s]	1.1

given by:

$$E_{listen} = 2t_B\lambda_{data}\overline{E_{listen}^{SCP}} = t_{listen}i_{rx}V_{cc}$$

The energy consumed during the node transmission process is given by:

$$E_{tx} = (t_{tone} + (CW_1^{max} - \varphi_{slot}^{CW_1})t_B + (\varphi_{slot}^{CW_2} - 1)t_B + L_{data}t_B)\lambda_{data}$$
$$\overline{E_{tx}^{SCP}} = t_{tx}i_{tx}V_{cc} \tag{5.9}$$

For the node reception process, the energy consumption is given by:

$$E_{rx} = n(t_{tone} + (CW_1^{max} - \varphi_{slot}^{CW_1})t_B + (\varphi_{slot}^{CW_2} - 1)t_B + L_{data}t_B)\lambda_{data}$$
$$\overline{E_{rx}^{SCP}} = t_{rx}i_{rx}V_{cc} \tag{5.10}$$

The energy consumption during the node polling process is given by:

$$E_{poll} = t_{p1}/T_p\overline{E_{poll}^{SCP}} = t_{poll}i_{rx}V_{cc}$$

Finally, the energy consumed during the sleeping time of a sensor node is given by:

$$E_{sleep} = (1 - (2t_B\lambda_{data} + (t_{tone} + L_{data}t_B)\lambda_{data} + n(t_{tone} + \varphi_{slot}^{CW_2}t_B$$
$$+ L_{data}t_B)\lambda_{data} + t_{p1}/T_p))\overline{E_{sleep}^{SCP}} = t_{sleep}i_{sleep}V_{cc} \tag{5.11}$$

The lifetime of a sensor node, t_{node}, depends on the total power, $\overline{P_{cons}^{SCP}}$, and the capacity of the battery. Therefore, we can establish a boundary with respect to the available capacity of the battery:

$$t_{node[s]} = \frac{C_{batt} \times V_{cc}}{\overline{P_{cons}^{SCP}}} \times 60 \times 60 \tag{5.12}$$

This equation is considered when assuming an ideal battery, where there are no battery recovery phenomena or other phenomena related with battery discharge.

One of the simplest models that describes the battery discharge is given by the Peukert's equation, which takes into account the rate capacity effect but does not take into account the recovery effect. Peukert's equation numerically shows how discharging at higher rates removes more power from the battery, than discharging at lower rates. The Peukert's battery model is simple but it has some flaws, such as the overestimation of battery lifetime for most cases [JH09]. The original equation is written as follows:

$$I^{n\_peuk} T = C_{batt} \qquad (5.13)$$

where I is the discharge current, in ampere, T is the time in hour, C_{batt} is the capacity of the battery in ampere hour, and n_peuk is the Peukert's exponent for that particular battery type. Peukert's equation, as it is, has to be used on batteries specified as the "Peukert Capacity". It means the capacity of the battery when discharged at 1 A. Rarely the batteries are quoted like this. To account for the way the battery capacity is quoted we need to modify the formula transforming the original Peukert's equation into the following one:

$$T_h = \frac{C_{batt}}{\left(I/(\frac{C_{batt}}{R})\right)^{n\_peuk}} \times \left(\frac{R_{batt}}{C_{batt}}\right) \qquad (5.14)$$

where R is the battery hour rating.

The battery hour rating in 99 % of the batteries is equal to 20 h. Depending on the type of battery the hour rating changes, as stated by the authors in [SK11]. The Peukert's exponent is directly related with the internal resistance of the battery and varies between 1 and 1.4. The higher the internal resistance is, the higher the losses while charging and discharging are, especially at higher currents. Hence, the faster a battery is discharged, the lower its capacity is.

By solving Equations (5.8) to (5.11), and considering the parameters from Table 5.1, we can estimate the maximum energy needed for a given network configuration. Lifetime is going to be estimated based on the estimated overall energy consumption, considering a variable inter arrival time (from 50 s to 200 s) while varying the number of neighbour nodes. We consider a polling interval T_p equal to 62 s, because the data is sampled every minute. Therefore, there is no need to scan for data during a time interval lower than the 62 s. Setting the polling interval to value mentioned will decrease significantly the energy consumption of a node using the SCP MAC protocol, because of the low duty cycle that it will follow. We define that all synchronization information can be included in the data frame when $r_{data} = r_{sync}$.

By observing Figure 5.5, with an ideal battery, the expected lifetime is decreasing as the number of neighbouring nodes increases, as expected. Since this is an ideal battery, the discharge process removes the same power from the battery both at higher or lower rate of discharge. When comparing the upper graphic with the lower one that models the lifetime by the Peukert's

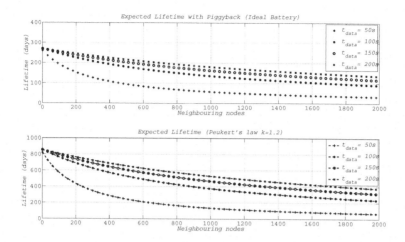

Figure 5.5 Expected lifetime with piggyback (higher bound), while considering $CW_1^{max} = 8$ and $CW_2^{max} = 16$ for different inter arrival periods, with $n\_peuk = 1.2$.

equation with $n\_peuk=1.2$, it is observed that the node will last three times more than using the ideal battery model. As the inter arrival time decreases the expected lifetime decreases, because of the low inter arrival time that causes the node to wake-up more times to send or receive data packets than with a higher inter arrival time where the node will wake-up fewer times to send or receive the data packets.

Comparing Figures 5.5 and 5.6, with the increase of the Peukert's exponent until it reaches the maximum value there is an increase of the expected lifetime.

The battery "dies" when it reaches a minimum voltage which does not allows to the radio transceiver (or other sensor node components) to work properly. We define that the critical component in our sensor node is the radio transceiver, because it needs a minimum voltage supply of 1.8 V. Although the radio transceiver stops working properly below 1.8 V, the microcontroller could continue to work. However, the communication capability from the sensor node will be turned off from the sensor node. To observe how the battery voltage drops, until it reaches the critical zone, we choose a fixed number of neighbour nodes, while varying the inter arrival time, the maximum size of the contention windows, and incrementing the time until it reaches the critical zone.

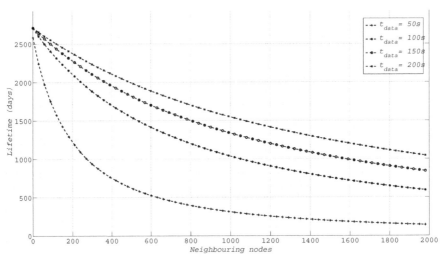

Figure 5.6 Expected lifetime with piggyback (higher bound), while considering $CW_1^{max} = 8$ and $CW_2^{max} = 16$ for different inter arrival periods, with $n\_peuk = 1.4$.

The following equation is considered to model the battery voltage dropping:

$$V_{drop} = V_{cc} - \frac{P_{node}}{\left(\frac{R(\frac{C_{batt}}{R})^{n\_peuk}}{T} \right)^{1/n\_peuk}}$$

(5.15)

where P_{node} is the power consumed with a fixed number of neighbouring nodes, a fixed maximum size for the contention windows, and a fixed inter arrival time.

By observing Figure 5.7, the battery voltage decreases faster when the inter arrival time is the shortest, and decreases slower when the inter arrival time is the largest. Therefore, the node that sends or receives data every 50 s will reach the critical zone after ≈ 3500 h. In turn, a node that receives data every 200 s will reach the same region after ≈ 6200 h.

By changing the number of neighbouring nodes to 91, one expects to increase the lifetime of the node. By observing Figure 5.8 and comparing it with Figure 5.7, we conclude that the node will reach the critical region only after ≈ 4800 h when considering an inter arrival time equal to 50 s (an increase in lifetime of ≈ 1300 h). If the inter arrival time is equal to 200 s, the node will reach the same region after ≈ 6500 h (a lifetime increase of ≈ 300 h). Another parameter that could lead to changes in the battery voltage

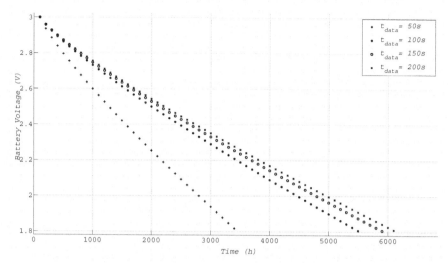

Figure 5.7 Battery voltage discharge, while considering $CW_1^{max} = 8$ and $CW_2^{max} = 16$ for different inter arrival periods, with $n\_peuk = 1.2$ and $n = 211$.

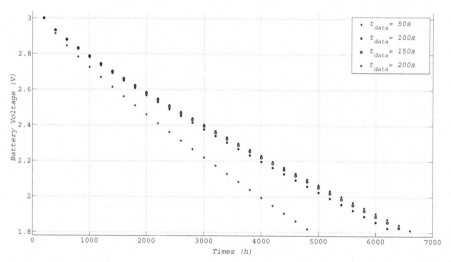

Figure 5.8 Battery voltage discharge, while considering $CW_1^{max} = 8$ and $CW_2^{max} = 16$ for different inter arrival periods, with $n\_peuk = 1.2$ and $n = 91$.

drop, and consequently on the expected lifetime, is the maximum size of the contention windows.

For the **first experiment**, we have set the $CW_1^{max} = CW_2^{max} = 208$, and have defined the number of neighbouring nodes equal to 211 nodes. When

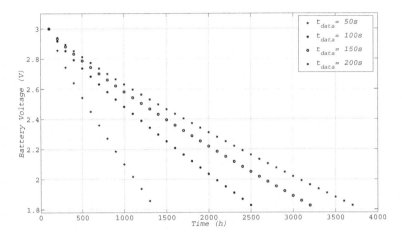

Figure 5.9   Battery voltage discharge, while considering $CW_1^{max} = CW_2^{max} = 208$ for different inter arrival periods, with $n\_peuk = 1.2$ and $n = 211$.

comparing this experiment with the one that employs the same number of neighbouring nodes ($n = 211$), but different maximum contention windows sizes, a lower lifetime is obtained for the node, because of the increase of the maximum contention time.

By observing Figure 5.9, we can conclude that the node that employs a larger maximum contention windows size presents a lower lifetime. When the inter arrival time is equal to 50 s, the node is expected to last ≈ 1300 h. For an inter arrival time equal to 200 s, the node is expected to last ≈ 3700 h.

For the **second experiment**, the 208 slots were maintained in each contention window, and the number of neighbouring nodes was defined to be 91 nodes. When comparing this experiment with the one that employs the same number of neighbouring nodes ($n = 91$), but with different maximum contention windows sizes it is expected to obtain a lower lifetime of the node, because the increase of the maximum contention time, even if the number of neighbouring nodes decrease.

By observing Figure 5.10, we conclude that the sensor node reaches the critical zone quicker for an inter arrival time equal to 50 s (than when considering the same inter arrival time but with a smaller maximum contention windows sizes). For an inter arrival time equal to 200 s the node will reach the critical zone ≈ 20 % faster when considering the 208 slots in each contention window.

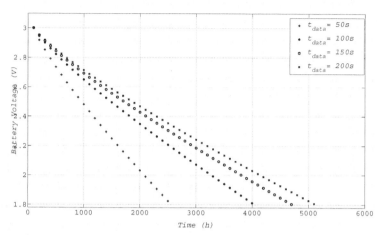

Figure 5.10 Battery voltage discharge, while considering $CW_1^{max} = CW_2^{max} = 208$ for different inter arrival periods, with $n\_peuk = 1.2$ and $n = 91$.

5.4 Performance Analysis of a Two-Phase Contention Scheme for Scheduled Channel Polling

5.4.1 Motivation for Using Two Contention Windows

As presented in Section 4.2, the two-phase contention scheme for scheduled channel polling (SCP) considers two contention windows, CW_1 and CW_2, divided into CW_1^{max} or CW_2^{max} time slots of equal duration. The choice of the slot to be used in the CW_1 and CW_2 follows a uniform distribution, as already mentioned in the aforementioned section. With two contention stages (CW_1 and CW_2) the collision probability decreases relatively to the case when a single contention window with equivalent size (sum of the sizes of CW_1 and CW_2) is adopted. Only the nodes succeeding in CW_1 enter CW_2. This procedure causes a decrease in the number of nodes effectively accessing the medium after the second contention stage. After the first phase of the contention procedure, only the successful nodes from CW_1 (the first one(s) that transmit the wake-up tone) contend in CW_2. With fewer contending nodes, the effective collision probability is drastically reduced. The motivation that drives us is that to the best of our knowledge, no work has addressed the analytical formulation for the collision probability in a two-phase contention window mechanism, which is characterized here for the collision avoidance mechanism employed by the SCP-MAC protocol [YSH06a].

5.4.2 Overview for the Saturated Regime

Usually, the performance analysis of wireless communication protocols is done in the saturated regime, since it significantly simplifies the analysis. In this case, the analysis is easier than in the unsaturated regime, in which the number of contending nodes is varying in time, depending on the packet arrival time instant at the application layer and the packet service time. Performing contention under saturated traffic conditions implies that nodes always have at least one data frame to be transmitted. An example of an application with saturated traffic conditions is a WSN with a high frequency event-based data dissemination scheme [YBO09]. This type of application is characterized by a set of nodes that simultaneously detect a high frequency phenomenon, which may drive the network to saturated traffic conditions for a given period of time. We analyze the impact of the two-phase contention window mechanism on the collision probabilities for different values of the size of the contention windows, CW_k^{max}, with $k \in \{1, 2\}$. This analysis is performed as a function of the number of contending nodes. In the remainder of this work we consider that the transmission period includes the first and second contention periods, although a frame transmission can be postponed to future transmission periods due to the need of retransmissions. This is mainly because the probability of collision is defined for a single transmission period, and retransmissions can be viewed as new transmissions originated by previous collisions.

For the saturated regime, the following assumptions are considered:

1. The sensor nodes only remove a data frame from the queue after gaining access to the medium and transmitting it;
2. Each transmitting node adopts the contention values within the interval $[1, CW_1^{max}]$ and $[1, CW_2^{max}]$, depending on the CW (CW_1 or CW_2, respectively);
3. If there is only one node which succeeds to pass from CW_1 to CW_2, the CW_2 collision probability is zero;
4. All the nodes have the same traffic load.

5.4.3 Overview for the Unsaturated Regime

The focus of the above description is the saturated regime, where every node in the network always has a frame ready for transmission. This means that the number of nodes contending for the channel is always the total number of nodes present in the network, n. However, from the application point of view, the saturated regime is unlikely to be realistic in real world applications, due to queue overflow and infinite queueing time [BGdMT06]. Hence, we also consider the unsaturated traffic condition of the network, which is more realistic, in practice.

In the unsaturated regime, nodes with an empty queue do not contend for the channel. Therefore, the average number of contending nodes varies between 1 and n, depending on the offered traffic load for each node. In the unsaturated regime we assume that the sensor nodes have one attempt to transmit each data frame. Otherwise, if the sensor node senses a busy channel, it discards the frame and waits for the next (generated) one.

For unsaturated traffic, we have tested CBR arrival traffic with a given inter-arrival time value, T_{tR}. The starting times for the CBR traffic are uniformly distributed over a certain percentage of the average inter-arrival time period (i.e., the drift rate, θ), for different simulation rounds. After the first uniformly generated inter-arrival time, the remaining ones are generated periodically during each traffic generation inter-arrival time.

Moreover, since Scheduled Channel Polling involves synchronization, the polling period time, T_{poll}, needs to be defined (in this work ot is set to 1 s), while the polling duration, T_p, is set to 2 ms. The sensor nodes wake-up each second in order to dispatch the frames, which can exist in the Tx queue, as quickly as possible. Since the nodes wake-up each second, the initial times are allocated uniformly over five possible time intervals (with duration of one second each), for $\theta = 0.1$. As the drift increases, the number of intervals also increases, while the probability of a node to generate frame(s) in one of the intervals decreases. The dependence of this probability decay on θ and T_{tR} is analyzed in detail in Section 5.4.5.

5.4.4 Stochastic Collision Probability Model for the Saturated Regime

In the stochastic model for the collision probability, n is the number of nodes simultaneously competing for time slots in CW_1, $(n > 0)$ and m is the number of successful nodes in CW_1 that enter CW_2, whilst simultaneously competing for the time slots in CW_2, $(m > 0)$. Considering CW_1^{max} and CW_2^{max} as the number of time slots in CW_1 and CW_2, respectively, the respective n or m nodes choose their time slot from the following set:

$$\varphi_{slot}^{(CW_k)} = \{a_i, i = 1, \ldots, CW_k^{max}\} \tag{5.16}$$

where a_i represents the slot and $k \in \{1, 2\}$.

The probability of success in CW_1 is the probability of a successful wake-up tone, i.e., it excludes the cases where two or more nodes choose the lowest order slot (although it still may lead to success in CW_2). Considering the slot selection events, A_i, which occur from the set of slots in CW_1, the set of possible slots is given by:

$$S_{CW_1} = A_1 \cup \ldots \cup A_{CW_1^{max}} \tag{5.17}$$

As the events are equiprobable, the probability of choosing a slot i, p_i, is given by:

$$p_i = \frac{1}{CW_1^{max}}, \ \forall \, i = 1, \ldots, CW_1^{max} \tag{5.18}$$

Firstly, the probability that l nodes choose the slot i, is expressed by the binomial distribution of l nodes choosing the slot i, with probability p_i in a sequence of n nodes, and is denoted as

$$P(X_i = l) = \frac{n!}{l!(n-l)!} p_i^l (1 - p_i)^{(n-l)}, \ \forall \, i = 1, \ldots, CW_1^{max} \tag{5.19}$$

where X_i represents a random variable which corresponds to the number of nodes that choose the slot i (to start the transmission of the wake-up tone).

The probability that one slot i in CW_1 is chosen by more than u ($u \geq 0$) nodes is given by the sum of the probabilities of having $u+1$ nodes choosing the slot i. The sum of the probabilities is therefore given by:

$$P(X_i > u) = \sum_{k=u+1}^{n} P(X_i = k), \ \forall \, i = 1, \ldots, CW_1^{max} \tag{5.20}$$

Since a collision occurs when two or more nodes access the medium in the same slot, and because the nodes choose a given slot with the same probability, the probability of collision in the first contention period, CW_1, is given by:

$$P_{c1} = P(X_i > 1) = \sum_{k=2}^{n} P(X_i = k), \ \forall \, i = 1, \ldots, CW_1^{max} \tag{5.21}$$

Figure 5.11 represents the CW_1 collision probability (P_{c1}) as a function of the number of contending nodes, n, with CW_1^{max} as a parameter. A

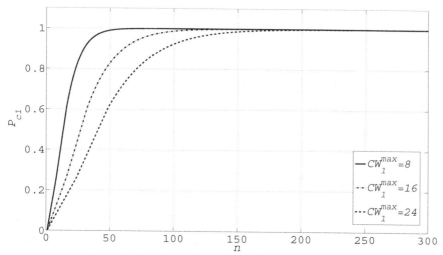

Figure 5.11 CW$_1$ collision probability as a function of n.

comparison between both traffic regimes is presented in section 5.4.5, where the differences are noticeable.

The expected number of contending nodes in CW$_2$, $E[S]$, depends on the probability that the i-th lowest order slot in CW$_1$ is chosen by one or more nodes. At least one node successfully passes from CW$_1$ to CW$_2$ if $n \geq 1$. The expected number of contending nodes in CW$_2$ therefore includes the nodes that had chosen the lowest order slot in CW$_1$ (slot i) and is given by:

$$E[S] = \sum_{k=1}^{n} kP(X_i = k), \ \forall \ i = 1, \ldots, CW_1^{max} \qquad (5.22)$$

Figure 5.12 plots $E[S]$ for different numbers of contending nodes and sizes of contention window CW$_1$. As the size of CW$_1$ increases, the expected number of nodes in CW$_2$, $E[S]$, decreases. This behaviour is due to the higher dilution of the nodes by the contention slots. By enlarging CW$_1$, for the same number of nodes, the expected number of nodes transmitting in each slot decreases, leading to fewer nodes that successfully pass from CW$_1$ to CW$_2$.

By comparing Figure 5.11 and Figure 5.12, we conclude that the CW$_1$ collision probability increases as the number of contending nodes increase. Two distinct zones can be identified in Figure 5.11:

i) a rising zone where $0 \leq P_{c1} \leq 1 - \epsilon, \forall n$;

ii) a flat zone where $1 - \epsilon < P_{c1} \leq 1, \forall n$.

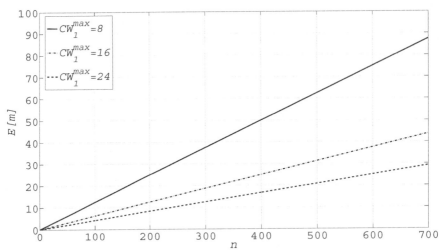

Figure 5.12 Expected number of nodes that successfully pass from CW_1 to CW_2.

The variable ϵ is such that the slope of the derivative of the collision probability curve from Figure 5.11 is close to zero in the flat zone. In the rising zone the expected number of nodes that pass from CW_1 to CW_2 is always smaller than the number of contending nodes. While in the flat one the CW_1 collision probability (P_{c1}) achieves the maximum collision ratio. From this point forward, as the number of contending nodes that compose the network increases, all nodes successfully pass to CW_2.

In this work, the numerical solution of the collision probability for the second contention window is defined by two different approaches: (i) individual slot state analysis approach, or (ii) a combinatorics approach. The former is based on the combination of the different slots' probabilities, namely the probability of finding a slot idle, or the probabilities of finding a slot busy due to a collision and full transmission success. The latter can be described as a "bins and balls occupancy" problem [Fel68], which is going to be further developed in this section. Both approaches are presented in this paper, to show the available possibilities to define the model for the CW_2 collision probability.

To define the collision probability in the second contention (CW_2), we must consider the slot selection events (denoted as C_i) which occur from the set of slots in CW_2, and are expressed by:

$$S_{CW_2} = C_1 \cup \ldots \cup C_{CW_2^{max}} \tag{5.23}$$

As the events are equiprobable (as in CW_1), the probability that a node chooses a given slot i is the same for all nodes and is given by:

$$\tau = P(Y_i = l) = \frac{1}{CW_2^{max}}, \ \forall \, i = 1, \ldots, CW_2^{max} \tag{5.24}$$

where Y_i represents a random variable which corresponds to the number of nodes that choose the slot i to start the transmission of the data frame. Assuming a network composed by n nodes, where $E[S]$ nodes successfully passed from CW_1 to CW_2, the probability that one node finds the channel idle (when performing a carrier sense in the chosen slot), is related with all the $E[S]$ nodes that do not start transmitting in the same slot. Consequently, the probability of a slot being idle is given by:

$$P_{Sidle} = (1 - \tau)^{\lceil E[S] \rceil} \tag{5.25}$$

where the function $\lceil x \rceil$ denotes the ceiling function. We note that the effective data frame collision in CW_2 only happens if the nodes that successfully pass from CW_1 to CW_2 randomly choose the same time slot in CW_2. This event corresponds to an effective data frame loss due to collision. Therefore, the successful slot probability in CW_2 depends on the number of nodes that choose a slot to transmit while the remaining $E[S] - 1$ nodes on its radio range do not start a transmission in the same time slot. Thus, the probability of occurring a successful transmission in a given slot of CW_2 is given by:

$$P_{Ssuc} = E[S]\tau(1 - \tau)^{\lceil (E[S]-1) \rceil} \tag{5.26}$$

By combining Equations (5.25) and (5.26), we are able to derive the probability of multiple access occurrences in the same slot (collision), which is expressed as follows:

$$P_{Scol} = 1 - P_{Sidle} - P_{Ssuc} \tag{5.27}$$

Finally, a collision can occur in any slot of CW_2. It occurs in the first frame's slot with probability P_{Scol}. It only occurs in the second slot if the first slot is idle. This probability of collision for the second slot is an intersection of the events of occurring a collision in the second slot (with probability P_{Scol}) and the event of not having nodes accessing in the first slot (with probability P_{Sidle}). Consequently, applying the same rationale for the remaining slots, the probability of occurring a collision in the second window CW_2 is expressed by:

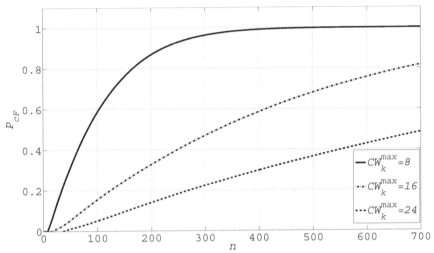

Figure 5.13 CW_2 collision probability under saturated regime considering a individual slot state analysis, $k \in \{1, 2\}$.

$$P_{cF} = \sum_{k=1}^{CW_2^{max}} P_{Scol} P_{Sidle}^{(k-1)} \qquad (5.28)$$

Figure 5.13 plots the probability of a collision occurring in CW_2. Given that a node broadcasts a frame, its success depends only on channel occupancy. Therefore, a collision occurs in CW_2 if more than one node choose the lowest order slot and transmit data frames simultaneously. Hence, the curves referred to the CW_2 collision probability are the numerical solutions for the P_{cF} expression.

P_{cF} increases as the number of contending nodes increases. As the contention window size increases, the maximum collision probability value is attained for higher number of contending nodes. Moreover, as the size of CW_2 increases (for the same number of contending nodes) the P_{cF} value sharply decreases. This is due to the two-phase contention window mechanism. Since the first contention window performs a first selection of nodes that pass to CW_2 (by means of contention), the second contention window handles fewer nodes. These nodes select a slot in CW_2 and try to transmit the data frame with success. In the flat zone, all the nodes that contend in CW_1 contend in CW_2, causing the CW_2 collision probability to reach the maximum collision probability value ($P_{cF} = 1$). This occurs because the nodes have a higher probability of accessing the channel when frames are broadcast (as plotted in Figure 5.13), which originates more collisions.

If we consider the combinatorics approach, the collision probability for the second contention window is partially based on the birthday paradox [Fel68]. In the *birthday paradox*, bins are distinguishable and balls are indistinguishable. However, in our model bins and balls are both indistinguishable. This means that besides the amount of balls that are inside of the lowest order bin, the number of the ball also matters. For the sake of simplicity, we assume that each of the m balls represents a node, while each of the p bins is a CW_2 time slot. Hence, the envisaged success probability may be defined as the probability that the lowest order bin contains more than one ball and the remaining balls should be distributed for the remaining (highest order) bins. The remaining bins may contain more than one ball, and no ball should remain outside a bin. If the lowest order (first) bin does not contain any balls, then the next order bin is treated as the lowest order bin. The expected number of contending nodes in CW_2 is given by Equation (5.22). Although the expected values of contending nodes in CW_2 from Equation (5.22) may not be an integer value, for our combinatorics solution, all the involved variables should be treated as integer values. Hence, the following rounding function must be applied to it:

$$E[m] = \lfloor E[S] \rfloor \tag{5.29}$$

The $\lfloor x \rfloor$ function is used to round down to the nearest integer value of $E[m]$, to avoid possible overestimation of probability values that may lead to higher deviation errors. To begin with, the number of different arrangements in CW_2 (with and without collision) for the assignment of nodes by the time slots is obtained by arrangements with repetition. As any of the CW_2^{max} slots often occur repeatedly (so that it can be chosen several times) one must consider arrangements with repetition.

The total number of possibilities, given by Equation 5.30, assumes that any number of nodes can be put in a slot:

$$A_t(m) = (CW_2^{max})^{E[m]} \tag{5.30}$$

The number of possibilities that exactly one node chooses the i-th lowest order slot and the remaining ones choose the subsequent slots is also given by arrangements with repetition. If the lowest order slot is selected by only one node then the number of different possibilities is given by an arbitrary assignment of the $E[m] - 1$ remaining nodes to the $CW_2^{max} - r$ remaining slots.

The variable r is the number of the lowest order slot that is selected by only one node and varies between 1 and CW_2^{max}. Therefore, the arrangements, A_s, corresponding to the choice of the lowest order slot by exactly one node is expressed by:

$$A_s(m) = E[m] \sum_{j=1}^{CW_2^{max}-1} (CW_2^{max} - j)^{(E[m]-1)} \qquad (5.31)$$

In turn, the number of possibilities for the lowest order slot to be chosen by more than one node is given by:

$$A_f(m) = (CW_2^{max})^{E[m]} - E[m] \sum_{j=1}^{CW_2^{max}-1} (CW_2^{max} - j)^{(E[m]-1)} \quad (5.32)$$

The probability of a collision occurring in CW_2 results from the ratio between the number of possibilities for the lowest order slot to be chosen by more than one node, given by Equation (5.32), and the total number of possibilities to distribute n nodes by CW_2^{max} slots, given by Equation (5.30). As a consequence, the CW_2 collision probability, P_{eF}, is given by:

$$P_{eF}(m) = \frac{A_f(m)}{A_t(m)} \qquad (5.33)$$

Figure 5.14 shows the CW_2 collision probability as a function of the number of contending nodes, with the contention window size as a parameter. By comparing the curves obtained for the CW_2 collision probability between each approach (as shown in Figures 5.13 and 5.14), the values are similar with a Mean Absolute Error (MAE) close to 0.5 %. The main difference between the two approaches is the slope at the rising zone. For a number of contending nodes higher than the ones that correspond to the beginning of the CW_1 collision probability flat zone, the CW_2 collision probabilities are almost identical. These differences between the different approaches is due to the rounding function that is applied to the expected number of nodes that pass from CW_1 to CW_2 (to compute the numerical solution for the CW_2 collision probability according to the second approach).

5.4.5 Stochastic Collision Probability Model for the Unsaturated Regime

In the saturated regime, all sensor nodes have frames ready for transmission. As a consequence, the number of contending nodes with non-empty queues

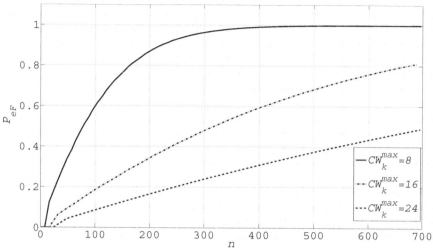

Figure 5.14 CW_2 collision probability considering saturated regime using a combinatorics approach, $k \in \{1, 2\}$.

in the network is constant and known by all nodes. This fact is advantageous for the analysis of the collision probability under saturation conditions. In the latter case, the number of contending nodes varies continuously based on the frame arrival rates. Although the unsaturated regime is more complex to analyze, for the sake of simplicity, we only analyze uniformly distributed generation arrival times. The generation of unsaturated traffic at node n_i, $i \in 0, \ldots, n$, is modelled by a uniformly distributed frame generation probability, P_{u_G}. The node n_i uniformly chooses the initial instant value for the first data frame generation, t_{θ_i}. The maximum drift rate, θ, varies between 0 and 1 ($\theta \in [0; 1]$). The frame generation initial instant value (given in seconds), t_{θ_i}, follows a uniform distribution, that is repeated at each inter-arrival time, as given by:

$$t_{\theta_i} \in [0; t_{\theta_{max}}], \ \forall \ i = 1, \ldots, n \qquad (5.34)$$

where $t_{\theta_{max}}$ is the maximum time of the interval when the frame generation times are chosen (in the experiments we considered $\theta_{max} = 5$ s). $t_{\theta_{max}}$ depends on the values chosen for the θ and T_{tR} parameters and is given by $t_{\theta_{max}} = \theta \cdot T_{tR}$. Since the nodes wake-up every second ($T_{poll} = 1$ s), the drift rate reflects the number of intervals the inter-arrival period is divided in. Figure 5.15 depicts how the time frame is subdivided into a uniform number of time intervals (of one second duration each). Figure 5.15 also presents,

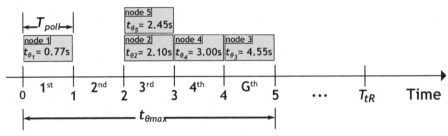

Figure 5.15   Uniform distribution example of nodes over the intervals.

an example of how the nodes are distributed among the generation intervals, depending on the chosen frame generation time, t_{θ_i}.

The interval order, G, is expressed as $G \in [1; \Pi_{max}]$, where Π_{max} is the highest interval order, given by the combination of values of θ, T_{tR}, and polling period time, T_{poll}, parameters. Hence, $\Pi_{max} = \left\lfloor \frac{t_{\theta_{max}}}{T_{poll}} \right\rfloor$. In Figure 5.15, we assume $\theta = 0.1$, $T_{tR} = 50$ s, and $T_{poll} = 1$ s, which, in turn, leads to $\Pi_{max} = 5$ (by dividing $t_{\theta_{max}}$ by T_{poll}).

From the uniform distribution, the probability of a given node to choose one of the G intervals is expressed by:

$$P_{u\_G} = \frac{T_{poll}}{t_{\theta_{max}}}, G \in [1; \Pi_{max}] \qquad (5.35)$$

Figure 5.16 shows the probability of a node to choose one of the G possible intervals as a function of the drift rate per node, θ, and of the network size n, for $T_{tR} = 50$ s. The surface shows two distinct generating probability zones. For values of θ higher or equal to 0.2, the generating probability decreases as θ increases. Hence, the probability P_{u_G} is independent of the number of contending nodes for a given value of inter-arrival time, T_{tR}. When the drift rate decreases ($\theta < 0.2$), P_{u_G} sharply increases. This sharp increase occurs because of the reduced number of possible intervals that leads to a higher probability of a node to fall into an interval.

Considering the unsaturated regime and a deployment of n nodes in a given area, the expected number of nodes starting a data frame generation in the interval G, $\widehat{E}[n]$, is represented by:

$$\widehat{E}[n] = \lceil P_{u\_G} \cdot n \rceil, G \in [1; \Pi_{max}] \qquad (5.36)$$

$\widehat{E}[n]$ is given by the ceiling integer value, since the expected number of nodes should not be decimal. Figure 5.17 shows the respective expected

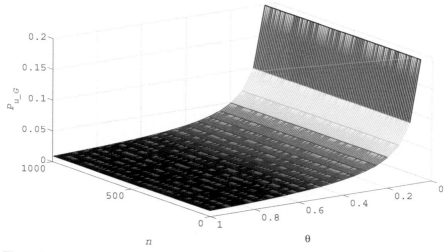

Figure 5.16 Probability of a node to choose one of the G possible intervals as a function of θ and n, for $T_{tR}= 50$ s.

number of nodes that choose one of the G possible intervals, $\widehat{E}[n]$. Unlike Figure 5.16, in Figure 5.17, $\widehat{E}[n]$ is not independent from the number of contending nodes. Since the generating probability varies only when θ increases or decreases, for a pre-set θ value, $\widehat{E}[n]$ increases as the number of possible generating nodes (n) increases. This proportional behaviour is depicted in the surface presented in Figure 5.17, where two distinct zones are identified. The first one ($\theta \geq 0.2$) presents an expected number of contending nodes varying between 0 and 25 nodes. The smallest $\widehat{E}[n]$ values match with the highest values of θ. The second one ($\theta < 0.2$) corresponds to the highest values of $\widehat{E}[n]$, where the $\widehat{E}[n]$ varies between 0 and 200 nodes. The highest values of $\widehat{E}[n]$ correspond to the reduced number of possible intervals that a node may choose to start the frame generation.

For the unsaturated regime, Equation 5.21 is rewritten as follows:

$$\widehat{P_{c1}} = \sum_{k=2}^{\widehat{E}[n]} \binom{\widehat{E}[n]}{k} p_i^k (1 - p_i)^{(\widehat{E}[n]-k)},$$

$$\forall\, i = 1, \ldots, CW_1^{max} \qquad (5.37)$$

which defines the probability that at least two nodes choose the same time slot in CW_1, for the expected number of nodes that choose one of the G possible intervals ($\widehat{E}[n]$).

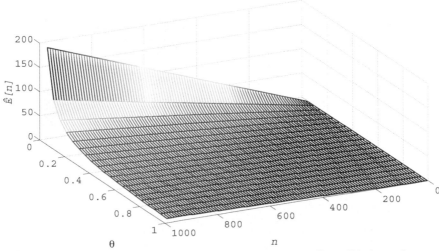

Figure 5.17 Expected number of nodes to choose one of the G possible intervals as a function of θ and n, for $T_{tR} = 50$ s.

Moreover, for the unsaturated case, Equation 5.22 is rewritten as follows:

$$\widehat{E}[S] = \sum_{k=1}^{\widehat{E}[n]} k \binom{\widehat{E}[n]}{k} p_i^l (1 - p_i)^{(\widehat{E}[n]-k)},$$

$$\forall \, i = 1, \ldots, CW_1^{max} \qquad (5.38)$$

where $\widehat{E}[S]$ is the expected number of contending nodes in CW_2 under the unsaturated regime. As in the saturated case, two distinct zones can be identified in Figure 5.18:

i) a rising zone where $0 \leq \widehat{P_{c1}} \leq 1 - \epsilon, \, \forall \, n$;

ii) a flat zone where $1 - \epsilon < \widehat{P_{c1}} \leq 1, \, \forall \, n$.

For the rising and flat zones the variable ϵ is characterized in the same way as in the saturated regime. Therefore, by following the first approach to obtain the collision probability in CW_2 (individual slot state), $E[S]$ in Equations 5.25, 5.26 and 5.27 is now equivalent to $\widehat{E}[S]$. Besides, if the second approach (combinatorics) is considered in Equations 5.30, 5.31 and 5.32, $E[m]$ is now equivalent to $\widehat{E}[m]$. Finally, the collision probability in CW_2 is obtained by the first approach (see Equation 5.28) under the unsaturated regime, and is represented by $\widehat{P_{cF}}$ as follows:

$$\widehat{P_{cF}} = \sum_{k=1}^{CW_2^{max}} \widehat{P_{Scol}} \widehat{P_{Sidle}}^{(k-1)} \tag{5.39}$$

For the combinatorics approach the CW_2 collision probability under the unsaturated regime, given by Equation 5.33, is expressed by

$$\widehat{P_{eF}}(m) = \frac{\widehat{A_f}(m)}{\widehat{A_t}(m)} \tag{5.40}$$

Figure 5.18 shows the characterization of the collision probability in CW_1, for different number of contending nodes, n, and for different CW_1^{max} values. The saturated and unsaturated regimes have been considered. The purpose of this analysis is to verify the validity of the model for different number of contending nodes, contention window sizes and levels of traffic load (saturated and unsaturated). In both regimes, one can observe that, for a given value of CW_1^{max}, the collision probability increases as the number of contending nodes increases, i.e., the ratio of non-interfering time slots per node decreases. Moreover, as the contention window size increases, the collision probability decreases, as expected.

However, for the unsaturated case, the CW_1 collision probability does not increase as fast as for the saturated one. The P_{u_G} values for the unsaturated case explains why the CW_1 collision probability reaches a value of 1 for a higher number of possible contending nodes.

Figure 5.19 presents the variation of the expected number of successful nodes that pass from CW_1 to CW_2 as a function of n. Both the saturated and unsaturated regimes are considered. By comparing these results with the ones in the saturated conditions, we conclude that the number of successful nodes in CW_1 increases, in saturated regime.

For different contention windows sizes, the expected number of success-ful nodes in CW_1, $E[m]$ for the saturated regime and $\widehat{E}[m]$ for the unsaturated one, increases linearly. By comparing both traffic regimes with respect to the contention window size, the expected number of successful nodes that pass to CW_2, $\widehat{E}[S]$, is lower in the unsaturated case. This is due to the low number of nodes that will choose one of the G-th intervals and contend for the channel.

Figure 5.20 shows the variation for the CW_2 collision probability with n (which, in turn, is related with the number of contending nodes in the CW_1 that succeed into reaching the CW_2) for the saturated and unsaturated

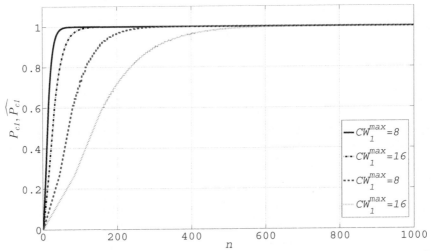

Figure 5.18 CW_1 collision probability for the saturated (black solid and dash-dot lines) and unsaturated (blue dashed and dotted lines) regimes as a function of n, for $\theta = 0.1$.

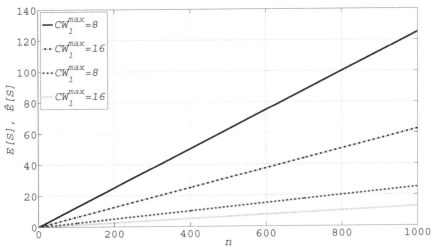

Figure 5.19 Expected number of successful nodes from CW_1 to CW_2 for the saturated (black solid and dash-dot lines) and unsaturated (blue dashed and dotted lines) regimes as a function of n, for $\theta = 0.1$.

regimes. The results are presented for different values of the contention window sizes, CW_1^{max} and CW_2^{max}.

When only one successful node enters into CW_2 the collision event never happens. Moreover, since $\widehat{E}[S]$ is lower in the unsaturated case. Then, the

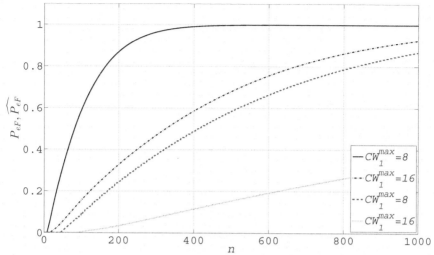

Figure 5.20 CW_2 collision probability for the saturated (black lines) and unsaturated (blue dashed and dotted lines) regimes as a function of n for $\theta = 0.1$, $k \in \{1, 2\}$.

CW_2 collision probability in the latter regime is always lower than in the saturated regime, as shown in Figure 5.20.

5.4.6 Simulation Scenario

In this section, we evaluate the impact of the number of contending nodes on the frame collision of the two-phase contention window mechanism, using both analytical and simulation approaches, under saturated traffic conditions. We evaluate the proposed model through OMNeT++ [Var10] simulations. The Mobility Framework from [REHD08] is used with our SCP protocol implementation, whilst considering different model parameters like n, CW_1^{max} and CW_2^{max}, as well as different values for the traffic generation rate, λ. In these simulations, we consider an ideal wireless channel, i.e., the network performance is not degraded by the physical layer impairments such as fast fading and shadowing, as free-space propagation mainly in line-of-sight is assumed. The ideal wireless channel assumption may be adequate to capture the general behaviour of the mechanism. The node topology represents a tight cluster of dense deployed wireless motes, which is a worst-case situation in terms of collisions. We assume that the sensor nodes do not move (an example of node deployment is shown in Figure 5.21). The number of contending nodes is kept constant during the simulations, in order

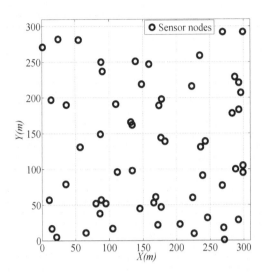

Figure 5.21   Homogeneous uniform WSN deployment simulation.

to properly evaluate the effective data frame collision. Nodes are deployed using a 2D Poisson process on a 300×300 m^2 area (i.e., x and y coordinates are randomly and uniformly chosen within $[0, 300]$). All sensor nodes are simultaneously sink and sources (whilst sending broadcast packets). Since broadcast frame transmission is performed by plain carrier sensing with duration of a time slot (in both contention windows), this setup is prone to the "hidden terminal" problem. To avoid this problem, we assure that all nodes can sense each other. To give a more realistic behaviour to our simulations, the accumulative SNIR interference model [KSEG+08] is considered. With this interference model, the SNIR of the given signals is recalculated each time the receiver predicts the change in the interference power. This model is frequently used in wireless communication simulations and is included in the Mobility Framework [REHD08]. For each deployment with n contending nodes, a set of 6 seeds (corresponding to 300 simulation rounds for each seed) is chosen for the random number generator used by SCP (which has six degrees of freedom). In our set of SCP simulations, we consider the Texas Instruments CC1100 low-power radio transceiver, available in the Mobility Framework [REHD08], with a transmission rate of 250 kbps. Frames are periodically generated at the application layer with a certain frame time interval parameter (50 or 100 s). System parameters are listed in Table 5.2.

Table 5.2 Simulation parameters.

Parameter	Value	Parameter	Value
TX/RX range	> 1 m	Frame length	50 Bytes
Data rate	250 kbps	Slot time	32 μs
Channel model	Free-space	$CW_k^{max}, k \in \{1, 2\}$	[8; 16]
Node number	2~440	Poll period	$\{1, 5\}$ s
Node placement	Uniform	Number of frames	50

5.4.7 χ^2 Test in the unsaturated traffic regime

The Pearson's chi-square test (χ^2) is used to assess the goodness-of-fit which allows to infer whether an observed distribution differs or not from the theoretical one. We have applied the Pearson's chi-square test to the simulation results obtained for the unsaturated regime. The procedure is as follows: the sample space is divided into the union of G disjoint intervals. The probability that a data frame is generated in the G-th interval under the postulated distribution, p_i, $i = 1, ..., G$, is calculated. The expected number of outcomes that fall in the G-th interval in n possible contending nodes of the network simulation is given by $e_i = n \cdot p_i$. The chi-square statistic is defined as the weighted difference between the observed number of generation times (f_i) in the G-th interval, and the expected theoretical frequency (e_i), given by:

$$Q^2 = \sum_{i=1}^{G} \frac{(f_i - e_i)^2}{e_i} \tag{5.41}$$

The hypothesis is rejected if $Q^2 \geq \chi^2$, where χ^2 is the critical point determined by a given level of significance. Otherwise, the fit is accepted. When expected frequencies are large, there is no problem with the assumption of normal distribution. However, the smaller the expected frequencies are, the less valid are the results of the chi-square test. Therefore, if the events show very low raw observed frequencies (five or below), the expected frequencies may also be too low for chi-square test to be appropriately used. To guarantee that the approximation to the chi-square is appropriate, the expected frequencies must not be too low. One must ensure that the following inequality is fulfilled [CL54]:

$$n \cdot p_i \geq 5, \qquad i = 1, ..., G \tag{5.42}$$

To validate the model for the collision probabilities in the unsaturated regime, it is fundamental to test if the times when the frames are generated can be modelled by a uniform distribution. We denote this hypothesis as H_0. In the simulations, we consider $\theta = 0.1$, which implies $G = 5$ possible disjoint intervals. A level of significance $\alpha = 0.05$ is assumed to ensure that condition in Equation 5.42 is verified. The condition in Equation (5.42) establishes that in a network that defines 5 possible disjoint intervals for frame generation, with a uniform distribution, the network must have at least 5 nodes. This way, each of the 5 intervals corresponds theoretically to one node. In order to fulfill the latter condition, the given inequality is used to calculate the minimum number of n possible contending nodes. Hence, the probability that a data frame is generated in the G-th interval under the uniform distribution is obtained by dividing one frame by the maximum number of possible intervals, $p_i = 1/5 = 0.2$. Manipulating the inequality, we obtain $n \geq 25$ for the minimum number of contending nodes, which verifies the condition in Equation 5.42. Since we consider network sizes lower than 25 nodes in our simulations, we can not rely on the chi-square table to obtain the critical point in these cases. To obtain the critical point we have performed Monte Carlo simulations with the IBM SPSS software, whilst considering $\alpha = 0.05$, for $n = \{5, 10, 20\}$ nodes. From the Monte Carlo simulations we have concluded that, for $n < 10$ nodes, it is not possible to adopt the χ^2 test. Table 5.3 shows the results for different seeds and $n = 10$ nodes. By analysing the Monte Carlo simulation results from Table 5.3 the hypothesis is accepted if the asymptotic significance value (Φ) is less than or equal to Monte Carlo significance level (Ω). Otherwise, H_0 is rejected.

For network sizes $n > 25$, the chi-square test was carried out to determine the goodness-of-fit between the H_0 and the simulation results. Since we consider $G = 5$, in our simulations, the critical point for $G - 1 = 4$ degrees of freedom (with a 5 % of significance level) is given by $\chi^2(\alpha = 0.05, V = 4) = 9.4877$. For $n > 25$, the goodness-of-fit of H_0 to the simulations results is guaranteed for all the seeds. For example, for a seed with a corresponding network size equal to $n = 30$ we have obtained $Q^2 = 6.6667$, which is well within the critical point, $Q^2 < \chi^2(\alpha = 0.05, V = 4)$. Hence, we conclude that the simulation results are in good agreement with Equation 5.39.

5.4.8 Discussion of the Results

In this section, the CW_1 and CW_2 collision probabilities are obtained as a function of the number of nodes, whilst considering the contention window size as a parameter.

Table 5.3 Monte Carlo simulation results for different seeds, $n = 10$.

f_i	27377	56190	63086	39755	53720	21803
1	3	2	3	2	4	4
2	4	2	2	3	3	2
3	1	3	2	2	1	2
4	1	2	2	2	1	1
5	1	1	1	1	1	1
χ^2	0.4	1	1	1	4	3
Φ	0.527	0.910	0.910	0.910	0.406	0.558
Ω	0.75	0.989	0.989	0.987	0.476	0.683

Six different random simulations are considered and the 95 % confidence intervals are represented. Figures 5.22 and 5.23 show the validation results for the expected number of nodes that pass from CW_1 to CW_2. Both traffic load regimes (saturated and unsaturated) are considered for two different contention window sizes, $CW_1^{max} = CW_2^{max} = \{8, 16\}$. The difference between the modelling and the simulation results "is insignificant" for all the range of n, which validates the model (for the expected number of nodes that pass to CW_2). Despite a smaller model accuracy for smaller number of nodes, as n increases, the curve for the simulation results converges to the curve plotted for the numerical solution of the expected number of nodes that pass from CW_1 to CW_2.

Figures 5.24 and 5.25 present the verification of the validity of the CW_1 collision probability, under the saturated and unsaturated traffic regimes. The difference between the simulation and modelling results is less noticeable for the saturated regime than for the unsaturated one. Nevertheless, as the number of contending nodes, n, increases the simulation curves converge to the model curves, which allows for verifying our model for the CW_1 collision probability. This initial deviation between the model and simulation results was already previously mentioned. We verify that, in all the cases presented in Figures 5.24 and 5.25, the collision probability increases with the increase of the number of contending nodes. Collisions become more frequent as the number of competing nodes increases. The results obtained for the CW_1 collision probability are consistent by the ones from [vHH09], which presents

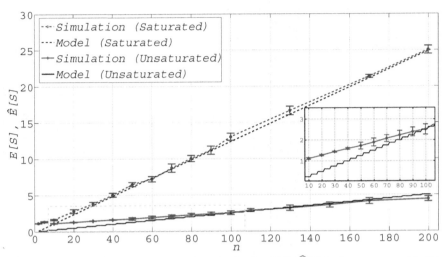

Figure 5.22 Simulation and analytical results for the $E[S]$, $\widehat{E}[S]$ in the unsaturated (solid line) and saturated (dashed line) regimes, for $CW_1^{max} = 8$.

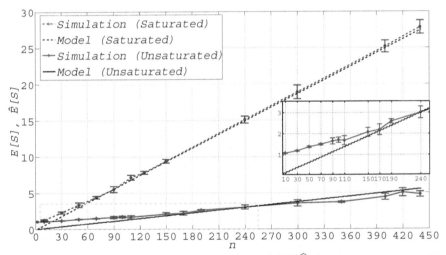

Figure 5.23 Simulation and analytical results for the $E[S]$, $\widehat{E}[S]$ in the unsaturated (solid line) and saturated (dashed line) regimes, for $CW_1^{max} = 16$.

a similar analysis for the collision probability (but for the case of a single contention window).

The deviation that is observed in Figures 5.24 and 5.25 for the unsaturated regime can be explained by the number of nodes that are present in the

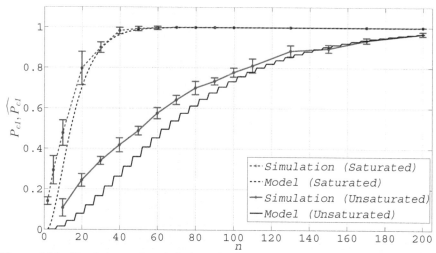

Figure 5.24 Simulation and analytical results for the CW_1 collision probability in the unsaturated (solid line) and saturated (dashed line) regimes, for $CW_1^{max} = 8$.

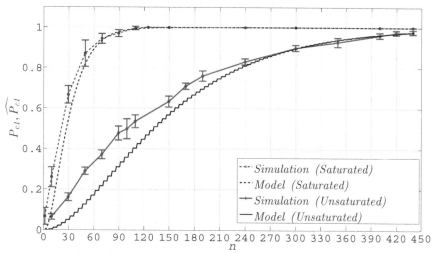

Figure 5.25 Simulation and analytical results for the CW_1 collision probability for $CW_1^{max} = 16$ in unsaturated (solid line) and saturated (dashed line) regimes.

lowest order slot in the CW_1. In each one of the aforementioned figures, a close-up of the chart area is presented, showing the deviation between the model and simulation results. For the case of $CW_k^{max} = 8$, $k \in \{1, 2\}$, the model underestimates the number of nodes that are able to pass from CW_1

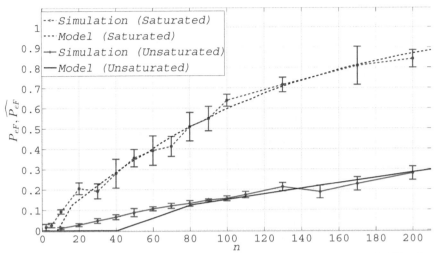

Figure 5.26 Simulation and analytical results for the CW_2 collision probability for $CW_k^{max} = 8, k \in \{1, 2\}$ in unsaturated (solid line) and saturated (dashed line) regimes.

to CW_2, i.e. the number of nodes that have chosen the lowest order slot in CW_1. The deviation observed for $\widehat{E}[S] < 2.5$ nodes, which corresponds to a network with 100 nodes. For $CW_k^{max} = 16$, $k \in \{1, 2\}$, the close-up presented in Figure 5.25 shows that the model underestimates the value of $\widehat{E}[S]$ as well. For this value of CW_k^{max}, a deviation is also observed for an expected number of nodes, $\widehat{E}[S]$, up to 2.5. This value matches to a network size of $n = 240$ nodes. Numerically, the CW_1 collision probability is only meaningful when two or more nodes pass from CW_1 to the CW_2. While this condition is not verified, the model presents this kind of deviations due to the decimal to integer rounding procedure. Hence, due to the underestimation of the expected number of nodes that pass from CW_1 to CW_2 this leads to errors that can reach a maximum of 12 %. With our model and the presented values, we obtain MAE values of 5.19 % and 5.62 % for the unsaturated regime and of 3.60 % and 2.76 % in the saturated case, for $CW_k^{max} = 8$ and 16, $k \in \{1, 2\}$, respectively.

Figures 5.26 and 5.27 present a comparison between simulation and analytical results for the collision probability in CW_2 as a function of n for two different values of the contention window sizes $CW_k^{max} = \{8, 16\}$, $k \in \{1, 2\}$. The saturated and unsaturated regimes are considered. For comparison purposes, the numerical model for the CW_2 collision probability that uses the combinatorics approach is considered. Figure 5.26 considers

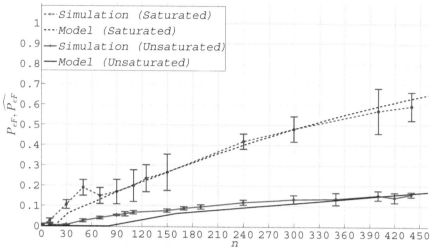

Figure 5.27 Simulation and analytical results for the CW_2 collision probability for $CW_k^{max} = 16, k \in \{1, 2\}$ in unsaturated (solid line) and saturated (dashed line) regimes.

$CW_k^{max} = 8$, $k \in \{1, 2\}$. We observe that the analytical curves are similar to the simulation ones, except for $n \leq 20$ and $n \leq 70$, for the saturated and unsaturated regimes, respectively, where the range of the 95 % confidence interval does not superimpose the analytical curves. For higher values of n, the numerical model for the CW_2 collision probability is very accurate, compared with the simulation values.

Figure 5.27 shows the results for $CW_k^{max} = 16$, $k \in \{1, 2\}$. One also verifies that the results for the saturated regime are more accurate than the ones for the unsaturated one. Furthermore, we observe that the analytical results are similar to the simulation ones, except for $n \leq 40$ in the saturated regime. For larger values of the number of contending nodes, the simulation curves are well fitted by the analytical ones. However, for the unsaturated regime, the analytical results present some deviation from the simulated ones for larger number of contending nodes. Nevertheless, the analytical curve fits the simulation results, for the largest network sizes. The analytical model underestimates the simulated collision probability in CW_2, presenting a maximum deviation of approximately 0.05 from the simulated value.

The similarity between the two curves for the CW_1 and CW_2 collision probabilities and the expected number of nodes that pass from CW_1 to CW_2 verifies the validity of the proposed models for the CW_1 and CW_2 collision probabilities in the SCP two-phase contention window mechanism.

5.5 Service Time and Throughput Theoretical Model for a Two- Phase Contention Window Mechanism

5.5.1 Stochastic Service Time and Throughput Model for the Saturated Regime

In this section, we apply the proposed model for the collision probability in the two-phase contention window mechanism to derive the average MAC service time ($\overline{T_{sv}}$) for a successful transmission as well as the achieved throughput ($\overline{\vartheta_s}$). We provide numerical results for both service time and throughput, verified by simulation. The simulations mimic the two-phase contention window mechanism implemented in the Mobility Framework [REHD08], which provides the required traces for the service time and throughput.

We define the MAC service time as the time needed by the MAC protocol operation time between the instant when a frame is withdrawn from the MAC queue to be transmitted until the end of the transmission process, i.e., until the instant the MAC sub-layer is ready to withdraw another frame from the MAC queue. A data frame is removed from the queue when one of the following events occurs: after being successfully received at the destination node or after being transmitted but involved in a collision. To investigate the asymptotic behaviour of the service time and throughput unlimited retransmissions attempts (ζ), $\zeta \to \infty$, are assumed. Even though we consider unlimited retransmissions attempts, a frame transmitted by a node is always bounded by a finite number ζ of retransmissions attempts.

We consider the following two events in the definition of the service time, $\overline{T_{sv}}$:

- The event of dropping data frames due to a transmission of a frame involved in a collision at the receiver, is given by P_{cF} defined in Equation 5.26;
- The event of a successful data frame transmission after i-th transmission attempts after $i - 1$ consecutive transmission failures, for $i = 1, ..., \zeta, \forall \zeta \to \infty$, is given by P_{sF} as follows.

During channel carrier sensing, if a node detects an occupied channel during CW_1 or CW_2, it keeps the frame in the queue and tries to send it in the next frame. To define the probability of no collision occurrence in the second contention window CW_2, the same rationale from the Equation 5.26 is applied. As a consequence, the probability of a successfull data frame transmission after it, P_{sF}, is given by:

$$P_{sF} = 1 - \sum_{k=1}^{CW_2^{max}} P_{Scol} P_{Sidle}^{(k-1)}$$ (5.43)

The average service time $\overline{T_{sv}}$ is given by the ratio between the initial size of the network, n, and the total number of nodes per slot, n_{tx}, that effectively transmit the data frame (with and without collisions at the receiver node). Here, it is assumed that the two-phase contention mechanism is repeated after T_{poll} seconds. Recall that T_{poll} is the length of the cycle between two consecutive frames and is referred as the polling or renewal interval. A more general analysis of the service time can be achieved by considering T_{poll}. Finally, the dependence of the average service time on n, n_{tx} and T_{poll} is given by:

$$\overline{T_{sv}} = \frac{n}{n_{tx}} \cdot T_{poll}$$ (5.44)

The number of nodes per slot that effectively transmit the data frame is obtained by computing the following expected number:

$$n_{tx\_S} = \frac{E[S] \cdot P_{sF}}{CW_2^{max}}$$ (5.45)

$$n_{tx\_F} = \frac{E[S] \cdot P_{cF}}{CW_2^{max}}$$ (5.46)

where n_{tx_S} and n_{tx_F} are the number of nodes per slot that transmit a data frame and experienced a successful reception or the number of nodes transmitting a frame involved in a collision, respectively. The total number of nodes per slot, n_{tx}, is given by:

$$n_{tx} = n_{tx\_S} + n_{tx\_F}$$ (5.47)

The throughput $\overline{\vartheta_s}$ is defined as the network capability of processing successful frame transmissions. Regarding the proposed model, a frame is successfully transmitted if the node detects the medium idle during the contention window CW_1 and CW_2. The frame transmission can either result into a successful reception or collision at the receiver node. The throughput is expressed by

$$\overline{\vartheta_s} = \frac{L_{data}}{\overline{T_{sv}}}$$ (5.48)

Note that the network throughput $\overline{\vartheta_s}$ is meaningful only when the WSN is under saturated conditions.

5.5.2 Simulation and Analytical Results Comparison

Figures 5.28 and 5.29 show the dependence of the average service time and throughput on the number of nodes with the contention window maximum sizes (varying from 4 up to 16 slots).

When the number of nodes is small, e.g., $n = 20$, there are some gaps between our analytical and simulation results for the average service time, $\overline{T_{sv}}$, shown in Figure 5.28. For instance, numerical results for the service time $\overline{T_{sv}}$ when $CW_k^{max} = 8, k \in \{1, 2\}$ are deviated from the simulation until it reaches a network size of 64 nodes. However, for network sizes that vary from 2 up to a maximum size of 20 nodes and larger than 70 nodes both numerical and simulation results are well matched for the service time. The reason for these deviations is the number of nodes that pass from CW_1 to CW_2. The CW_1 collision probability is only meaningful when two or more nodes pass from CW_1 to CW_2. If less than two nodes pass from CW_1 to CW_2, these deviations are observed. They become more significant as the contention windows sizes increases.

Figure 5.29 presents the network throughput $\overline{\vartheta_s}$. Similarly to the average service time from Figure 5.28, the model matches the simulation results although some deviations are observed. It can be observed from the figures that the average service time increases, while the network throughput decreases significantly as the number of sensor nodes and the contention

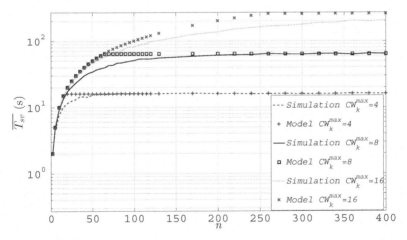

Figure 5.28   Simulation and analytical results for the service time for $CW_k^{max} = \{4; 8; 16\}$, $k \in \{1, 2\}$ in the saturated regime.

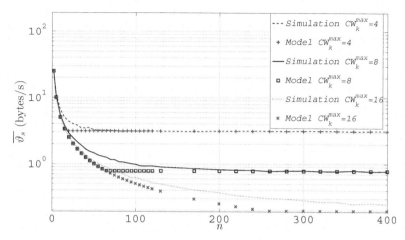

Figure 5.29 Simulation and analytical results for the throughput for $CW_k^{max} = \{4; 8; 16\}$, $k \in \{1, 2\}$ in the saturated regime.

windows sizes increase. The average service time and throughput are bounded (both reach the saturation) by an asymptotic value when the collision probability in the contention window CW_1 reaches the maximum value.

5.5.3 Summary and Conclusions

This Section focuses on the collision probability of a MAC protocol with a two-phase contention mechanism. Following the analytical study from Section 5.4 on the collision probability in the first contention window (CW_1) and in the second one (CW_2) along with the expected number of nodes that pass from CW_1 to CW_2 (under saturated and unsaturated traffic loads), the simulation results from this Section verify the analytical models (individual slot state analysis approach and combinatorics one). The collision probabilities of SCP two-phase contention mechanism are investigated whilst considering saturated and unsaturated traffic load regimes. We show that the CW_2 collision probability is directly related to the expected number of nodes that pass to CW_2. For the expected number of nodes that pass from CW_1 to CW_2, simulation and analytical results are well fitted. The results from the model under the unsaturated regime present some deviation relatively to the simulation results, although they follow them for larger number of nodes. For $CW_k^{max} = 8$, $k \in \{1, 2\}$, the CW_2 collision probability analytical curves are similar to the simulation ones, except for $n \leq 20$ and $n \leq 70$, for the saturated and unsaturated regimes, respectively. For $CW_k^{max} = 16$, $k \in \{1, 2\}$, the analytical results match the simulation ones, except for $n \leq 40$ in the saturated regime. In the unsaturated case, some deviation from the simulated results occurs for larger number of contending nodes. The analytical model underestimates the simulated collision probability in CW_2, presenting a maximum deviation of approximately 5 % from the simulated value. Once again, the observed deviations are due to the rounding operations of the expected number of nodes that pass from CW_1 to CW_2. From our analytical model for the collision probability in the two-phase contention window mechanism we derive the average MAC service time for a successful transmission and the achieved throughput. The average service time increases while the network throughput decreases significantly as the number of sensor nodes and the contention windows sizes increase. The average service time and throughput are bounded by an asymptotic value when the collision probability in the contention window CW_1 reaches the maximum value.

5.6 Simulation of SCP in the Context of IEEE 802.15.4

5.6.1 IEEE 802.15.4 Compliant

The SCP has been first implemented by us in the OMNeT++ simulator [Var10], using the Mobility Framework initially introduced by the authors from [REHD08]. Then, the SCP code has been ported to the new MiXiM Framework [KSW+08]. The MiXiM Framework supports CC1100 [CC107] and CC2420 [Ins07] radio energy consumption models, as well as several propagation models. Moreover, we have extended it to work with AT86RF231 [AT809] radio transceiver. The latter is the best radio available in the market in terms of energy consumption. From all the radio transceivers the CC1100 is the only one that is non-compliant IEEE 802.15.4. We intend to compare the performance of SCP with IEEE 802.15.4 compliant and non-compliant transceivers. IEEE 802.15.4 is a double standard: PHY and MAC layers. By definition, if we program a board with our own MAC, it cannot be IEEE 802.15.4 compliant. The reason for using an IEEE 802.15.4 compliant radio is that hundreds of millions of 15.4 chips have been sold in the world and it is by the far the *de facto* standard for those low power networks. The key aspects of the IEEE 802.15.4 standard are the following: the transmitter output power can be tuned from approximately -30 dBm to +3 dBm on the majority of the radios; the sensitivity has to be <-85 dBm, but the sensitivity for some radios can go as low as -101 dBm; there are 16 different frequency channels (2.405 GHz+n×5 MHz), which are orthogonal (i.e., adjacent channels do not interfere); a radio chip consumes power when it is on (about the same power when transmitting, receiving or idle listening), and no power when the radio is off; it takes 192 μs for a radio chip to perform the transition from Transmission (TX) to Reception (RX) mode, and vice-versa; The maximum length of the IEEE 802.15.4 frame should be 127 Bytes. We consider an IEEE 802.15.4 non-beacon scheme in our simulations.

Our simulator considers single and multi-hop network topologies. The nodes may communicate only with reachable neighbour nodes. The deployment area defined in the configuration file for the single-hop scenario is 400×400 m^2, while for the multi-hop scenario it is 800×100 m^2. For each experiment, a set of six seeds is chosen for the random generator of SCP.

In our simulator protocol stack, each layer is connected with the higher and lower layers. There is an application module layer connected to the MAC and PHY layer modules. Besides these modules, a battery module and statistics battery module are included in order to simulate the network or node lifetime, while varying the capacity of the batteries. In the end of the simulation the statistics battery module presents the battery statistics collected during the simulation. Nodes are assumed static.

The typical values for the current consumption, data rate, and sensitivity of the radios are also defined in the simulation configuration file, as presented in Table 5.4. The sensor nodes battery voltage is 3.3 V. The following channel parameters are also defined in the MiXiM simulation Framework: $max\_TXPower$ corresponds to the maximum transmission power allowed in the channel (1.99 mW); $txPower$ is the transmission power (1 mW), while $thermalNoise$ is the electronic noise generated by the thermal agitation (-110 dBm). The Analogue Model Type parameter is the propagation model used in the channel ($SimplePathlossModel$), while $alpha$ is the propagation exponent for the free-space path loss formula (i.e., 2). The $berLowerBound$ parameter is the lower bound of the bit-error-rate (1×10^{-8}). Modulation type is the modulation that the transceiver works with (MSK for IEEE 802.15.4 non-compliant radio transceivers and OQPSK for IEEE 802.15.4 compliant ones). The $queueLength$ parameter is the maximum number of frames waiting to be transmitted allowed in the queue (i.e., 50). In our simulations, the three radio models (from MiXiM) do not include the poll state [Var10]: when a node wakes-up and does not have data frames in its MAC queue, it polls the channel for 2 ms, while the power consumption is the same as in the receiving state.

Table 5.4 Typical values for CC1100, CC2420 and AT86RF231 radios (P_{tx}=1 mW).

Symbol	Meaning	CC1100	CC2420	AT86RF231
I_{tx}	Current in transmitting	16.9 mA	17.4 mA	11.6 mA
$I_{rx}=I_{listen}$	Current in receiving/listening	16.9 mA	18.8 mA	12.3 mA
I_{poll}	Current in polling	16.4 mA	18.8 mA	12.3 mA
I_{sleep}	Current in sleep	0.02 mA	0.000021 mA	0.00002 mA
r_B	Bit rate	250 kbps	250 kbps	250 kbps
S_{min}	Sensitivity	-111 dBm	-94 dBm	-101 dBm
f_b	Carrier Frequency	868 MHz	2.4 GHz	2.4 GHz

5.6.2 Comparison between IEEE 802.15.4 Compliant and IEEE 802.15.4 Absence

Power Consumption without Piggyback (periodic traffic)

It is worthwhile to compare simulation and analytical results for the average power consumption for each node. When no piggybacking is considered, we define the "adapted" analytical model (lower and upper bounds) from the original SCP work [YSH06a] and ran the simulator to extract the values.

The employed topology considers five nodes in star topology. In this simulations, the *sync slave* mode is enabled, the synchronization time is defined as t_{sync}= 60 ms and the inter arrival period is variable. Moreover the simulator also considers, t_{sync}, the synchronization period, which depends on the inter-arrival period (already defined by the authors from [YSH06a]), T_{sync}. As the synchronization period increases the wake-up tone duration also increases. We consider for the maximum contention window size CW_1^{max}= 8 and CW_2^{max}= 16. For this experiment all five nodes from the network form in a single-hop configuration, where each node generates and broadcasts 50 data frames (with 50 Bytes each). Since the inter-arrival period varies between 50 and 300 s, the traffic load is considered to be light. Simulations are run five times for each of the six seeds, generating a total of thirty experiments. For each experiment the average power consumption of each node is obtained, as shown in Figure 5.30. The achieved 95 % confidence intervals are negligible.

Figures 5.30 and 5.31 presents a power consumption comparison between simulation and analytical results (lower and upper bound), as a function of the inter-arrival period for three different radio transceivers. The CC1100 is the

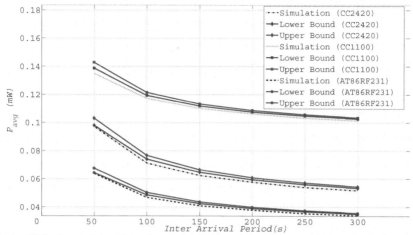

Figure 5.30 Lower/upper bound and simulation results with no piggyback for the power consumption per node of CC1100, CC2420 and AT86RF231.

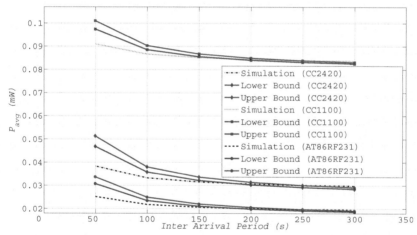

Figure 5.31 Lower/upper bound and simulation results with piggyback for the power consumption per node of CC1100, CC2420 and AT86RF231.

one that presents the highest power consumption when is compared with the CC2420 and AT86RF231 ones.

The CC2420 transceiver consumes approximately 0.04 mW less than the CC1100 one, while AT86RF231 consumes around 0.08 mW less. This is due to the low power consumption that the AT86RF231 presents combined with the long times the node remains in sleep mode. Besides, the MSE is calculated with respect to the analytical lower bound curve, and a value of 5.1319×10^{-7} mW for CC1100, a value of 1.4213×10^{-6} mW for CC2420 and a value of 6.3434×10^{-7} mW for AT86RF231 is obtained. This shows a high similarity between the simulation and analytical results.

If the t_{sync} parameter is changed from 60 to 50 ms while maintaining the remaining parameters, the power consumption becomes less similar to the lower bound (in comparison with the previous one). From these results, one may conclude that the power consumption with the *sync slave* mode enabled presents better results when t_{sync}= 60 ms. However, for values of t_{sync} lower than 50 ms it was not possible to achieve stable results with the SCP simulations. Moreover, for values of t_{sync} higher than 60 ms, higher values for power consumption are achieved.

Power Consumption with Piggyback (periodic traffic)

The lower and upper bound for the "adapted" analytical model are also defined for the case when synchronization schedules are piggybacked, whilst maintaining the simulator topology. However, since all the exchanged traffic is of the broadcast type, all the SYNC frames can be suppressed and the schedule synchronization is performed by using the piggyback technique. Hence, the piggyback mode is enabled for this experiment. There is no need to define the values for the parameter t_{sync}, as the schedule information is already piggybacked into the data frames, leading to a consequent decrease of node energy consumption. For this set of simulations, the wake-up tone duration is always equal to 2 ms. The only parameter that varies is $T_{data} \in$ {50; 100; 150; 200; 250; 300}.

The dependence of the average power consumption of each node on the inter-arrival period is shown in Figures 5.30 and 5.31, where the simulations and analytical results (upper and lower bounds) are also presented for comparison purposes. The achieved 95 % confidence intervals are negligible.

For CW_1^{max}= 8 and CW_2^{max}= 16, the obtained simulation results are similar to the lower bound curve defined by the analytical model until the inter-arrival period achieves 200 s. For an inter-arrival period higher than 200 s, the simulation results are similar to the upper bound curve and even superimposes the upper bound when it achieves 300 s. The MSE values obtained for this experiment with respect to the lower bound curve are equal to 7.9825×10^{-6} mW, 1.3524×10^{-5} mW and 5.8479×10^{-6} mW, for the CC1100, CC2420 and AT86RF231 transceivers, respectively.

By comparing the power consumptions with explicit SYNC frames and with piggyback synchronization, we observe that the piggyback technique is truly efficient, since the power consumption with piggybacking achieves lower values than the one with explicit SYNC frames. In this set of experiments the poll period is set to 5 s, which is more than enough for example for a healthcare monitoring application reporting sensors values. Therefore, in a healthcare application, if we use a sensor node with the AT86RF231 radio transceiver whilst enabling the piggyback synchronization, the node lifetime will increase significantly due to the low associated power consumption.

Throughput Performance with Heavy Traffic Load

In real world applications, WSNs do not have the capability to predict what type of traffic load is being exchanged with high accuracy. For example, in a healthcare scenario, the sensor nodes deliver the monitoring data from the vital signs of the patients at a lower data rate (in usual situations). However, when a patient presents anomalies in the vital signs the sensor node may deliver data at a higher rate, leading to the need for higher network throughput.

For this set of experiments, we have enabled the *Throughput* mode at the application layer, in order to test the network's response capability as the number of transmitting nodes increases. By varying the number of transmitting nodes the contention time of the network will increase, leading to an increase of the collision probability. Each node generates 20 frames (with 100 Bytes each). A five nodes in star topology is considered. In these experiences the nodes poll the channel each second. Besides varying the number of transmitting nodes, the CW_1 and CW_2 contention window sizes also vary.

The throughput is the ratio between the amount of data frames received with success by the destination node and the time taken to arrive at this node. By analysing Figure 5.32, all the results for the throughput (for all the three radio transceivers) match while the contention window sizes vary. The throughput is maximum (for all contending time configurations) when there is only one node transmitting, as expected, since there is no more nodes competing for the channel to send frames. When two nodes are competing for the channel, the throughput drops around 25 % of the maximum achievable throughput, because there is more than one node trying to transmit the data frame. As the number of transmitting nodes increases, the contending time will also increase, leading to a collision probability increase.

In terms of energy consumption when considering the IEEE 802.15.4 compliant radios (with high duty cycles) the CC2420 energy consumption is higher than the CC1100 one, because the CC2420 achieves lower power

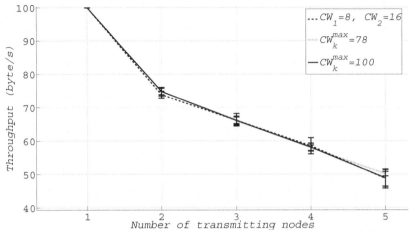

Figure 5.32   SCP throughput as a function of the number of transmitting nodes and maximum contention windows sizes, $k \in \{1, 2\}$.

consumption when the radio transceiver remains more time in sleep mode. Since the *Throughput* mode enables the node to wakeup every second to poll the channel, the energy consumption will increase when compared with the CC1100. The AT86RF231 radio transceiver is the one that presents the lowest power consumption (from all the three radio transceivers), mainly due to the lowest values of the power consumptions it achieves whatever the operating mode is.

Node energy consumption for Linear Chain (multi-hop)

By considering experiences with multi-hop topology in a linear chain, we intend to evaluate the SCP energy and latency performance. We enabled the *multi_hop* mode, where only the application layer from node 1 generates data frames to be sent to the other nodes. Node n receives data frames and answers with ACK frames when a data packet is received correctly, as presented in Figure 4.7. The wake-up tone duration is always set to be equal to 2 ms, whilst each node polls every second. In our topology, we have deployed $n - 2$ forwarders nodes, one sink node, and one source node. For the multi-hop evaluation the acknowledgement mode may be enabled/disabled.

In multi-hop networks, there is no optimal working point and network performance depends on what application the network supports. The topology used in this set of experiences is the one presented in Figure 4.7, with a network composed by a 9-hop linear topology with ten nodes (each one separated by 50 m from each other).

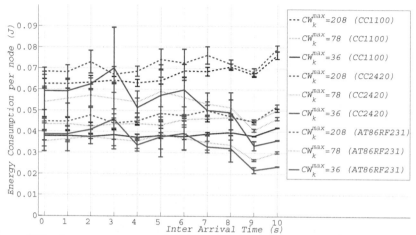

Figure 5.33 Average energy consumption as a function of the inter-arrival time and different contention windows sizes for the multi-hop experiments, $k \in \{1, 2\}$.

The source node generates 20 frames (each with 50 Bytes), with an inter-arrival period that varies between 0 and 10 s. Nodes will rebroadcast the data frames along the linear chain until the packet reaches the most distant node from the chain. Considering an inter-arrival time equal to zero means the application layer generates data frames and sends it to the lower layers the fastest way possible, as previously explained in the $Throughput$ mode.

For a contention window sizes configuration of $CW_1^{max} = CW_2^{max} = 208$, this combination is the one that presents the highest energy consumption per node for all the three radio transceivers, Figure 5.33, due to the maximum contention time nodes use to avoid the frames collisions. Comparing the energy consumptions between the three radio transceivers, the one that presents the lowest mean energy consumption is the AT86RF231 one. The CC2420 radio transceiver presents energy consumptions similar to the CC1100 radio transceiver. The energy consumptions are lower when comparing with the energy consumptions for multi-hop experiments in the original paper [YSH06a].

For all the configurations the energy consumption behaviour is approximately stable as the inter-arrival time is increasing.

Node latency (multi-hop)

This section addresses the SCP latency over a 9-hop network. We alternate the polling period between 0.15 s and 1s, enable or disable the

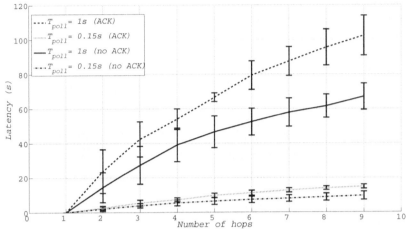

Figure 5.34 Mean frame latency over 9 hops as a function of number of hops, for CW_1^{max} = CW_2^{max} = 16, while enabling/disabling the ACK frames feature.

acknowledgement mode, and set the contention windows maximum sizes to CW_1^{max}= CW_2^{max}= 16 slots. When the inter-arrival time is equal or higher than 1 s (light traffic) it is expected that only one frame is sent through one hop, during one polling interval. When the polling period is changed to 0.15 s we expect that the latency decreases significantly, since the node will wake-up almost seven times in one second, and will schedule and send more data frames in less time.

Regarding Figure 5.34, we may conclude that the latency curve when SCP polls each second (with the ACK frames feature enabled) is the one that presents the highest delay, due to the data frames retransmissions. When the polling period is set to 0.15 s, the maximum data a frame latency is less than 20 s, but the energy consumption increases due to the number of times the node wakes-up. The other two latency curves are related with the same experiments, but the ACK frames feature is disabled. With the ACK frames feature disabled for all three radio transceivers, we observe that the maximum data frame latency is less than 70 s (at the end of the linear chain), when polling each second. Moreover, when polling at each 0.15 s (with the ACK frames feature disabled), the latency is less than 10 s.

Node data a frame success rate (multi-hop)

To observe if there is any tradeoff when the ACK frames feature is enabled or disabled, we have plotted the DPSR for all four configurations described previously. Figure 5.35 shows that, with the ACK feature enabled for both

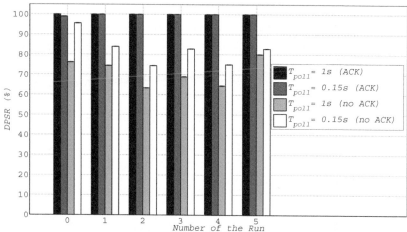

Figure 5.35   DPSR as a function of the simulator run over 9 hops, for $CW_1^{max} = CW_2^{max} = 16$, while enabling/disabling the ACK feature.

polling periods, there is around 100 % frame delivery success for all the three radio transceivers. When the ACK feature is disabled for both polling periods, the frame delivery in both experiments varies around 75 and 89 % for all the considered radio transceivers. Regarding these results, enabling ACK frames feature will result into an increase of data frame delivery. However, the tradeoff is the increase of the latency and energy consumption.

5.6.3 Summary and Conclusions

In this Section, we provide a detailed analysis of the mechanism used in SCP, learned from our own experience while implementing the SCP protocol in the MiXiM Framework. Our results help to clarify some missing aspects in the original SCP protocol description, as well as providing an evaluation of performance metrics (for single and multi-hop topology) by means of simulation. By comparing the power consumptions in the presence and absence of a fixed wake-up tone duration, the energy savings obtained with a fixed wake-up tone duration are adequate (more or less 0.005 mW). The use of IEEE 802.15.4 compliant radio transceivers leads to a clear decrease in the energy consumption in the presence and absence of piggybacking. Considering a multi-hop linear chain of sensor nodes, the SCP energy consumption is stable as the inter-arrival time varies. However, only the AT86RF231 transceiver presents lower mean energy consumption compared with the radio transceiver that does not complies with the IEEE 802.15.4 standard. Besides, the values of the achieved throughput are similar for both types of transceiver. In terms of delay, all the three radio transceivers present the same data frame latency, since the transmission data rate is the same and the specific time parameters of all these radio transceivers are approximately the same. Considering a healthcare application where the nodes report periodically to a sink node the values from the sensors, the SCP can be selected has the MAC protocol that can deliver data efficiently to the sink with low power consumption cost. However, some tradeoff exists when choosing the reporting period of the node. The latency, DPSR and energy consumption (considering a linear chain of sensor nodes in the Hospital environment) should be well balanced in order to fulfil the minimum requirements for the application.

5.7 Final Remarks

This chapter addressed four main issues: i) simulation of SCP-MAC in single-hop and multi-hop scenarios, ii) the proposal of a stochastic model for the collision probability, iii) the application of the model to derive the average MAC service time for a successful transmission as well as the achieved throughput, and iv) the comparison of the performance metrics of the SCP while considering an IEEE 802.15.4 compliant and IEEE 802.15.4 agnostic physical layer.

The SCP performance evaluation, which was implemented in the simulator, included the analysis of different performance metrics (in single-hop and multi-hop scenarios), such as power consumption (with and with no Piggyback), energy consumption and throughput, considering heavy and light periodic traffic patterns. It has been shown in Section 5.1 the adequacy between the model(s) and simulation results. Besides, the use of piggybacking technique has led to a gain of 10-30 % in the power consumption compared to explicit SYNC frames (piggybacking is not enabled). In addition, has also been investigated how some values of the parameters from SCP may cause changes in the performance. By setting the duration of the wake-up tone to a fixed value of 2 ms (minimum duration of the wake-up tone in SCP) an effective decrease of the energy consumption is achieved.

The comparison between different CW sizes has shown that, for a $CW_1^{max} = CW_2^{max} = 78$ and a $CW_1^{max} = CW_2^{max} = 100$ configurations, the achieved throughput are higher and more stable (80 B/s) than with $CW_1^{max} = CW_2^{max} = 16$ slots. It can also be concluded that the node that employs a larger maximum contention windows size presents a shorter lifetime.

A stochastic model for the collision probability in the collision avoidance mechanism with two contention windows from the SCP protocol has been proposed, whilst considering the saturated and unsaturated regimes. Results have shown that the CW_2 collision probability is directly related to the expected number of nodes that pass to CW_2. It has also been shown that, for the expected number of nodes that pass from CW_1 to CW_2, simulation and analytical results fit well.

The use of IEEE 802.15.4 compliant radio transceivers leads to a clear decrease in the energy consumption in the presence and absence of piggybacking. In terms of delay, all the three radio transceivers present the same data frame latency, since the transmission data rate is the same and the specific time parameters of all these radio transceivers are approximately the same.

6

MAC Sub-layer Protocols Employing RTS/CTS with Frame Concatenation

6.1 Introduction

Frame concatenation facilitates the aggregation of several consecutive frames by means of channel reservation and different types of acknowledgment and/or Network Allocation Vector (NAV) procedures. In this context, the RTS/CTS mechanism enables to reserve the channel and avoids to repeat the backoff phase for every consecutive transmitted frame, probably reducing overhead. In the presence of RTS/CTS two solution are considered, one with DATA/ACK handshake and other without ACKs, simply relying in the establishment of the NAV.

In particular, the Sensor Block Acknowledgement MAC (SBACK-MAC) protocol allows the aggregation of several acknowledgement responses into one special frame *BACK Response* being compliant with the IEEE 802.15.4 standard. Two different solutions are addressed. The first one considers the SBACK-MAC protocol in the presence of *BACK Request* (concatenation mechanism), while the second one considers SBACK-MAC in the absence of *BACK Request* (the so-called piggyback mechanism).

Therefore, we reduce the overhead, which is one of the fundamental problems of MAC inefficiency. The proposed protocol uses detailed information from the PHY (collision detection, data frame synchronization, CCA and state of the radio transceiver) and data link layers (carrier sense control, radio duty cycle control, use of fragmentation and retransmissions). The throughput and delay performance is mathematically derived under ideal conditions (a channel environment without transmission errors) and non ideal conditions (presence of interference). The SBACK-MAC will also be evaluated in terms of bandwidth efficiency and energy consumption, and the proposed schemes are compared against IEEE 802.15.4 with and without RTS/CTS by means of extensive simulations by employing the OMNeT++ simulator.

6.2 Motivation

The design of WSNs MAC protocols envisages satisfying application-specific QoS requirements, whilst maximizing the network lifetime. Considerable energy savings can be achieved by placing nodes into the sleep state during the periods when they are not receiving or transmitting. These nodes opportunistically enter into the sleep mode after successfully receiving RTS/CTS control frame reservation messages. In WSNs, the use of ACK frames introduces overhead. Since the length of these frames may be similar to the one of the data frames, they will increase the collision probability inside a given cluster. Moreover, nodes are battery operated; therefore the transmission of such frames also leads to energy decrease whilst reducing the number of data frames transmitted containing useful information. Based on Sensor-MAC (S-MAC) [YHE04b], an innovative MAC protocol that uses a BACK mechanism is proposed, i.e., the SBACK-MAC protocol. The main difference from SBACK-MAC relatively to S-MAC is the way it treats ACK control frames and the fact of being totally compliant with IEEE 802.15.4, e.g., in terms of interframe spacing. By using a BACK mechanism we improve channel efficiency by aggregating several ACK frames into one special ACK frame (*BACK Response*). This way, energy consumption may be significantly reduced when a series of data messages needs to be transmitted, because it is not necessary to transmit and receive several ACK control frames (one for each data frame) which would lead to an extra energy waste while increasing the delay and decreasing the throughput as well as bandwidth efficiency. Moreover, these extra control signalling frames are considered to be overhead and do not not directly result in the communication of information.

Likewise other MAC protocols [WLA05, WPA11, KA13], SBACK-MAC is a contention-based MAC protocol that considers a duty cycle scheme in order to reduce the energy consumption. It considers the IEEE 802.15.4 nonbeacon-enabled mode and uses unslotted carrier sense multiple access with CSMA-CA for transmitting frames. When nodes have data to send, they will contend for the wireless channel, whereas if a collision occurs, or the medium is found to be busy, nodes will perform backoff for a random duration of time before attempting again to access the channel later on.

6.3 Design Considerations for IEEE 802.15.4 Non-beacon-Enabled Mode with Frame Concatenation

The IEEE 802.15.4 standard has been widely accepted as the *de facto* standard for WSNs, enabling to provide ultra-low complexity, cost and power for low-data rate wireless connectivity for wireless sensors. Due to its low power, it has been used as a basis for ZigBee®, WirelessHart and MiWi applications. Moreover, it represents a significant breakthrough from the "bigger and faster" standards that the IEEE 802 organization continues to develop and improve: instead of higher data rates and more functionality, this standard addresses the simple and low-data universe, in terms of control and sensor networks, which existed without global standardization through a series of proprietary methods and protocols. The IEEE 802.15.4 standard can operate in the beacon-enabled and non-beacon-enabled modes, which is presented in more detail in Chapter 3. In this report, we assume that the IEEE 802.15.4 non-beacon-enabled CSMA-CA algorithm for the basic access mode is employed by the SBACK-MAC protocol. Next we address the design considerations of the PHY and MAC layer aspects for IEEE 802.15.4 and SBACK-MAC.

6.3.1 PHY Layer

The PHY layer is responsible for providing status information for the MAC layer, switch between different radio states (i.e., RX, TX and SLEEP), send/receive/listen frames to/from channel, provide hooks for statistical information and configurable settings. The information can be passively (e.g., current channel state) or actively provided to the MAC layer (e.g., frame transmission complete) based on events. Some information must be provided on demand, as follows:

- Channel state (busy/idle) and Received Signal Strength Indicator (RSSI);
- Current radio state (e.g., RX, TX and SLEEP);
- Control information (e.g., transmission over and send).

Figure 6.1 presents the state transition diagram for switching between the different radio states. The time for switching mainly depends of the IEEE 802.15.4 compliant transceivers.

In order to maintain a simple interface, both MAC and PHY layers share a common frame structure, as presented in Figure 6.2. This type of frame is known as the PPDU, being responsible for the encapsulation of all data structures from the higher layers of the protocol. The frame is divided into three basic components: Synchronisation header (SHR), PHY header (PHR) and variable length payload, which contains the PSDU as follows:

- The SHR consists basically of two fields, a preamble sequence and a start of frame delimiter. The preamble field has a length of 4 bytes, allowing

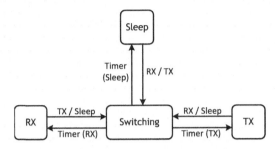

Figure 6.1 State transition diagram for the switching between different radio states.

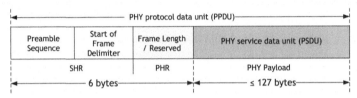

Figure 6.2 IEEE 802.15.4 PHY protocol data unit.

for achieving chip and bit synchronisation. The Start of frame delimiter (SFD) has a length of 1 byte and allows a receiver to set up the beginning of a frame;

- The PHR field has a length of 1 byte, where the Most significant bit (MSB) is reserved and the other 7 bits are used to specify the frame length information, enabling to have frames with a total length of 127 bytes;
- The PHY payload contains one field called PSDU that carries the data payload from the PPDU.

The IEEE 802.15.4 [WPA11] PHY layer, (implemented in the OMNeT++ Network Simulation Framework [OMN13]) operates in the 2.4 GHz band and considers an O-QPSK modulation, offering extremely low BER performance, in the presence of a low SNR. The O-QPSK modulation is presented in detail in Appendix C.

The BER is addressed by considering an interference signal similar to Additive white gaussian noise (AWGN). The radio transceivers operating in the 2.4 GHz band employ a DSSS spreading technique. As a consequence, the transmitter signal takes up more bandwidth than required to transmit the information signal being modulated. The name "spread spectrum" derives from the fact that the carrier signals occur over the full bandwidth (spectrum) of the sensor nodes transmitting frequency band.

The modulation and spreading technique uses a 16-ary quasi-orthogonal modulation, where four information bits are used to select one of 16 nearly orthogonal Pseudo-random noise (PN) sequences to be transmitted. Therefore, we use a power-efficient modulation method that achieves low SNR and Signal-to-interference ratio (SIR) requirements.

The O-QPSK modulation uses a chip rate, R_c, of 2 Mchip/s and a bit rate, R_b, of 250 kb/s, by using a codebook of $M = 16$ symbols. The conversion from SNR to the noise density, (E_b/N_0) assumes a matched filtering and a half-sine pulse shaping, as shown in Equation 6.1:

$$\frac{E_b}{N_0} = \frac{0.625 \cdot R_c}{R_b} SNR = \frac{0.625 \times 2000000}{250000} SNR = 5 \cdot SNR \quad (6.1)$$

where SNR is the Signal-to-noise ratio.

Equation (6.2) presents the conversion of these ratios from bit noise density, E_b/N_0, to symbol noise density, E_s/N_0:

$$\frac{E_s}{N_0} = log_2(M)\frac{E_b}{N_0} = 4 \cdot \frac{E_b}{N_0} \quad (6.2)$$

As shown in [Ber01], the probability of Symbol error rate (SER), P_s, is computed by using the following equation:

$$P_s = \frac{1}{M}\sum_{j=2}^{M}(-1)^j \binom{M}{j} e^{\left(\frac{E_s}{N_0}\left(\frac{1}{j}-1\right)\right)} \quad (6.3)$$

Finally, for an M-ary orthogonal signal, the conversion from P_s to the probability of bit error, P_b, is given by:

$$P_b = P_s \left(\frac{M/2}{M-1}\right) = P_s \left(\frac{8}{15}\right) \quad (6.4)$$

By combining Equations (6.1) to (6.4), one obtains the BER function:

$$BER = \left(\frac{8}{15}\right)\left(\frac{1}{16}\right)\sum_{j=2}^{M}(-1)^j \binom{M}{j} e^{\left(\frac{E_s}{N_0}\left(\frac{1}{j}-1\right)\right)} \quad (6.5)$$

Since the modulation and spreading technique consists of using a 16-ary quasi-orthogonal modulation technique, one obtains the following equation for BER.

$$BER = \left(\frac{8}{15}\right)\left(\frac{1}{16}\right)\sum_{j=2}^{M}(-1)^j \binom{16}{j} e^{\left(20 \cdot SNR\left(\frac{1}{j}-1\right)\right)} \quad (6.6)$$

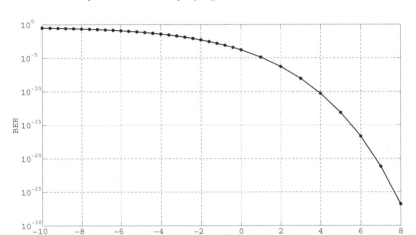

Figure 6.3   BER as a function of the SNR for IEEE 802.15.4 and SBACK-MAC.

Figure 6.3 presents the BER as a function of the SNR for both SBACK-MAC and IEEE 802.15.4 (by considering the 2.4 GHz band). The Signal-to-interference-plus-noise ratio (SNIR) values correspond to the BER lower bound, and are lower than 1×10^{-8}. In our simulator, this value is used as a parameter for the network. The Packet success rate (PSR), is defined in [KW05a] as the probability that the frame has no errors, and is given by:

$$PSR = (1 - BER)^{(nBits-1)} \qquad (6.7)$$

where $nBits$ is the length of the frame, in bits.

The Packet error rate (PER) for a given frame is given by:

$$PER = (1 - PSR) \qquad (6.8)$$

Figure 6.4 shows the PERs for different SNRs by considering 2 different data frames lengths with data payload of 118 bytes and 18 bytes respectively, and an ACK control frame with length of 11 bytes.

By analyzing Equation (6.8) and Figure 6.4, one concludes that, by increasing the SNR, the ratio between the received power and noise is increased, so that the PER decreases. Moreover, for longer frame lengths (i.e., for data frames) the PER is higher. In [BVBL14] the authors have proposed a reliability based decider to improve the frame reception rate at the PHY layer which is characterized by metrics like the SNIR, BER, SER, PER, and outage probability, whose implementation in the MiXiM framework of the OMNeT++ simulator is considered here.

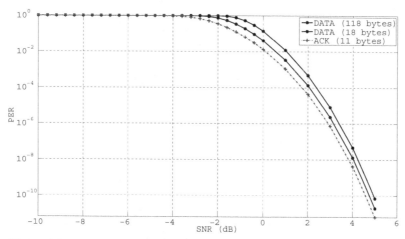

Figure 6.4   BER as a function of the SNR for IEEE 802.15.4 and SBACK-MAC.

6.3.2 MAC Sub-layer

In the IEEE 802.15.4 [WPA11] basic access mode, nodes use a non-beacon-enabled CSMA-CA algorithm for accessing the channel and transmit their frames. The backoff phase (N.B., this time period is not generally called contention window in IEEE 802.15.4) algorithm is implemented by considering basic units of time called backoff periods. The backoff period duration is equal to $T_{BO} = 20 \times T_{symbol}$ (i.e., 0.32 ms), where $T_{symbol} = 16\mu$s is the symbol time [WPA11]. Before performing CCA, a device shall wait for a random number of backoff periods, determined by the backoff exponent (BE). Then, the transmitter randomly selects a backoff time period uniformly distributed in the range $[0, 2^{BE} - 1]$. Therefore, it is worthwhile to mention that even if there is only one transmitter and one receiver, the transmitter will always choose a random backoff time period within $[0, 2^{BE} - 1]$. Initially, each device sets the BE equal to $macMinBE$, before starting a new transmission and increments it, after every failure to access the channel. In this work we assume that the BE will not be incremented since we are assuming ideal conditions.

Table 6.1 summarizes the key parameters for both the IEEE 802.15.4 standard and the SBACK-MAC protocol for the 2.4 GHz band, by considering the DSSS PHY layer with the O-QPSK modulation, which is described in detail in Chapter 3.

IEEE 802.15.4 [WPA11] nodes support a maximum over-the-air data rate of 250 kb/s. However, in practice, the effective data rate is lower due to the

Table 6.1   Parameters, symbols and values for the IEEE 802.15.4 standard and the proposed MAC sub-layer protocols employing RTS/CTS with frame concatenation for the 2.4 GHz band, by considering the DSSS PHY layer with the O-QPSK modulation.

Description	Symbol	Value
Symbols per octet for the current PHY	$phySymbolsPerOctet$	2 symbols
PHY SHR length	L_{SHR}	10 symbols
RX/TX or TX/RX maximum turnaround length	$aTurnaroundTime$	12 symbols
Backoff period length	$aUnitBackoffPeriod$	20 symbols
Symbol period	T_{symbol}	$16\mu s$
Backoff period duration	T_{BO}	$320\mu s$
PHY SHR duration	T_{SHR}	$160\mu s$
CCA detection time	T_{CCA}	$128\mu s$
Setup radio to RX or TX states [RDA$^+$09]	$rxSetupTime$	$1720\mu s$
Time delay due to CCA	$ccaTime$	$1920\mu s$
TX/RX or RX/TX switching time	T_{TA}	$192\mu s$
PHY length overhead	L_{H_PHY}	6 bytes
MAC overhead	L_{H_MAC}	9 bytes
DATA payload	L_{DATA}	3 bytes
DATA frame length	L_{FL}	18 bytes
ACK frame length	L_{ACK}	11 bytes
Short interframe spacing (SIFS) time	T_{SIFS}	$192\mu s$
Long interframe spacing (LIFS) time	T_{LIFS}	$640\mu s$
ACK transmission time	T_{ACK}	$352\mu s$
Request-to-send (RTS) transmission time	T_{RTS}	$352\mu s$
Clear-to-send (CTS) transmission time	T_{CTS}	$352\mu s$
RTS ADDBA transmission time	T_{RTS_ADDBA}	$352\mu s$
CTS ADDBA transmission time	T_{CTS_ADDBA}	$352\mu s$
BACK Request transmission time	$T_{BRequest}$	$352\mu s$
BACK Response transmission time	$T_{BResponse}$	$352\mu s$
ACK wait duration time	T_{AW}	$560\mu s$
Number of TX frames	n_{frames}	1 to 112
Data Rate	R	250 kb/s

protocol timing specifications, [WPA11]. This can also be explained by the various mechanisms that are employed to ensure robust data transmission, including channel access algorithms, data verification and frame acknowledgment. In this work we address unicast data transmissions with ACKs, being the channel access time a dominant factor in the overall performance of the network. The regular procedure of the IEEE 802.15.4 non-beacon-enabled mode is presented in Figure 6.5. When a device wishes to transfer data, it simply transmits its data frame, using unslotted CSMA-CA, to the

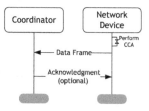

Figure 6.5 IEEE 802.15.4 - Communication to a coordinator in a non-beacon-enabled PAN.

coordinator. The coordinator acknowledges the successful reception of the data by transmitting an ACK control frame.

In this research work, we only consider the non-beacon-enabled mode. The beacon-enabled mode is not considered because collisions can occur between beacons or between beacons and data or control frames, making a multi-hop beacon-based network difficult to be built and maintained [SSZ09]. Another important attribute is scalability, due to changes in terms of network size, node density and topology. Nodes may die over time. Other nodes may be added later and some may move to different locations. Therefore, for such kind of networks, the non-beacon-enabled mode seems to be more adapted to the scalability requirement than the beacon-enabled mode. In the former case, all nodes are independent from the PAN coordinator and the communication is completely decentralized.

Moreover, for beacon-enabled networks [WPA11], there is an additional timing requirement for sending two consecutive frames, so that the ACK frame transmission should be started between the T_{TA} and $T_{TA}+T_{BO}$ time periods, and there is time remaining in the CAP, for the message, appropriate, Interframe space (IFS) and ACK. Figure 6.6 presents the timing requirements for transmitting a frame and receiving an ACK for the beacon and non-beacon-enabled modes, respectively.

In IEEE 802.15.4 [WPA07a, WPA11], the CSMA-CA algorithm is significantly different from the one used in IEEE 802.11e [WLA05]. The main differences are related to the backoff algorithm. In IEEE 802.11e [WLA05], the value of the Contention Window (CW) depends on the number of failed retransmissions for the frame, whereas, in the basic access mode for IEEE 802.15.4, this value (denoted as backoff phase) depends on the Backoff Exponent (BE), and Number of Backoffs (NB). Moreover, in IEEE 802.11e, the backoff time counter (BO_c) is decreased as long as the channel is sensed idle and is frozen when a transmission occurs. In the IEEE 802.15.4 basic access mode, nodes do not continuously monitor the channel during the

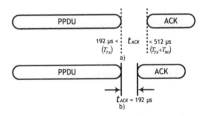

Figure 6.6   IEEE 802.15.4 acknowledgement frame timing: a) beacon and b) non-beacon-enabled modes.

backoff phase and the sensing phase (i.e., CCA) only occurs at the end of the backoff phase.

According to the IEEE 802.15.4 standard [WPA07a], a sensor node that sends a data or a MAC command frame with its ACK Request subfield set to one shall wait for at most an ACK wait duration period, T_{AW}, for the corresponding ACK frame to be received. The T_{AW}, already includes the time for the ACK frame itself. The transmission of an ACK frame in a non-beacon-enabled PAN or in the CFP shall start $aTurnaroundTime$ symbols (i.e., 192 μs) after the reception of the last symbol of the DATA or MAC command frame ([WPA07a], Section 7.5.6.4.2).

The ACK wait duration period, T_{AW}, is calculated as follows:

$$T_{AW} = T_{Symbol} + T_{TA} + T_{SHR} + [6 \times T_{symbol} \times phySymbolsPerOctet] \quad (6.9)$$

By considering the DSSS PHY layer for the 2.4 GHz band, the maximum ACK wait duration period, T_{AW}, is given by:

$$T_{AW} = 16\,\mu s + 192\,\mu s + 160\,\mu s + 192\,\mu s = 560\,\mu s \qquad (6.10)$$

Figure 6.7 presents the ACK timing required for the IEEE 802.15.4 standard, by considering the DSSS PHY layer for the 2.4 GHz band at 250 kb/s. The receivers start transmitting the ACK, 192 μs (i.e., T_{TA}) after the reception of the DATA frame. By assuming a DATA and an ACK frame with 18 and 11 bytes, respectively (including the PHY and MAC overhead), the transmission time is 576 μs and 352 μs, respectively. As reference, Figure 6.7 also includes the ACK wait duration period, T_{AW}.

For every DATA frame transmitted, there is a random deferral time period, D_T, before transmitting, which is given by:

$$D_T = InitialbackoffPeriod + ccaTime + T_{TA} \qquad (6.11)$$

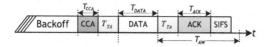

Figure 6.7 Acknowledgement process timing.

Figure 6.8 IEEE 802.15.4 basic access mode (the number of frames, n_{frames}, is represented by n).

The initial backoff period, $Initialbackoff Period$, is given as follows:

$$Initialbackoff Period = CW_{NB} = (2^{BE} - 1) \times T_{BO} \qquad (6.12)$$

The time delay, due to CCA, is given by:

$$ccaTime = rxSetupTime + T_{CCA} \qquad (6.13)$$

The $rxSetupTime$ is the time to setup the radio from a previous state to the transmission or reception states and it mainly depends on the radio transceiver used. During the T_{CCA}, the radio transceiver must determine the channel state within a 8 symbol duration (i.e., 128 μs, which corresponds to one symbol duration of 16 μs). In a normal transmission, for every DATA frame sent an ACK must be received, as shown in Figure 6.8.

6.3.3 Analytical Model for the Maximum Throughput and Minimum Delay

One of the fundamental reasons for the IEEE 802.15.4 standard MAC inefficiency is overhead. In this work, we consider the unslotted (non-beacon-enabled mode) version of the protocol for the 2.4 GHz band. The 2.4 GHz band allows having more channels at the highest data rate. Besides, the unslotted version is the one that has the least overhead as explained in the previous Section (Figure 6.6). The IEEE 802.15.4 sources of overhead are as follows:

- **Interframe Spaces** - For every IEEE 802.15.4 transmission, there is an idle period before accessing the medium. This idle period is called IFS. In IEEE 802.15.4 in the basic access mode for IFS, SIFS is used when MPDU (i.e., $F_{H\_MAC} + L_{DATA}$) is less or equal than 18 bytes; otherwise, LIFS is considered. The purpose of IFS is to regulate the data exchange flow and provide priority for certain types of transmissions;

- **Backoff Period** - When IEEE 802.15.4 nodes contend for accessing the wireless medium, they use a backoff algorithm. This process ultimately reduces collisions and allows for achieving QoS prioritization. The random backoff time represents a number of "slots" (i.e., periods of time) when the wireless medium must be idle;
- **PHY and MAC Headers** - In IEEE 802.15.4 in the basic access mode the PPDU must contain the SHR and the PHR fields in order to achieve reliable reception of frames. The MAC header contains information about how to coordinate nodes, and provide a fair mechanism to share the medium access among other nodes, being responsible for how and when it should use the PHY functions for accessing the shared physical medium by the nodes. Although the described headers are necessary, they introduce overhead and are responsible for decreasing the throughput;
- **Acknowledgements** - The lossy and inherently unreliable wireless medium imposes the use of ACK control frames in order to confirm that frames have successfully reached the destination. However, ACK frames are also considered as overhead, since from the point of view of the communication, they do not contain any useful information;
- **Interference** - In a very generic sense, all sources of interference create overhead, since IEEE 802.15.4 nodes must perform a CCA procedure to determine whether the wireless medium is busy or idle;
- **Retransmissions** - When a transmitted frame is not received by the intended node, i.e., there is no ACK response, retransmissions are required. Transmitting a frame more than once creates more overhead, e.g., repetition of the backoff phase, IFS, etc. Therefore, retransmissions are one of the most "problematic" sources of overhead, which are mitigated by using efficient retransmission mechanisms.

Next, we will analyse the maximum average throughput (S_{max}) and the minimum average delay (D_{min}) for the IEEE 802.15.4 standard. S_{max} is defined as the number of data bits generated from the MAC layer that can be transmitted per second on average to its destination including the ACK reception, on average. D_{min} is the time needed to transmit a frame and the successfully reception of the ACK or *BACK Response*, on average. In this analytical model, we will only consider the non-beacon-enabled version of the protocol. Although, we are considering the 2.4 GHz band, the proposed formulation is also valid for other different frequency bands as well as different PHY layer.

As explained before, initially, nodes uses the same backoff procedure like IEEE 802.15.4 basic access mode. Therefore, BE is set to $macMinBE$ (which corresponds to the minimum value of the backoff exponent, in the CSMA-CA algorithm). By considering the default value $BE = 3$ for $macMinBE$, and assuming that the channel is free, the worst-case channel access time that corresponds to the maximum backoff windows is given by Equation (6.12).

The average backoff window is given by:

$$\overline{CW} = \left(\frac{2^{BE} - 1}{2}\right) \times T_{BO} \qquad (6.14)$$

S_{max} and the D_{min} can be determined for the best scenario, i.e., the channel is an ideal channel without errors. During one transmission cycle, there is only one active node that has always a frame to be send, whereas the other neighbouring nodes can only accept frames and provide ACKs. An analytical model to evaluate S_{max} and D_{min}, is proposed. Table 6.1 presents the key parameters, symbols and values. Hence, there is no need to redefine every parameter after every equation again. The transmission times for the DATA and ACK frames are given as follows:

$$T_{DATA} = 8 \times \frac{L_{H\_PHY} + L_{H\_MAC} + L_{DATA}}{R} \qquad (6.15)$$

$$T_{ACK} = 8 \times \frac{L_{H\_PHY} + L_{ACK}}{R} \qquad (6.16)$$

The minimum average delay, D_{min}, in seconds, is given by:

$$D_{min} = (\overline{CW} + ccaTime + T_{TA} + T_{DATA} + T_{TA} + T_{ACK} + T_{IFS}) \qquad (6.17)$$

For IFS, SIFS is considered when the MAC protocol data unit, MPDU ($= L_{H\_MAC} + L_{DATA}$), is less or equal than 18 bytes; otherwise LIFS is considered.

In IEEE 802.15.4 basic access mode, the maximum average throughput, S_{max}, in bits per second, is given by:

$$S_{max} = \frac{8L_{DATA}}{D_{min}} \qquad (6.18)$$

By analysing Equations (6.17) and (6.18), we conclude that, if a short frame is transmitted, then the data transmission time is relatively short when compared to the associated overhead time, resulting into relatively low throughput. When a long frame is transmitted (by increasing the payload) data transmission time increases. This way, IEEE 802.15.4 is capable of achieving a much higher throughput.

Previously we have presented results for the channel with no errors for IEEE 802.15.4 basic access mode. However, data collisions may occur between neighbouring nodes if two or more nodes during the CSMA-CA perform CCA simultaneously, the channel is found to be idle and frame transmissions occur at the same time. Therefore, there is a need to study the impact of retransmissions in IEEE 802.15.4 networks by considering an erroneous channel. The flowchart of the CSMA-CA algorithm for the non-beacon-enabled mode is presented in Figure 6.9. The CSMA-CA algorithm requires listening to the channel before transmitting in order to reduce the collision probability (which is used before the transmission of data frames within the CAP) unless the frame can be quickly transmitted due to an ACK of a data request command.

The IEEE 802.15.4 non-beacon-enabled CSMA-CA algorithm in the basic access mode, maintains two variables for each frame as follows:

1. **Number of Backoffs** (NB): number of times the CSMA-CA algorithm was required to experience backoff due to unavailability. It is initialised to zero before each new transmission attempt.

2. **Backoff Exponent** (BE): enables the computation of the backoff delay, and represents the number of backoff periods that need to be clear of channel activity before a transmission can occur. The backoff delay is a random variable in the range $[0, 2^{BE} - 1]$.

The CSMA-CA algorithm can be summarised in four steps as follows:

1. In the first step, after initialization of the NB and BE, the MAC layer will delay the activity based on a backoff time counter (BO_c) uniformly distributed in the range $[0, CW_{NB}]$, where $CW_{NB} = [2^{BE} - 1]$ is the dimension of the initial contention window and NB ranges from $[0, NB_{max} = macMaxCSMABackoffs]$. The $macMaxCSMABackoffs$ variable [WPA07a] represents the maximum number of backoffs the CSMA-CA algorithm will experience before declaring a channel access failure.

2. In the second step, the node will perform CCA during the $ccaTime$ time period.

3. In the third step, if the channel is found to be busy, both the NB and BE (BE is less or equal than BE_{max}) will be incremented by one unit. If

the value of NB is larger than NB_{max}, then the CSMA-CA algorithm will finish by entering in the "Failure" state, which means that the node does not succeed in accessing the channel.

4. In the fourth step, if the channel is found to be idle during the CCA procedure, the MAC layer starts immediately transmitting its current frame, and the algorithm will finish by entering in the "Success" state, which means that the node has success in accessing the channel.

By considering the same assumptions from the model presented in [BV⁺09], one can model the backoff procedure by a dimensional process

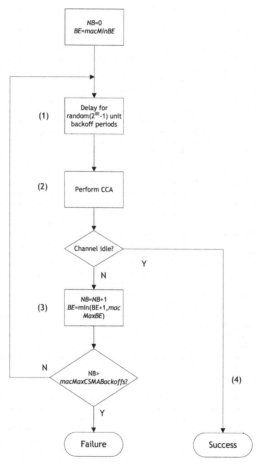

Figure 6.9 IEEE 802.15.4 and SBACK-MAC CSMA-CA algorithm for the non-beacon-enabled mode.

Table 6.2   Backoff stages for IEEE 802.15.4 and SBACK-MAC.

NB	BO_s	BE_i	$CW_{NB} = [2^{BE_i} - 1]$	$\overline{CW}_{NB}$
0	0	3	$CW_0 = 7$	3.5
1	1	4	$CW_1 = 15$	7.5
2	2	5	$CW_2 = 31$	15.5
3	3	5	$CW_3 = CW_2 = 31$	15.5
4	4	5	$CW_4 = CW_2 = 31$	15.5

$Q(t) = \{BO_c(t), BO_s(t)\}$, where according to [BV$^+$09], t is an integer, which represents the time slot. More precisely, the jth slot (varying from $j \cdot T_{BO}$ to $(j + 1) \cdot T_{BO}$) is denoted by $t = j$. The variables $BO_c(t)$ and $BO_s(t)$ represent the backoff time counter and the backoff stage at slot t, respectively.

Since the $BO_c(t)$ is not a memoryless process, the dimensional process given by the $BO_c(t)$ and $BO_s(t)$ cannot be derived by considering a Markovian chain [BV$^+$09]. Moreover, as explained before, the BE is dependent of the $BO_s(t)$. By analysing the possible combinations between the pair (NB, BE) one concludes that there are $NB_{max} + 1$ different backoff stages, where NB_{max} represents the maximum number of backoffs allowed by the CSMA-CA algorithm.

Table 6.2 presents the different values for the backoff stage and the corresponding CW_{NB} by assuming the combination pair $(NB_{max} = 4, BE_{max} = 5)$.

As shown in Figure 6.10(a), for each transmission attempt in the basic access mode of IEEE 802.15.4, nodes perform a random backoff from a uniform distribution over $[0, CW_{NB}]$. The $CW_{NB} = [2^{BE_i} - 1]$ represents the backoff delay before performing CCA. The BE_i value depends on the $BO_s(t)$. After performing CCA, if the channel is found to be idle there will be a data transmission. Otherwise, if the channel is found to be busy, nodes will defer the data transmission for a random backoff period defined by BE_i in the next backoff stage.

The backoff contention window is doubled whenever the channel is determined busy during CCA until BE_i reaches its maximum value and it cannot be increased any further. Thus, the backoff window remains the same until the maximum retry limit (e.g., $NB_{max} = 4$) is reached [WPA11]. In the last retransmission case, if the channel is found to be busy, and nodes cannot succeed in sending the frame, the frame will be deleted from the MAC queue. Then, a new transmission cycle will start again, with a backoff delay defined by (e.g., $CW_0 = 7$) for the next data frame to be transmitted, and the CSMA-CA algorithm is repeated.

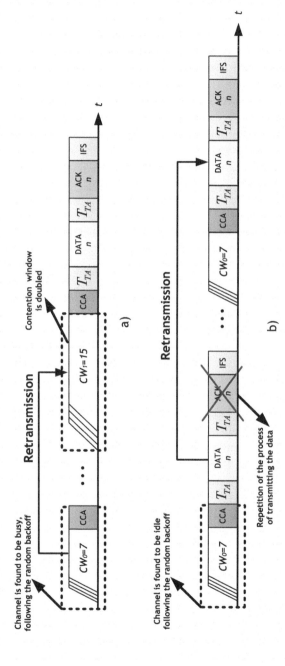

Figure 6.10 IEEE 802.15.4 basic access mode with retransmissions: a) channel is found to be busy and b) channel is found to be idle (the number of frames, n_{frames}, is represented by n).

In case the channel is found to idle during CCA following the backoff phase and an ACK is not received within the ACK wait duration period, T_{AW}, nodes shall conclude that the transmission attempt has failed. For this case, nodes shall repeat the DATA/ACK handshake process and the backoff window is given by the first backoff phase, i.e., first contention window, CW_0, as shown in Figure 6.10(b).

To determine the maximum average throughput, S_{max}, in an erroneous channel, first one needs to determine the minimum average delay, D_{min}. D_{min} is obtained by taking into account, the channel state (i.e., idle or busy) during CCA and the number of retransmissions for the non-received ACK control frames, by considering frame collisions and channel errors.

Details on the analytical model are presented in [NBC20].

6.4 IEEE 802.15.4 with Frame Concatenation

The BACK mechanism was previously introduced in the IEEE 802.11e standard [WLA05]. It improves channel efficiency by aggregating several ACK control frame responses into one special frame, the *BlockAck*. Hence, an ACK control frame will not be received in response to every data frame sent. The IEEE 802.11e standard defines two types of BACK mechanisms: immediate and delayed, depending on whether a *BlockAck* frame is transmitted, immediately after a *BlockAckReq* frame reception or not. Immediate BACK is suitable for high-bandwidth and low-latency traffic, while the delayed BACK is suitable for applications that tolerate moderate latency.

Since we intend to investigate the impact of the BACK mechanism in the context of IEEE 802.15.4 standard, defined for Low-rate wireless personal area network (LR-WPAN), some modifications were made to accommodate the energy, throughput and end-to-end delay requirements for these networks. Therefore, the proposed solutions are totaly compliant with the standard.

Since WSN are composed by a large number of battery-operated nodes, one of the most important issues is energy conservation. In order to reduce the energy consumption due to the protocol overhead, the SBACK-MAC protocol is proposed, which combines contention-based, scheduling-based and BACK-based schemes to achieve energy efficiency. As explained before, the SBACK-MAC protocol considers the use of the RTS/CTS mechanism, in order to avoid the hidden terminal problem of non-beacon-enabled multi-hop wireless networks, which may significantly degrade their performance [SSZ09]. Although, the RTS/CTS scheme is often employed to improve performance as it shortens frame collision duration and addresses the hidden station problem, until now, RTS/CTS is not proposed in any of the IEEE 802.15.4 standards. This reservation scheme actually involves the transmission of the short RTS and CTS control frames prior to the transmission of the actual data frame.

There is a number of studies in the literature on the performance of the RTS/CTS reservation mechanism in either IEEE 802.15.4 or 802.11 wireless networks [MMU11, DZM12, CBV04a, CBV04b]. The authors from [CBV04a] study proposed a broadcasting method based on RTS/CTS for IEEE 802.15.4 networks, in which the sender node uses a RTS frame to select the receiver node from neighboring nodes. The work from [DZM12] proposes an adaptive mechanism based on RTS/CTS in order to solve the hidden terminal problem in an IEEE 802.15.4 non-beacon-enabled multi-hop network. Authors in [CBV04a] studied the effectiveness of the RTS/CTS scheme in reducing the collision duration for IEEE 802.11 Wireless local area network (WLAN)s under certain scenarios. In a later work, authors in [CBV04b] tried to optimally employ the RTS/CTS scheme by deriving an all-purpose expression for the RTS threshold, a manageable parameter that indicates the data length under which the data frames should be sent without RTS/CTS.

In the framework of this work, we also propose the use of RTS/CTS for IEEE 802.15.4 by considering the non-beacon-enable mode combined with frame concatenation. The proposed solution shows that by considering the RTS/CTS mechanism combined with frame concatenation we improve the network performance in terms of maximum average throughput, minimum average delay and bandwidth efficiency.

6.4.1 IEEE 802.15.4 with RTS/CTS Combined with Frame Concatenation

The main reasons why IEEE 802.15.4 basic access mode does not consider the adoption of the RTS/CTS handshake mechanism are the following ones. (i) The introduction of RTS/CTS frames adds additional protocol overhead and, in a situation with low traffic load, short frame sizes could have the same order of magnitude of a RTS/CTS frame, (ii) The absence of a RTS/CTS handshake mechanism allows to reduce the system complexity.

Although these assumptions are true for some particular cases, we argue that in the presence of link layer errors the additional protocol overhead due to the use of RTS/CTS frames is mitigated by our concatenation mechanism. So, there is no need to repeat the backoff procedure for each data frame sent, but only once for each RTS/CTS set. Moreover, by using the proposed RTS/CTS mechanism frame collisions between the hidden nodes are avoided, allowing for decreasing the number of retransmitted frames and, thus, increasing the network performance.

In our proposal, we also assume that both the RTS and CTS frames have the structure of an ACK frame, which is assumed to have a limited size of 11 bytes, as shown in Table 6.1. The maximum data payload for IEEE 802.15.4 depends on the application (maximum payload could range between 102 and 118 bytes). Therefore, the length of the data frames could be approximately ten times greater than the length of the control frame.

IEEE 802.15.4 employing RTS/CTS with frame concatenation is composed by the following time periods:

- Backoff phase;
- CCA mechanism;
- Time needed for switching from receiving to transmitting;
- RTS transmission time;
- Time needed for switching from transmitting to receiving;
- CTS reception time.

Figure 6.11 presents the IEEE 802.15.4 RTS/CTS handshake mechanism.

As shown in Figure 6.11, in IEEE 802.15.4 employing RTS/CTS, nodes will use the same backoff procedure like IEEE 802.15.4 basic access mode.

However, this process is not repeated for each data frame sent, but only for each RTS/CTS set. Therefore, the channel utilization is maximized by decreasing the deferral time period before transmitting a data frame, as shown in Figure 6.8. The proposed solution shows that, by considering the RTS/CTS mechanism combined with frame aggregation, the network performance is improved in terms of maximum throughput, minimum delay and bandwidth efficiency.

This report also introduces an analytical model capable of accounting the retransmission delay and the maximum number of backoff stages. The successful validation of our analytical model will be carried out by comparison

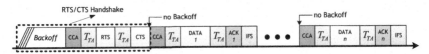

Figure 6.11   IEEE 802.15.4 frame sequence with RTS/CTS.

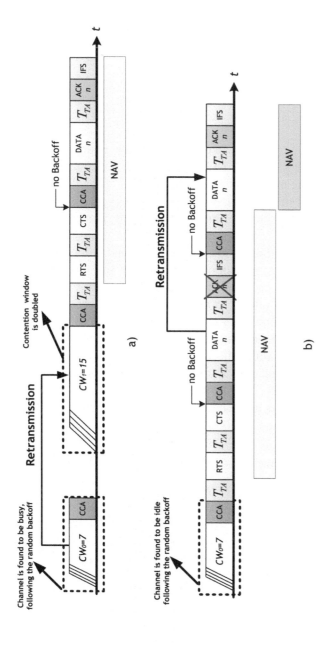

Figure 6.12 IEEE 802.15.4 with RTS/CTS with retransmissions: a) channel is found to be busy and b) channel is found to be idle (the number of frames, n_{frames}, is represented by n).

against simulation results by using the MiXiM framework of the OMNeT++ simulator [OMN13].

The minimum delay due to CCA, $D_{min\_CCA\_RTS}$, to determine if the channel is found to be busy or idle, after the backoff phase, and before each RTS/CTS set, is given by:

$$D_{min\_CCA\_RTS} = \sum_{i=1}^{n/N_{agg}} \sum_{k=0}^{k \leqslant NB} \left(\overline{CW_k} + ccaTime \right), \quad NB \in [0, NB_{max}]$$

(6.19)

By analysing Equation (6.19), it can be concluded that nodes only determine the channel state once per RTS/CTS. Therefore, if a node has $n = 100$ data frames to send and the number of aggregated frames is equal to $N_{agg} = 10$, nodes only determine the channel state 10 times ($n/N_{agg} = 100/10 = 10$) plus the time need for transmitting the frames (until the maximum retry limit, $NB_{max} = 4$, is reached).

If the channel is found to be idle during CCA and, after sending a data frame an ACK is not received within a duration of T_{AW}, the retransmission process will not consider the use of the backoff phase between two consecutive data frames, which allows to decrease the total overhead (as shown in Figure 6.12(b). Since any other stations can receive both the RTS, CTS, DATA or ACK frames, in the first transmission attempt they will set an internal timer called Network Allocation Vector (NAV) that is responsible for defining the time period a node will defer the channel access in order to avoid collisions.

Details on the analytical model for RTS/CTS combined with Frame Concatenation are presented in [NBC22].

6.4.2 Proposed Scheme Design with *Block ACK Request*

In the previous Section we have presented the introduction of an RTS/CTS mechanism combined with frame concatenation to improve the IEEE 802.15.4 MAC layer performance. Next, SBACK-MAC protocol proposed, which combines contention-based, scheduling-based and BACK-based schemes to reduce the energy consumption due to the protocol overhead.

The version of the SBACK-MAC protocol without piggyback considers the exchange of two special frames: *RTS ADDBA* and *CTS ADDBA*, the structure of the frame is presented in Figure 6.13(a). *ADDBA* stands for "Add Block Acknowledgement". The structure of these frames is presented in Figure 6.13(a). After this successfully exchange, the data frames are transmitted

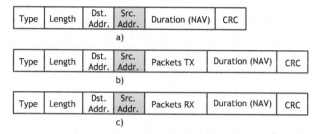

Figure 6.13 a) *RTS ADDBA Request* and *CTS ADDBA Response*, b) *BACK Request* and c) *BACK Response* frames format.

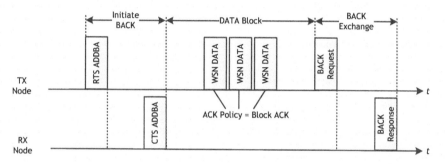

Figure 6.14 SBACK-MAC protocol - BACK mechanism with *BACK Request*.

from the transmitter to the receiver. Afterwards, by using the *BACK Request* primitive, the transmitter inquires the receiver about the total number of data frames that successfully reach the destination. In response, the receiver will send a special data frame called *BACK Response* identifying the frames that require retransmission, and the BACK mechanism finishes. The structure of these frames is shown in Figure 6.13(b) and (c).

Figure 6.14 presents the message sequence chart for the BACK mechanism with *BACK Request*, based on [WLA05]. The exchange of two special control frames, used in the beginning and at the end of the BACK mechanism, allows for mitigating the hidden-terminal and exposed-terminal problems like in IEEE 802.11e [WLA05] (by using a RTS/CTS handshake).

The BACK mechanism aims at reducing the power consumption by transmitting less ACK control frames whilst decreasing the time periods the transceivers should switch between different states. By using the BACK, there is no need to receive an ACK for every DATA frame sent, as shown in Figure 6.15. However, during the data transmission there is no way to know

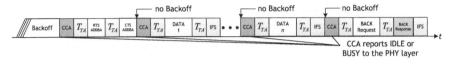

Figure 6.15 Timing relationship in the SBACK-MAC protocol with *BACK Request*, i.e., concatenation (the number of frames, n_{frames}, is represented by n).

how many frames have successfully reached the destination, except at the end of communication by using the *BACK Request/BACK Response*. Like in [WPA⁺07c], we set BE equal to 0, as if there is no congestion. Therefore, the channel utilization is maximized by decreasing the deferral time before transmitting (i.e., the $Initial backoff Period$ will be 0).

As presented in Figure 6.15, to overcome the overhead of the IEEE 802.15.4 MAC, several efficient MAC enhancements are proposed in which the frame concatenation concept is adopted. The idea is to transmit multiple MAC/PHY frames by using the BACK mechanism. In this work, a distributed scenario is considered, with single-destination and single-rate frame aggregation. In single-destination approaches, frames can be aggregated if they are available, and have the same source and destination address. Moreover, we also assume that the payload of the MAC frames cannot be modified.

Details on the analytical model for SBACK MAC with block acknowlege-ment request are presented in [NBC20], and shown in Figures 6.16(a) and (b).

6.4.3 Proposed Scheme Design without *Block ACK Request*

The version of the SBACK-MAC protocol whitout *BACK Request* ("piggyback mechanism") also considers the exchange of the *RTS ADDBA* and *CTS ADDBA* frames at the beginning of the communication. However, at the end of the communication the *BACK Request* primitive is not transmitted. Therefore, the last aggregated data frame, must include the information about the frames previously transmitted. However, with this scheme, the system becomes less robust. If the last aggregated data frame is lost, the destination does not know that an ACK needs to be sent back, Figure 6.17.

As presented in Figure 6.17, the SBACK-MAC version with piggyback does not consider the use of the *BACK Request* primitive. Therefore, the control overhead and the delay is reduced whilst increasing the throughput. By "piggybacking" the BACK information into the last data fragment, however, with this scheme, the system becomes less robust. If the last aggregated frame

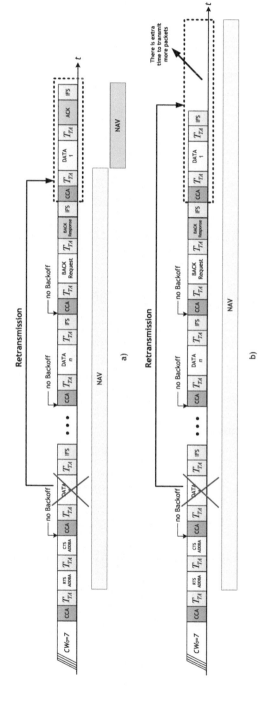

Figure 6.16 IEEE 802.15.4 with *BACK Request* (concatenation): a) Retransmissions by using DATA/ACK handshake and b) Retransmissions by using an NAV extra time (the number of frames, n_{frames}, is represented by n).

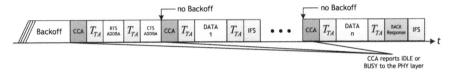

Figure 6.17   Timing relationship in the SBACK-MAC protocol without *BACK Request* (piggyback).

(DATA frame n) is lost, the destination does not know that an ACK needs to be sent back.

The version of SBACK-MAC without *BACK Request*, has the same retransmission scheme like in the SBACK-MAC with *BACK Request*, Figure 6.18(a) and (b).

Details on the analytical model for SBACK MAC without block acknowledgement request are presented in [NBC20].

6.5 State Transition Diagram for IEEE 802.15.4 and SBACK-MAC

Figure 6.18 presents the state transition diagram for the SBACK-MAC and IEEE 802.15.4. There are 8 states been shared by both IEEE 802.15.4 and SBACK-MAC.The BACK mechanism is activated in the **TYPE ACK** state. The characterisation of possible states for the state machine is the following one:

States associated to transitions

- **IDLE_1**: The node is "ON", and there is no scheduled activity to take place;
- **BACKOFF_2**: The node will delay any activities for a random number of backoff periods;
- **CCA_3**: The node will use the PHY layer to determine the channel occupancy;
- **TRANSMIT_4**: The node transmit RTS, *RTS ADDBA*, DATA and *BACK Request* frames;
- **WAIT_ACK_5**: The node waits for an ACK frame;
- **WAIT_CTS_6**: The node waits for CTS or *CTS ADDBA* frames;
- **WAIT_BACK_RESPONSE_7**: The node waits for a *BACK Response* frame;

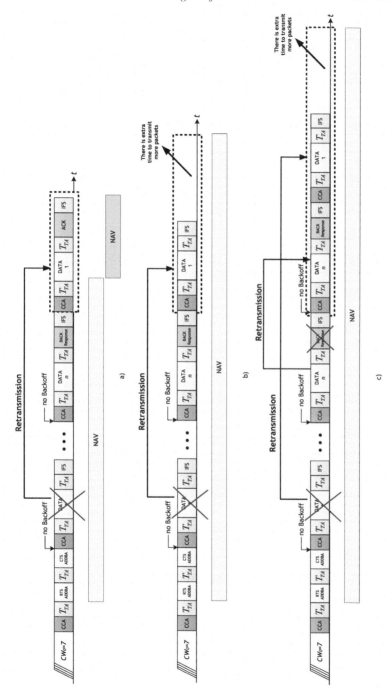

Figure 6.18 IEEE 802.15.4 without *BACK Request* (piggyback): a) Retransmissions by using DATA/ACK handshake and b) Retransmissions by using an NAV extra time c) Retransmission of the last aggregated frame (the number of frames, n_{frames}, is represented by n).

- **WAIT_SIFS_8**: The time between the reception of RTS, *RTS ADDBA, BACK Request* and DATA frames and the transmission of CTS, *CTS ADDBA, BACK Response* and ACK frames;
- **TRANSMIT_ACK_9**: The node transmits ACK frames;
- **TRANSMIT_CTS_10**: The node transmits *CTS or CTS ADDBA* frames;
- **TRANSMIT_*BACK Response*_11**: The node transmits *BACK Response* frames;
- **TRANSMIT_SYNCH_12**: the node transmits SYNCH frames;
- **ACK_POLICY_13**: The node will choose the use of normal ACK or BACK;
- **SLEEP_14**: The node will "turn OFF" the radio.

States of the frame

- **Type**: The type of the frame can be RTS, CTS, *RTS ADDBA, CTS ADDBA*, DATA, ACK, *BACK Request, BACK Response*, SYNCH;
- $N_{collisions}$: Number of collisions the frame as suffered;
- **Payload**;
- **Time of generation**;
- **Fragmentation**: If fragmentation is used;
- **Origin**;
- **Destination**;
- **First RTS**: Before sending frame put an RTS in first place on the queue to reserve the wireless medium;
- **Backoff value**.

Medium Access States

- **Free**: Medium is free, there is no transmission ongoing;
- **Busy**: Medium not free, there is a transmission ongoing.

Queue States

- **Not empty**: N frames are waiting for transmission;
- **Busy**: The buffer is empty.

Events that change the state of the machine

- **SLEEP_TIMEOUT**: Wake up from periodically sleep;
- **RETURN_TO_IDLE**: If the node receive has no frame to transmit, and have just choose the ACK Policy;
- **NAV_SLEEP_SCHEDULE**: Shutdown radio if a communication is sensed;

- **PERFORM BACKOFF**: Start the backoff procedure;
- **PERFORM CCA**: Start the CCA procedure;
- **CCA COMPLETE**: Channel is free, start transmission.;
- **DATA/RTS/*RTS ADDBA/BACK Request* TRANSMISSION COM-PLETE**: The node has successfully transmitted DATA, RTS, *RTS ADDBA and BACK Request* frames;
- **RETRANSMIT DATA/RTS/*RTS ADDBA/BACK Request* FRAME**: The node has successfully retransmitted DATA, RTS, RTS *ADDBA and BACK Request* frames;
- **ACK/CTS/*CTS ADDBA/BACK Response* RECEIVED**: The node has correctly received ACK, CTS, *CTS Response* and *BACK Response* frames;
- **PERFORM SIFS**: Start the SIFS procedure;
- **ACK/CTS/*CTS ADDBA/BACK Response* SIFS TIMEOUT**: Transmit ACK, CTS, *CTS Response* and *BACK Response* frames;
- **TYPE_OF_ACK_REQUEST**: ACK or BACK procedure;
- **TYPE_OF_ACK_REQUEST**: ACK or BACK procedure.

The state transition diagram for the SBACK-MAC and IEEE 802.15.4 was considered in the simulations by considering the WSN communication stack based on the MiXiM simulation framework [MiX13] from the OMNeT++ [OMN13] simulation engine. Figure 6.19 presents the main components that are responsible for simulating the channel access.

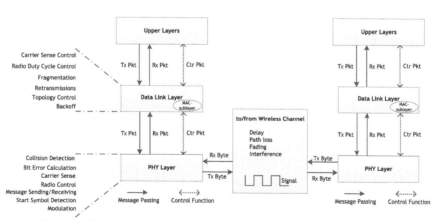

Figure 6.19   Layered model used by the IEEE 802.15.4 and SBACK-MAC

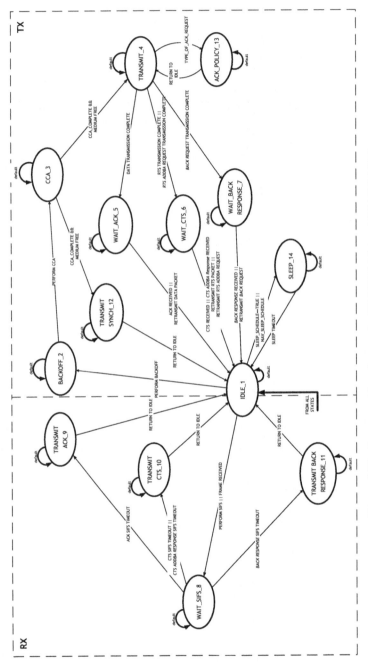

Figure 6.20   State transition diagram for IEEE 802.15.4 and SBACK-MAC.

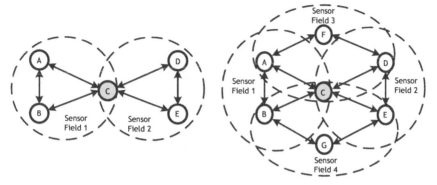

Figure 6.21 Multi-hop star topology simulation scenario: a) without interferers and b) with two interferers.

6.6 Model for Energy Estimation

In order to know how much energy is spent by the SBACK-MAC protocol in each state an analytical model was conceived. A two-hop network, with two sources, one relay and two sinks, has been considered. Figure 6.21 shows the OMNeT++ [OMN13] multi-hop star topology simulation setup. The frames from source node **A** flow, through node **C**, to sink node **D** while the frames originated by source node **B** flow, through node **C**, to reach sink node **E**.

The star network topology is challenging because there are abundant overhearing opportunities between neighbouring nodes. A node acting as coordinator may therefore take advantage of these opportunities to seek network optimisation. The star topology may also be viewed as a part of a larger network.Therefore, this type of network can be viewed as a building block for larger scale wireless networks. The performance metrics considered for a specific evaluation of the number of state transition of the SBACK-MAC protocol are the following ones:

- **Energy to Transmit**: is the amount of energy spent to transmit a frame;
- **Energy to Receive/Listen**: is the amount of energy spent to receive a frame or listen to the medium;
- **Energy to Sleep**: is the amount energy spent by a node during the time of inactivity (sleep state);
- **Energy Waste**: is the amount of energy spent by a node during the retransmission of a frame;

Table 6.3 Specification of the CC2420 radio transceiver.

Parameter	CC2420
Frequency [GHz]	2.4
Modulation	O-QPSK
P_{Sleep} [μW]	0.06
P_{RX} [mW]	56.4
P_{TX} [mW]	52.2
Data rate [kb/s]	250

- **Total energy consumption**: is the total energy spent per node. The total energy consumption metric incorporates all previous metrics together into a single one. The intention is to have a joint perspective of all the factors that optimise the total energy in a global way.

The analysis of the sensor nodes performance is obtained through simulation by considering the CC2420, radio transceiver operating in the 2.4 GHz band. The reason for choosing the CC2420 radio transceiver from Chipcon is related to the fact that it currently the most popular radio chip on wireless sensor nodes [HNL07].

Table 6.1 shows the specifications of the radio transceiver. Where P indicates the power spent by each state by considering a supply voltage of 3V.

The energy consumption of a given node over a period of time t is given as follows:

$$E(t) = (t_{tx} \times P_{TX}) + (t_{rx} \times P_{RX}) + (t_{sleep} \times P_{Sleep}) + (t_{idle} \times P_I) \quad (6.20)$$

The mean of each variable is presented in Table 6.4.

Table 6.4 Notations for energy estimation.

Notation	Parameter
t_{tx}	Time on TX state
P_{TX}	Power consumption in the transmitting state
t_{rx}	Time on RX state
P_{RX}	Power consumption in the receiving state
t_{sleep}	Time on SLEEP state
P_{Sleep}	Power consumption in the sleep state
t_{idle}	Time on IDLE state
P_I	Power consumption in the idle state

6.7 Performance Evaluation for IEEE 802.15.4 in the Presence/Absence of RTS/CTS Combined with Frame Concatenation

In this Section, we evaluate IEEE 802.15.4 while applying IEEE 802.15.4 without RTS/CTS, The MiXiM simulation framework [MiX13] from the OMNeT++ [OMN13] simulator is used. Figure 6.21.a) presents a two-hop network topology, with two sources, one relay, two sinks.

Figure 6.21.b) considers the same topology but two interferers are considered. Frames flow from source node **A**, through node **C**, to sink node **D**, while the frames originated by source node **B** flow, through node **C**, to reach sink node **E**. This topology is a simple star topology where node **C** acts as a central node.

The interferer nodes **F** and **G** are responsible for sending broadcast frames that will collide with the frames being sent by both the sources and the central node (i.e., in the case of interference). The level of interference of nodes nodes **F** and **G** imposes the retransmission on average of 20 % of the total number of frames being exchanged in the network.

The maximum average throughput, S_{max}, the minimum average delay, D_{min}, and the bandwidth efficiency, η, have been obtained through simulation by considering five different seeds, and a 95% confidence interval. Table 6.1 presents the MAC parameters considered for the network in our simulations. The performance analysis of the proposed schemes has been conducted for both the best-case scenario (no errors) and by assuming an erroneous channel.

6.7.1 Minimum Average Delay in the Presence and Absence of RTS/CTS

A. Dependence on the number of TX frames for a fixed payload of 3 bytes

Figure 6.22 presents the analytical and simulation results for the minimum average delay, D_{min}, versus the number of transmitted frames with and without retransmissions, by considering IEEE 802.15.4 with and without RTS/CTS, as presented in Figures 6.10 and 6.12. A fixed payload size of 3 bytes (i.e., $L_{DATA}= 3$ bytes) is considered.

In [CUK⁺10], the authors have show that for shorter frame sizes IEEE 802.15.4 results in lower bandwidth efficiency, delay and throughput. Therefore, we will simulate the worst case scenario for wireless sensor networks,

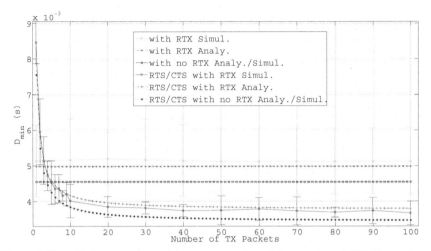

Figure 6.22 Minimum average delay as a function of the number of TX frames for a fixed payload size of 3 bytes for IEEE 802.15.4 with an without RTS/CTS (no RTS/CTS is represented by the absence of the label RTS/CTS in the legend).

showing that our proposed mechanism can improve channel efficiency even under this circumstances.

By analysing Figure 6.22 we conclude that by varying the number of TX frames, n, between 1 and 100, IEEE 802.15.4 basic access mode without RTS/CTS globally presents the worst performance in terms of minimum average delay, D_{min}, for both the cases with and without retransmissions. By comparing the case without retransmissions with the one presented in [CUK+10], we conclude that the differences in terms of minimum average delay, D_{min}, are justified by the fact that the authors in [CUK+10] have not considered the time needed to switch the radio between the different states (i.e., $rxSetupTime$) [RDA+09].

The size of the 95 % confidence intervals shown in the simulations is explained by the fact that nodes will backoff for a random duration of time before attempting to access the channel later on.If the channel is found to be busy for each TX frame, nodes will double the backoff time counter during the backoff phase and the process is repeated until the maximum contention window is reached.

By considering IEEE 802.15.4 basic access mode with RTS/CTS combined with the frame concatenation feature, if the number of aggregated frames is higher than 5, IEEE 802.15.4 employing RTS/CTS outperforms IEEE 802.15.4 without RTS/CTS in terms of delay (and also throughput). This is explained by the fact that the backoff phase is not repeated for each transmitted data frame but only for each RTS/CTS set. Then, there is no backoff phase between two consecutive data frames, which allows for decreasing the total overhead, as shown in Figures 6.12.a) and b). Note that in Figure 6.22 the absence of RTS/CTS is represented by the absence of the label RTS/CTS in the legend.

For the case with retransmissions, since all the frames include the NAV duration, when nodes receive a frame that is not intended for them, they will avoid sending frames as long as this time has not expired. The retransmission process does not include the backoff phase like in the IEEE 802.15.4 basic access mode (i.e., $BE = 0$). By comparing the analytical and simulation results, we conclude that the theoretical results are similar and this actually verifies the accuracy of our proposed retransmission model. Performance results for D_{min} as a function of the number of TX frames show that, by using IEEE 802.15.4 with RTS/CTS with frame concatenation, for 5, 7 and 10 aggregated frames, D_{min} decreases 8 %, 14 % and 18 %, respectively. For more than 28 aggregated frames, D_{min} decreases approximately 30 %.

B. Dependence on the payload size for a number of transmitted frames equal to 10

Figure 6.23 presents the analytical results for the minimum average delay, D_{min}, as a function of the payload size by considering the four different scenarios presented on Figures 6.10 and 6.12. For this case the number of transmitted frames (aggregated) is equal to 10. The discontinuity around 18 bytes is due to the use of SIFS and LIFS (i.e., MPDU less of equal than 18 bytes must be followed by a SIFS, whilst MPDU longer than 18 bytes must be followed by a LIFS).

The results show that, IEEE 802.15.4 basic access mode without RTS/CTS globally presents the worst performance in terms of minimum average delay, D_{min}, for both the cases with and without retransmissions. By comparing IEEE 802.15.4 without RTS/CTS with IEEE 802.15.4 with RTS/CTS and with retransmissions for small frames sizes (i.e., data payload less than 18 bytes), D_{min} is reduced by 25 %. For larger frame sizes, by considering the IEEE 802.15.4 standard with RTS/CTS, D_{min}, decreases 9 %. By comparing IEEE 802.15.4 without RTS/CTS with IEEE 802.15.4 with RTS/CTS without retransmissions for small frames sizes (i.e., data payload less than 18 bytes), D_{min} is reduced by 16 %. For larger frame sizes, by considering the IEEE 802.15.4 standard with RTS/CTS, D_{min}, decreases 9% .

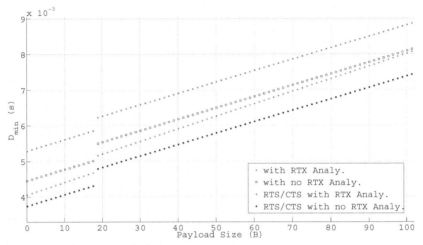

Figure 6.23   Minimum average delay as a function of the payload size for a number of TX frames equal to 10 for IEEE 802.15.4 with an without RTS/CTS (no RTS/CTS is represented by the absence of the label RTS/CTS in the legend).

6.7.2 Maximum Average Throughput in the Presence and Absence of RTS/CTS

A. Dependence on the number of TX frames for a fixed payload of 3 bytes

Figure 6.24 compares the analytical and simulation results for the maximum average throughput, S_{max}, as a function of the number of TX frames, with and without RTX between the cases of presence and absence of RTS/CTS.

By analysing Figure 6.24, we conclude that for the shortest payload sizes (i.e., $L_{DATA} = 3$), it is possible to improve the network performance by using the proposed RTS/CTS mechanism with frame concatenation. When the number of TX frames is higher than 5, IEEE 802.15.4 with RTS/CTS (with and without retransmissions) achieves higher throughput in comparison to IEEE 802.15.4 without RTS/CTS.

Moreover, we observe that by considering IEEE 802.15.4 in the basic access mode, S_{max} does not depend on the number of TX frames, and achieves the maximum value of 5.2 kb/s. For IEEE 802.15.4 with RTS/CTS combined with the frame concatenation feature, the maximum achievable throughput is 6.3 kb/s. Results for S_{max} as a function of the number of TX frames in IEEE 802.15.4 with RTS/CTS with frame concatenation show that, for 5, 7 and 10 aggregated frames, S_{max} increases 8 %, 14% and 18

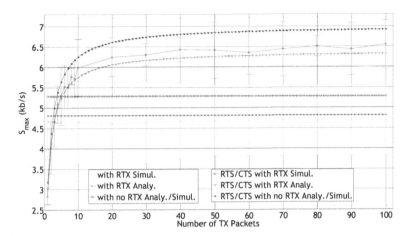

Figure 6.24 Maximum average throughput as a function of the number of TX frames for a fixed payload size of 3 bytes for IEEE 802.15.4 with an without RTS/CTS (no RTS/CTS is represented by the absence of the label RTS/CTS in the legend).

%, respectively. For a number of aggregated frames higher than 28, S_{max} increases approximately 30 %.

B. Dependence on the payload size for a number of transmitted frames equal to 10

Figure 6.25 presents the analytical results for the maximum average throughput, S_{max}, as a function of the payload size by considering the four different scenarios presented on Figures 6.10 and 6.12.

The results show that IEEE 802.15.4 basic access mode without RTS/CTS globally presents the worst performance in terms of maximum average throughput, S_{max}, for both the cases with and without retransmissions. By comparing IEEE 802.15.4 not employing RTS/CTS with IEEE 802.15.4 employing RTS/CTS with retransmissions, for short frames sizes (i.e., data payload less than 18 bytes), S_{max}, increases 25%. For longer frame sizes, by considering the IEEE 802.15.4 standard employing RTS/CTS, S_{max}, increases 9%. By comparing IEEE 802.15.4 not employing RTS/CTS with IEEE 802.15.4 employing RTS/CTS without retransmissions, for small frames sizes (i.e., data payload less than 18 bytes), S_{max} is increased 16%. For longer frame sizes, by considering the IEEE 802.15.4 standard with RTS/CTS, S_{max}, increases 9%.

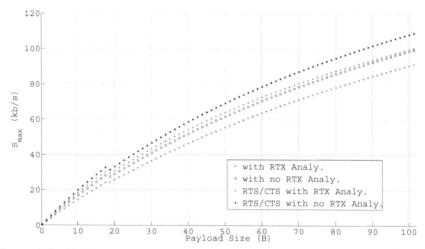

Figure 6.25　Maximum average throughput as a function of the of the payload size for a number of TX frames equal to 10 for IEEE 802.15.4 with an without RTS/CTS (no RTS/CTS is represented by the absence of the label RTS/CTS in the legend).

6.7.3 Bandwidth Efficiency in the Presence and Absence of RTS/CTS

The bandwidth efficiency, η, of IEEE 802.15.4 with and without RTS/CTS, suggested by the authors from [LDMM+05], is obtained by the following equation:

$$\eta = \frac{S_{max}}{R} \tag{6.21}$$

where R is the maximum data rate.

Figure 6.26 presents the bandwidth efficiency as a function of the number of TX frames, for a payload size of 3 bytes, again.

By analysing Figure 6.26 it is observable that, for IEEE 802.15.4 in the absence of RTS/CTS and with and without retransmissions, the bandwidth efficiency is 1.9% and 2.1%, respectively. It can also be observed that, by varying the number of TX frames, the channel efficiency remains constant. The results show that, in the case of no retransmissions, the obtained results are very similar to the ones obtained in [CUK+10], [LDMM+05], which again verifies the appropriateness of the proposed analytical formulation.

For IEEE 802.15.4 employing RTS/CTS and with and without retransmissions, the bandwidth efficiency is dependent on the number of TX frame (by considering aggregation). The results show that by aggregating more

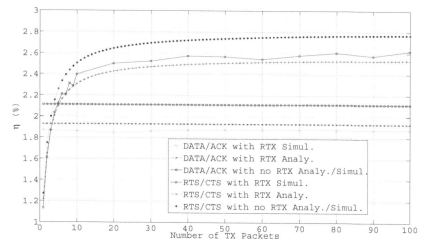

Figure 6.26 Bandwidth efficiency as a function of the number of TX frames for a fixed payload size of 3 bytes for IEEE 802.15.4 with an without RTS/CTS (no RTS/CTS represents the absence of the label RTS/CTS in the legend).

than 5 frames IEEE 802.15.4 with RTS/CTS outperforms IEEE 802.15.4 without RTS/CTS, in terms of bandwidth efficiency, η, where the maximum achievable value is 2.8% for payload size of 3 bytes.

Based on the results, we can conclude that even for short frame sizes (i.e., $L_{DATA} = 3$), by using the proposed RTS/CTS mechanism with frame concatenation, it is possible to improve channel efficiency.

6.8 Performance Evaluation for SBACK-MAC with and without BACK Request by using a NAV Extra Time for Retransmissions

As in their current solutions for medium access control (MAC) sub-layer protocols, channel efficiency has a margin for improvement, in [NBC20], the authors evaluate the IEEE 802.15.4 MAC sub-layer performance by proposing to use the request-/clear-to-send (RTS/CTS) combined with frame concatenation and block acknowledgment (BACK) mechanism to optimize the channel use. The proposed solutions are studied in a distributed scenario with single-destination and single-rate frame aggregation. The throughput and delay performance is mathematically derived under channel environments without/with transmission errors for both the chirp spread spectrum and direct sequence spread spectrum physical layers for the 2.4 GHz Industrial, Scientific and Medical band. Simulation results successfully verify the authors proposed analytical model. In this Chapter, we present additional results we have obtained while performing the research.

In this Section we present the numerical results for the SBACK-MAC with and without *BACK Request* by considering an NAV extra time for retransmissions. A comparison with IEEE 802.15.4 with and without RTS/CTS is also performed.

Figure 6.21.a) presents a two-hop network topology, with two sources, one relay, two sinks. Figure 6.21.b) considers the same topology but two interferers are considered. Frames flow from source node **A**, through node **C**, to sink node **D**, while the frames originated by source node **B** flow, through node **C**, to reach sink node **E**. This topology is a simple star topology where node **C** acts as a central node.

The interferer nodes **F** and **G** are responsible for sending broadcast frames that will collide with the frames being sent by both the sources and the central node (i.e., in the case of interference). The level of interference of nodes nodes **F** and **G** imposes the retransmission on average of 20% of the total number of frames being exchanged in the network.

The maximum average throughput, S_{max}, the minimum average delay, D_{min}, and the bandwidth efficiency, η, have been obtained through simulation by considering five different seeds, and a 95% confidence interval. Table 6.1 presents the MAC parameters considered for the network in our simulations. The performance analysis of the proposed schemes has been conducted for both the best-case scenario (no errors) and by assuming an erroneous channel.

In the following Section we present results for the SBACK-MAC protocol with and without *BACK Request* by considering both the numerical and simulation results by using the MiXiM simulation framework [MiX13] from the OMNeT++ [OMN13], enabling to verify the validity of our analytical model presented in Section 6.4.3. These results include both the cases with retransmissions by using DATA/ACK handshake and the NAV extra time.

6.9 Minimum Average Delay

A. Dependence on the number of TX frames for a fixed payload of 3 bytes

Figure 6.27 presents the analytical and simulation results for the minimum average delay, D_{min}, as a function of the number of TX frames without retransmissions (RTX) for a fixed payload size of 3 bytes.

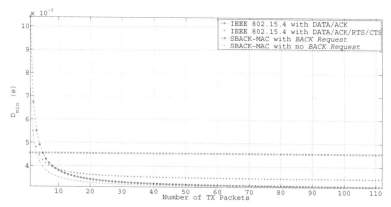

Figure 6.27 Minimum average delay as a function of the number of TX frames for a fixed payload size of 3 bytes without RTX fot IEEE 802.15.4.

By considering a number of aggregated frames equal to 10 we have the following observations:

- For the IEEE 802.15.4 standard with RTS/CTS, D_{min}, is decreased approximately 18% when compared with IEEE 802.15.4 in the basic access mode;
- For the SBACK-MAC protocol with and with no *BACK Request*, D_{min}, is decreased by 18% and 30%, respectively when compared with IEEE 802.15.4 in the basic access mode;
- The minimum average delay, D_{min}, for the IEEE 802.15.4 with and without RTS/CTS is 0.0045 s and 0.0038 s, respectively;
- The minimum average delay, D_{min}, for the SBACK-MAC with and without *BACK Request* is 0.0038 s and 0.0034 s, respectively.

By considering a number of aggregated frames equal to 80 we observe the following behaviour:

- For the IEEE 802.15.4 standard with RTS/CTS, D_{min}, is decreased approximately 28% when compared with IEEE 802.15.4 in the basic access mode;
- For the SBACK-MAC protocol with and with no *BACK Request*, D_{min}, is decreased by 40% and 45%, respectively when compared with IEEE 802.15.4 in the basic access mode;
- The minimum average delay, D_{min}, for the IEEE 802.15.4 with and without RTS/CTS is 0.0045 s and 0.0035 s, respectively;

- The minimum average delay, D_{min}, for the SBACK-MAC with and without *BACK Request* is 0.0032 s and 0.0031 s, respectively.

B. Dependence on the payload size for a number of transmitted frames equal to 10

Figure 6.28 presents the analytical and simulation results for the minimum average delay, D_{min}, as a function of the payload size without retransmissions (RTX), for a number of transmitted frames equal to 10.

By considering small frame sizes (i.e., data payload less than 18 bytes), for a payload of 17 bytes we observe that:

- By considering the IEEE 802.15.4 standard with RTS/CTS, D_{min}, is decreased approximately 17% when compared with IEEE 802.15.4 in the basic access mode;
- For the SBACK-MAC protocol with and with no *BACK Request*, D_{min}, is decreased by 17% and 25%, respectively, when compared with IEEE 802.15.4 in the basic access mode;
- The minimum average delay, D_{min}, for IEEE 802.15.4 with and with no RTS/CTS is 0.005 s and 0.0043 s, respectively;
- The minimum average delay, D_{min}, for SBACK-MAC with and with no *BACK Request* is 0.0043 s and 0.004 s, respectively.

For longer frame sizes (i.e., data payload longer than 18 bytes), and for a data payload equal to 118 bytes we observe that:

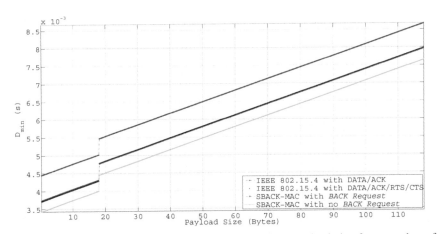

Figure 6.28 Minimum average delay as a function of the payload size for a number of transmitted frames equal to 10, without RTX.

- By considering the IEEE 802.15.4 standard with RTS/CTS, D_{min}, is decreased approximately 8.8% when compared with IEEE 802.15.4 in the basic access mode;
- For the SBACK-MAC protocol with and with no *BACK Request*, D_{min} is decreased by 8.6% and 13%, respectively when compared with IEEE 802.15.4 in the basic access mode;
- The minimum average delay, D_{min} , for IEEE 802.15.4 with and with no RTS/CTS is 0.0087 s and 0.0079 s, respectively;
- The minimum average delay, D_{min} , for the SBACK-MAC with and without *BACK Request* is 0.008 s and 0.0076 s, respectively.

6.10 Maximum Average Throughput

A. Dependence on the number of TX frames for a fixed payload size of 3 bytes

Figure 6.29 presents the analytical and simulation results for the maximum average throughput, S_{max}, as a function of the number of TX frames by considering a fixed payload size, of $L_{DATA} = 3$ bytes. When the number of TX frames (formed by the physical and MAC header plus the data payload, with a total number of bytes given by: $L_{PHY} + L_{MAC} + L_{DATA}$) is less than 3 the IEEE 802.15.4 standard in the basic access achieves higher throughput than SBACK-MAC with and without *BACK Request*. Moreover, by using the IEEE 802.15.4 standard in the basic access mode S_{max}, does not depend on the number of TX frames, attaining a value equal to 5.2 kb/s, whereas for SBACK-MAC with and with no *BACK Request* (i.e., concatenation and piggyback) S_{max} increases by increasing the number of TX frames (i.e., the number of aggregated frames).

For a number of aggregated frames equal to 10:

- By using the IEEE 802.15.4 standard with RTS/CTS, S_{max}, is increased approximately 20% when compared with IEEE 802.15.4 in the basic access mode;
- For the SBACK-MAC protocol with and with no *BACK Request*, S_{max}, is increased by 21% and 30%, respectively when compared with IEEE 802.15.4 in the basic access mode;
- The maximum average throughput, S_{max}, for IEEE 802.15.4 with and without RTS/CTS is 5.2 kb/s and 6.25 kb/s, respectively;
- The maximum average throughput, S_{max}, for SBACK-MAC with and with no *BACK Request* is 6.3 kb/s and 6.8 kb/s, respectively.

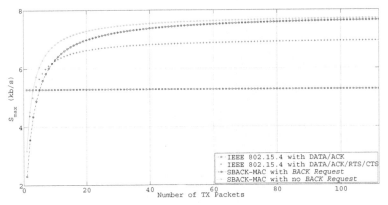

Figure 6.29 Maximum average throughput as a function of the number of TX frames for a fixed payload size of 3 bytes without RTX, for IEEE 802.15.4.

For a number of aggregated frames equal to 80:

- By considering the IEEE 802.15.4 standard with RTS/CTS, S_{max}, is increased approximately 30% when compared with IEEE 802.15.4 in the basic access mode;
- For the SBACK-MAC protocol with and with no *BACK Request*, S_{max}, is increased by 43% and 45%, respectively when compared with IEEE 802.15.4 in the basic access mode;
- The maximum average throughput, S_{max}, for IEEE 802.15.4 with and without RTS/CTS is 5.2 kb/s and 6.9 kb/s, respectively;
- The maximum average throughput, S_{max}, for SBACK-MAC with and with no *BACK Request* is 7.58 kb/s and 7.67 kb/s, respectively.

B. Dependence on the payload size for a number of transmitted frames equal to 10

Figure 6.30 illustrates the analytical and simulation results for maximum average throughput, S_{max}.

By considering small frame sizes (i.e., data payload less than 18 bytes), for a payload of 17 bytes:

- By considering the IEEE 802.15.4 standard with RTS/CTS, S_{max}, is increased approximately 17% when compared with IEEE 802.15.4 in the basic access mode;
- For the SBACK-MAC protocol with and with no *BACK Request*, S_{max}, is increased by 17% and 25%, respectively when compared with IEEE 802.15.4 in the basic access mode;

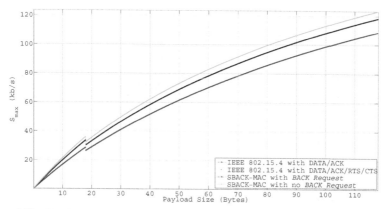

Figure 6.30 Maximum average throughput as a function of the number of the payload size for a number of transmitted frames equal to 10 without RTX.

- The maximum average throughput, S_{max}, for IEEE 802.15.4 with and without RTS/CTS is 28.66 kb/s and 33.53 kb/s, respectively;
- The maximum average throughput, S_{max}, for SBACK-MAC with and with no *BACK Request* is 33.6 kb/s and 36 kb/s, respectively.

By using longer frame sizes (i.e., data payload longer than 18 bytes), for a data payload equal to 118 bytes:

- By considering the IEEE 802.15.4 standard with RTS/CTS, S_{max}, is increased approximately 8.8% when compared with IEEE 802.15.4 in the basic access mode;
- For the SBACK-MAC protocol with and with no *BACK Request*, S_{max}, is increased by 8.6% and 13%, respectively when compared with IEEE 802.15.4 in the basic access mode;
- The maximum average throughput, S_{max}, for IEEE 802.15.4 with and without RTS/CTS is 108.9 kb/s and 118.5 kb/s, respectively;
- The maximum average throughput, S_{max}, for SBACK-MAC with and with no *BACK Request* is 118.3 kb/s and 123.4 kb/s, respectively.

6.11 Summary

In this Section we have shown a new innovative MAC sub-layer employing RTS/CTS with frame concatenation. The SBACK-MAC with and with no *BACK Request* was compared against IEEE 802.15.4 with and with no RTS/CTS. Lessons learned are discussed in the next Section, where we also extract conclusions.

6.12 Summary and Conclusions

In this Chapter, we have proposed an IEEE 802.15.4 MAC layer enhancement by employing RTS/CTS combined with the frame concatenation feature for WSNs. The use of the RTS/CTS mechanism improves channel efficiency by decreasing the deferral time before transmitting a data frame. The proposed solution have shown that for the shortest payload sizes (i.e., $L_{DATA} = 3$) for the case with retransmissions, if the number of TX frames is lower than 5 (i.e., the number of aggregated frames), IEEE 802.15.4 with RTS/CTS and the application of frame concatenation achieves higher throughput values in comparison to IEEE 802.15.4 without RTS/CTS even for shorter frame sizes. The advantage comes from not including the backoff phase into the retransmission process like IEEE 802.15.4 basic access mode (i.e., $BE = 0$). By comparing the analytical and simulation results, we conclude that the theoretical results are very similar and this actually verifies the accuracy of our proposed retransmission model.

Performance results for D_{min} as a function of TX frames show that, by using IEEE 802.15.4 with RTS/CTS with frame concatenation, for 5, 7 and 10 aggregated frames, D_{min} decreases by 8%, 14% and 18%, respectively. For more than 28 aggregated frames, D_{min} decreases approximately 30%.

Figure 6.31 presents the maximum throughput, S_{max}, as a function of the payload size. For small frame sizes, S_{max}, is increased by 17% for IEEE 802.15.4 with RTS/CTS and SBACK-MAC with *BACK Request* and 25% for the SBACK-MAC with no *BACK Request*. By considering longer frame sizes, S_{max}, is increased by 8.8% for IEEE 802.15.4 with RTS/CTS, 8.6% for SBACK-MAC with *BACK Request* and 25% for SBACK-MAC with no *BACK Request*.

Figure 6.32 presents the minimum average delay, D_{min}, as a function of the payload size. For small frame sizes, D_{min}, is decreased by 17% for IEEE 802.15.4 employing RTS/CTS and SBACK-MAC with *BACK Request* and 25% for the SBACK-MAC with no *BACK Request*. By considering longer frame sizes, D_{min}, is decreased by 8.8% for the IEEE 802.15.4 employing RTS/CTS, 8.6% for SBACK-MAC with *BACK Request* and 25% for SBACK-MAC without *BACK Request*.

Figure 6.33 presents the maximum average throughput, S_{max}, as a function of the number of TX frames (aggregated). For 10 aggregated frames, S_{max}, is increased by 20% for IEEE 802.15.4 employing RTS/CTS, 21% for SBACK-MAC with *BACK Request* and 30% for the SBACK-MAC with no *BACK Request*. By considering 80 aggregated frames, S_{max} is increased by

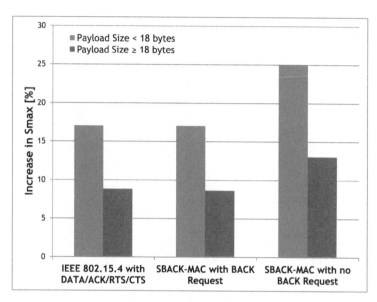

Figure 6.31 Increase of the maximum average throughput, S_{max}, as a function of the payload size for the presented MAC sub-layer mechanisms.

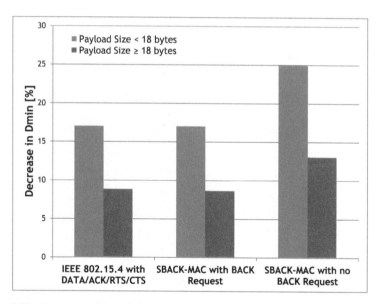

Figure 6.32 Decrease of the minimum average delay, D_{min}, as a function of the payload size for the presented MAC sub-layer mechanisms.

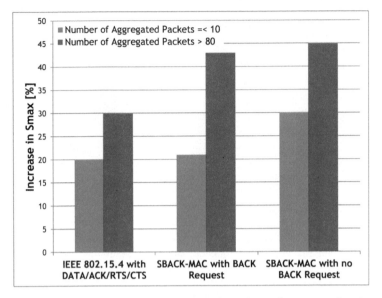

Figure 6.33 Increase of the maximum average throughput, S_{max}, as a function of the number of TX frames for the presented MAC mechanisms.

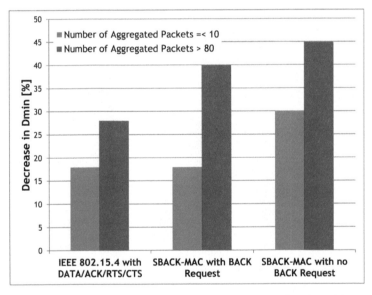

Figure 6.34 Decrease of the minimum average delay, D_{min}, as a function of the number of TX frames for the presented MAC mechanisms.

30% for IEEE 802.15.4 with RTS/CTS, 43% for SBACK-MAC with *BACK Request* and 45% for the SBACK-MAC without *BACK Request*.

Figure 6.34 presents the minimum average delay, D_{min}, as a function of the number of TX frames. For 10 aggregated frames, D_{min}, is decreased by 18% for the IEEE 802.15.4 with RTS/CTS and SBACK-MAC with *BACK Request* and 30% for the case we consider the SBACK-MAC with no *BACK Request*. By considering 80 aggregated frames, D_{min}, is decreased by 28% for the IEEE 802.15.4 with RTS/CTS, 40% for SBACK-MAC with *BACK Request* and 45% for SBACK-MAC without *BACK Request*.

7

Multi-Channel-Scheduled Channel Polling Protocol

7.1 Motivation

Multi-Channel-Scheduled Channel Polling (MC-SCP) is a scheduled and channel polling multi-channel MAC protocol. Its main rationale is based on the single-channel SCP-MAC [YSH06b] protocol, which is an energy-efficient medium access protocol designed for WSNs. The SCP protocol goes one step further in preamble sampling based protocols by combining scheduling with channel polling to minimize energy consumption. Compared to [HNS⁺06a, YV06b], the SCP's main aspects are:

- Two-phase collision avoidance mechanism employs a wake-up tone on a single channel between the two contention windows. The wake-up tone is similar to a busy tone and informs the other listening nodes of an ongoing transmission;
- It achieves synchronization by piggybacking the schedules in broadcast data frames;
- By default, nodes wake-up and poll the channel for activity. A sending node reduces the duration of the wake-up tone by starting it right before the receiver starts listening;
- Energy efficiency is also achieved, since channel polling together with the two-phase contention mechanism, mitigates the problems that may result from collisions and avoid the waste of energy.

Moreover, time-scheduled communication facilitates the coordination of multi-channel communication. Since the nodes (senders and receivers) have to switch their radio interfaces between different channels, coordination of channel switching is required. This channel switching mechanism facilitates that a sender and a receiver node to be simultaneously in the same channel, to exchange frames. Scheduled and synchronized access provides a way for the nodes to meet in the same channel.

In terms of the MAC contention mechanism, even tough single window slotted contention protocols may suffer collisions as the transmitter node fails to successfully allocate the medium, SCP protocol employs a double slotted contention mechanism whilst minimizing the collisions at expense of small overhead. SCP alleviates the strict behaviour of TDMA MAC protocols (e.g., LMAC), which may lead to high values for the delay and low values for the throughput while presenting the advantage of collision-free access [IVHJH11].

In scenarios of high scalability, the proposed MC-SCP protocol is considered as an Hybrid MAC protocol. This is due to the employment of a deterministic TDMA-based channel hopping mechanism and a slotted contention basis that relies on an enhanced two-phase contention window mechanism similar to the one from the SCP protocol. Besides the SCP protocol, other mechanisms from other recent MAC protocols are also being considered and adapted, in order to increase the performance of the proposed multi-channel-scheduled channel polling (MC-SCP) protocol.

7.2 Main States of the MC-SCP Protocol

Apart from the denial channel list state nodes can be in one of the five main states: startup, synchronization, discovery-addition, slot channel, and medium access, as shown in Figure 7.1. Further details are given in Section 7.4.

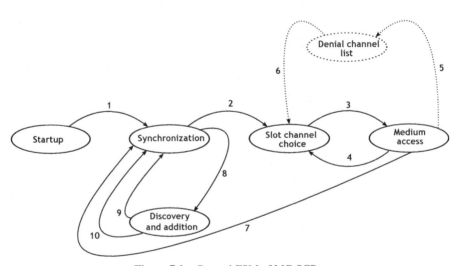

Figure 7.1 General FSM of MC-SCP.

7.2.1 Startup State

All the nodes that were recently deployed or rejoin the network, due to a reset or a battery replacement, are in the Startup state. In this state, the nodes sample the medium for an incoming SYNC frame sent by the sink node in the dedicated channel for control messages. After the node receives the SYNC frame it synchronizes with the network and enters into the Synchronization state. Here the child nodes are the Reduced Function Devices (RFD) nodes, while the Full Function Devices (FFD) are the remaining ones (coordinator and forwarders). FFDs are considered to be the parent nodes.

7.2.2 Synchronization State

The synchronization is performed by a hierarchical scheme similar to the one from the MC-LMAC protocol [IVHJH11]. Every node synchronizes with its parent. Every node chooses a parent node from the set of nodes that are closer to the sink node in terms of number of hops. In the MC-SCP protocol, a parent node is defined as a node closer to the sink node that is at one-hop distance from the sender node. This synchronization mechanism is similar to the one employed in the single channel SCP MAC protocol [YSH06b]. In the MC-SCP protocol, one of the channels is defined as the SYNC and control channel, which is dedicated to transmission and exchange of control messages, including SYNC frames. Therefore, each node has a single radio, that is used for both control and data frames [IVHJH11].

When the nodes are switched on, they go to the receive mode and the channel is, by default, the SYNC and control channel. Since the synchronization is initiated at the parent nodes, the remaining nodes (the RFD ones) are awaiting for an incoming SYNC frame. The first SYNC frame is always disseminated by the PAN coordinator [WPA07b]. When the neighbours from the coordinator receive the SYNC frame, they synchronize their clock with the coordinator's clock. The synchronization continues hop-by-hop while each node synchronizes with the associate forwarder node, as the RFD node does not has a direct link with the coordinator. Since the coordinator node cannot have a direct link with every node of the WSN, some FFD nodes have to assume the function of forwarder, in order to forward the received frames to the coordinator.

Synchronization Procedures

Every node that receives frames targeted to it, is considered as a parent node. It always forwards the frames to or not to the the lower layers (until it reaches the PAN coordinator). The PAN coordinator is the only one that does not forwards frames in the WSN. The remaining nodes are the child ones. The coordinator node has a field in its address that identifies it as a unique entity in the network. The node that receives a SYNC frame from a parent node (coordinator or forwarder) is locked onto this parent node. The sensor node performs the Carrier Sense (CS) of the radio channel for collision avoidance. After performing the CS for 1920 μs if the medium is free, the sender gets access to the medium and can start the SYNC frame transmission. This SYNC frame contains fields from the address of the sender and the time of its next wake-up. Immediately after the CS the sensor node transmits the SYNC frame (21 bytes). The parent node sends a SYNC frame to the child node. This SYNC frame containing the ID of the parent node, the time when the child nodes initiate the discovery-addition state and the time when SYNC frame was sent. This waiting time allows for the child nodes to stabilize the internal registers of the microprocessor and radio transceiver.

If the child node receives a SYNC frame from a parent node, it first verifies what type of frame has been received. If the type of received frame is a SYNC frame the child node extracts the parent node ID, the time when the SYNC frame was sent (t_{tx_sync}) and updates the schedule table. Besides, it also stores the time instant when the corresponding SYNC frame was received. This time instant is obtained from the clock of the child node, which is not synchronized with the clock of the parent node yet. Based on these time instants, the child node synchronizes its clock with the parent node clock. Some nodes may miss the synchronization with other nodes at beginning of the synchronization period, due to frame collisions that could occur on the dedicated radio channel. As a consequence, a single contention window is employed to mitigate the possible SYNC frame collisions. This scheme is further explained in Section 7.4. Periodically updating each of the schedule tables prevents long-time clock drift. This procedure of sending SYNC frames is repeated after a synchronization period, T_{sync}, but can be suppressed by piggybacking in the data frames schedules. This periodic synchronization is performed by the parent node by means of the control channel. After the first synchronization procedure the child nodes employ the piggybacking technique, in which the data frames sent to the parent node include the frame generation time. This piggybacked time instant is useful to the parent node be aware of possible child node clock drifts. Due to possible clock drifts, synchronization errors may occur from time to time. Besides the parent nodes, child ones also detect synchronization errors during their normal operation. This is achieved by using guard intervals that ensure receivers to be ready to listen before the senders start transmitting, leading to small timing differences. It has been shown in [YSH06b] that, if the nodes synchronize to every frame then a maximum drift of about 2 clock ticks is observed. Hence, in this work it is considered a guard interval of 2 clock ticks before and after the expected time of a message reception, whilst considering synchronization of nodes every frame by means of piggybacking. If the child nodes detect a time difference in the synchronization clock, the nodes transit back to the start-up state. The child nodes are not the only ones that are aware of clock drifts. The parent nodes may also enable a similar mechanism. This mechanism considers a "warning" frame that is sent immediately after receiving the data frame from the child node. In the slot structure from the child node there is a field named SYNC_EM, which is a SYNC frame that it is transmitted on the same channel that the child node used to transmit the data frame. In order to save energy, the child nodes after sending the data frame perform a CS with a duration of 9 Bytes, just to check if the parent node sends any frame to set the SYNC clock.

Based on the piggybacked time synchronization information in data frames, the MC-SCP uses a technique similar to the one from EM-MAC [TSGJ11], to keep a track of the clock drifts, after the first synchronization with his parent node. This technique is called Adaptive Time modelling that allows a sender and a receiver node to accurately predict the time of each node that sends/receives a frame. The child node keeps track of the drifts based on the modified ACK/NACK frames sent by the parent node after receiving correctly or not the data frames. With this ACK frames, there is no need to repeat SYNC frame transmission and reception for the nodes that are already in the WSN.

Implementation Details

The parent node (coordinator or FFD) broadcasts the SYNC frames only for new nodes that are in his range and have a direct link with the coordinator. This is possible because the PAN coordinator has extended processing capabilities and unlimited power supply. In MC-SCP, the child nodes model the time clock as a one-way message dissemination scheme [WCS11]. Parent nodes broadcast their timing information to various child nodes, while these nodes record the arrivals times of the broadcast message, as presented in Figure 7.2.

The dependence of the current time of the child node, $T_{2,n}$, on the current time of the parent node, $T_{1,n}$, is defined as follows [WCS11]:

$$T_{2,n} = T_{1,n} + \psi + \omega + X'_n \tag{7.1}$$

here, the clock offset, ω, and the delay, ψ, cannot be distinguished.

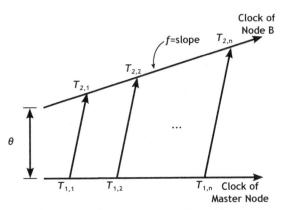

Figure 7.2   One way message dissemination.

Nevertheless, assuming that the fixed delay, ψ, is negligible, X_n' denotes the non-fixed delays in the transmission from the parent to the child node. The clock skew, f, is the slope that needs compensation and is approximately 1. More details about this assumption are given in [WCS11]. Hence, Equation 7.1 can be approximated by

$$T_{2,n} \approx f \cdot T_{1,n} + \omega + X_n' \tag{7.2}$$

After collecting all the time stamps and putting Equation 7.2 into the form of a matrix, the authors from [WCS11] state that the Least Square (LS) estimation for f and ω is possible to be obtained. However, this procedure can be computationally demanding for the sensor node. However, if a simple adaptive clock modelling is considered, based on the reception of a SYNC frame or an ACK/NACK frame by the child node, the procedure becomes less demanding. The clock offset ω is the initial time difference between the two nodes (child and coordinator nodes). When the child node receives the first frame it assumes that $f = 1$ and models the time of the parent node using the equation $T_{2,n} = T_{1,n} + \omega$. Hence, the child node assumes the new clock modelling as his clock reference for all the frames exchange. Depending on the reception of the SYNC information from his parent node by means of ACK/NACK or SYNC frames, the child node applies an algorithm to readjust its clock model. This is achieved by considering the last synchronization time related variables as well as the new ones. This can be translated into $T_{2,0} = f \cdot T_{1,0} + \omega$ and $T_{2,1} = f \cdot T_{1,1} + \omega$, in which $T_{2,0}$ is the previous received time sample of the parent node, while $T_{2,1}$ is the more recent received time sample from the parent node. The time variables $T_{1,0}$ and $T_{1,1}$ are the corresponding time stamps when the frames from the parent node are received by the child node. Based on these time samples, the child node is able to compute the values of f and ω, which allows for the node to be synchronized with the parent node. Theoretically, the time stamps added by the child node should be exactly the time when the frame was received by the child node. However, in real sensor hardware, a gap can be induced due to several factors, such as the operating system, carrier sensing, or radio backoff. Therefore, special attention must be paid to remove possible errors that may appear while applying this method in real sensor node hardware. The work of the authors from [TSGJ11] had the same problem and managed to remove these errors in the time stamps. A more detailed description of the EM-MAC protocol is presented in [TSGJ11]. Another important issue is the maximum allowable deviation time between the child and the parent nodes, ρ_{dev}. If the child node needs to send a frame to the parent node within the time window allocated for the chosen channel by the child node, the clock deviation time of the child node cannot be too large. Otherwise, the child node had changed to the same channel of the parent node either too early or too late. Based on [TSGJ11], the time needed to switch channel is 305 μs (10 ticks) for the considered TelosB platform. They considered a hopping period much larger than 305 μs and that synchronizing nodes within 6 ticks of deviation (i.e., ±183.1 μs) in 95 % of the time is sufficient for this purpose. The maximum deviation time considered by the authors from [TSGJ11] is about 60 % of the channel switch time. In our case, the considered channel switch time is 200 μs for evaluation purposes for the CC2420 transceiver. Therefore, the maximum considered deviation time considered in this work is 0.6×200 μs$= 120$ μs. A predictive wake-up mechanism based on a pseudo-random number generator is considered in order to properly wake-up a child node whilst turning its radio on right before the intended receiver (parent node) wakes-up the sender and the receiver for frames exchange.

Details on the node channel rendezvous scheduler (slot channel choice state, including as pects of the predicitive wake up mechanism) are addressed in [VBBC19].

7.2.3 Discovery-Addition State

In MC-SCP, the data frame includes the ID of the sender and the time when the frame was sent, along with the data collected from the sensors. Any node that receives this frame can adjust or set its internal clock and know the channel hopping sequence of the sender, based on the sender's ID, which is the seed of the LCG.

In this work, the sensor nodes that try to join the WSN for the first time must follow an algorithm that allows for joining the network and start sensing. There are two algorithms envisaged in MC-SCP-MAC protocol that allow the addition of nodes: first, the node uses the sensing node mechanism, in which it randomly chooses a channel and senses it for an initial period of 10 s. The channel choice is performed by considering a uniform distribution, as shown in Equation 7.3

$$\varphi_{ch} \in [12; N_{ch} - 1] \tag{7.3}$$

It is preferable the node performs a random channel choice based on uniform distribution than considering the LCG choice based on his ID, since it is less probable that any other node has chosen the same channel with a different seed.

When the new node detects a wake-up tone, a data frame or any frame being transmitted, it waits and receives it. With this procedure, the new node can synchronize his internal clock and assumes that the node that sends the received frame is the temporary parent node. The node is temporary because with the internal clock synchronized, the node can try to listen for the control and synchronization channel to check if it can directly communicate with the PAN coordinator. If so, it will keep the information of the 1-hop neighbour that it considers as a temporary parent node, just in case the PAN coordinator switch off or moves from the initial position (considering mobility). The time stamp stored in the frame must be taken after any queuing or any algorithm delay at the MAC layer.

If the new node does not senses any frame in the chosen channel, it will initiate a second mechanism called hopping sequencing, which hops throughout all the channels. However, it will sense each channel half or $1/4$ of the maximum time a node can use each channel to communicate. By decreasing the sensing time in each channel, the number of times a channel is sensed by the node in a frame duration (1 s) is two times (if it senses half of the time of a slot channel) or four times (if it senses a quarter of the time of the slot channel). This mechanism is the last resource of the new node to join the network.

Required number of slot channels per frame

In slot contention-based MAC protocols with single channel, the required number of time slots in the MAC frame is a parameter of paramount importance. Since these protocols rely on the slot choice to perform a CS and send a frame when an idle medium is sensed, the number of time slots affects the probability of collision (i.e., characterizing nodes that choose the same slot) and the service time. The majority of single-channel slot contention-based MAC protocols (e.g., SCP-MAC [YSH06b]) consider a fixed number of slots in the the MAC frame. However, some MAC protocols already consider a dynamic slot allocation mechanism. The required number of slots in a MAC frame depends on the expected number of nodes that compose the WSN. Moreover, tradeoffs must be established in the definition of the number of slots in the MAC frame by relating the number of slots, collision probability, and number of nodes. On the one hand, the choice of few slots in a high node density leads to high collision probability. On the other, the choice of a large number of slots in a low node density leads to low collision probability. However, the energy costs are higher due to the energy wasted while waiting for frame transmission.

In the single-channel SCP-MAC protocol, within the 2nd-hop neighbourhood a node should not use the same slot from the contention windows 1 or 2. Otherwise, if the sensor nodes sense the channel as idle and decide to transmit frames simultaneously a frame collision will occur.

Since MC-SCP-MAC is a multi-channel protocol, the nodes can choose one channel from the available ones to transmit their frames. As a consequence, in this case a node can use the same slot as the 2nd-hop neighbours but on a different channel. In this case co-channel interference is negligible [SWM+07]. The number of required slot channels dedicated to data exchange is calculated in the startup phase by considering the node density information. At the startup phase, since only a maximum of 15 channels can be used for a maximum number of nodes, n_{max}, deployed in the WSN, the following relation is considered:

$$N = \begin{cases} \left\lceil \frac{n_{max}}{2} \right\rceil & 0 < n_{max} \leq 2N_{ch} \\ 15 & n_{max} > 2N_{ch} \end{cases} \tag{7.4}$$

Only the default channel (slot channel 11, for synchronization purposes) is known and common to all the nodes. During the network startup the node knows the node density and assigns the number of channels to be considered by the channel choice mechanism. Here is assumed that a channel supports at most four nodes contending in the same channel. If new nodes are later added to the network, the parent node (PAN coordinator) can decide whether to restart the network or continue with the chosen number of channels. The associated collision probability is decreased by using an enhanced-two phase contention window mechanism, which considerably reduces the number of collisions compared to a single-contention window protocol (e.g., multi-channel MAC, (McMAC) [SWM⁺07]).

7.2.4 Medium Access State and Algorithm

After the node starts up and synchronizes with the network while selecting the first slot channel, it enters into the enhanced-two phase contention window mechanism, corresponding to the medium access phase. In the medium access phase, the node follows the state transition diagram presented in Figure 7.3. For every frame the node wants to transmit, the node repeats the choice of the slot channel given by the LCG and the choice of the slot for each enhanced two-phase contention window mechanism phases, as further explained in Section 7.3.1 (N.B., this is one of the key mechanisms to reduce potential collisions in the same channel).

After the reception and transmission of the frames, the nodes will check if any frame is stored in the queue. If so, the child node uses the remaining time of the slot channel to send the frames without the need to perform the choice of the slot in the two contention windows. This greatly reduces the time needed to transmit all the remaining frames.

It is therefore important to explain the frame and channel assignment structure, i.e., the frame structure and the enhanced-two phase contention window mechanism involved in the message exchange procedure.

Frame and channel structure

A frame corresponds to a hopping sequence of the available channels, in which the parent node senses the data and the child nodes choose a channel to transmit their data frames. Each channel is assumed as a slot channel, S_c, as shown in Figure 7.3. The enhanced-two phase contention window mechanism is employed in each slot channel, in order to facilitate that nodes contend for the channel. This time period is denoted as the node-contention (NC) phase. After the NC phase, the slot channel has an extra resolution (ER) phase, which can be used by the sensor nodes to transmit more frames that are stored in their queue, after winning the two phase contention window mechanism while having already sent the first data frame with no collisions. If the node needs to send more than one frame per slot channel, it must enable the "more bit" and include the number of remaining frames from the queue to be included in the control information from the first data frame sent to the parent node. This allows for the parent node to wait for the remaining frames, suppressing the need of performing all the two phase contention

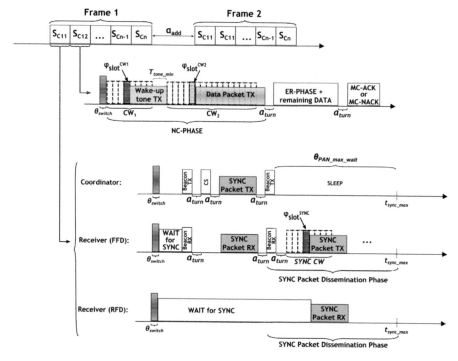

Figure 7.3 Frame structure for the MC-SCP-MAC protocol.

window mechanism when the sender needs to transmit consecutive frames to the parent node, in the current slot channel. This mechanism is already used by the WiseMAC [EHD04b] protocol, in which a high performance is attained when consecutive transmissions are envisaged. After the ER phase, the node (child or parent node) will enter into the inactive phase, in which nodes go to sleep mode to save energy. The slot channel structure presented in Figure 7.3 is considered in all the available channels from the IEEE 802.15.4 standard, except for channel 11, which has a different purpose (SYNC and control frames exchange) and, consequently, a different slot channel structure. The slot channel structure for the SYNC and control channel is also presented in Figure 7.3.

The number of slot channels, S_c, is equal to the number of channels, and each slot channel is indexed by a channel number. The slot channel S_{c11} is dedicated just for SYNC and control frames exchange, while the remaining ones ($S_{c12} \dots S_{c26}$) are dedicated to the node contention and data frames exchange. The number of slot channels on which the radio transceiver can be adjusted to, depends on the nodes density during the deployment, as expressed by Equation 7.4. However, there is the option to disable this dynamic channel adjustment, allowing for considering a fixed number of channels, independently of the nodes' density. This dynamic mechanism is more efficient for networks with low node density, since it uses less channels. The PAN coordinator will be more energy efficient, as the channel hopping sequence is performed for less slot channels, allowing for the nodes to consume less energy. In Figure 7.3, each frame has a duration of $t_F = 1$ s. The duration of each slot channel is $\Delta t_{SC} = t_F/N_{ch}$, with $N_{ch} = 16$. This means that the duration of each slot channel is $\Delta t_{SC} = 62.5$ ms. Figure 7.3 also depicts the different slot channels that are reserved for the senders of the frames whose number given by the LCG generator, is equal to the slot channel index (the first slot channel is reserved for the SYNC and control frames exchange, the second slot channel is for channel 12, and so on, up to channel 26). In the SYNC and control channel, S_{c11}, during the first time the child nodes are switched on, they initiate the sensing mode awaiting for the SYNC frame. Hence, the direct (1-hop) communication links between child and parent nodes can be established. The nodes (child and parent nodes) need to switch to the channel 11 before performing the synchronization procedure, as shown in Figure 7.3. This switching channel time in our protocol is 200 μ s. However, depending on the radio transceiver this switching channel time may vary. After switching to the slot channel S_{c11}, the child nodes wait for a beacon frame sent by the parent node, in order to initiate the synchronization of the nodes. As soon as the beacon frame is completely sent, the parent node initiates the transmission of the SYNC frame to the child nodes. This SYNC frame has a length of 18 Bytes. The control information in these 18 Bytes includes the time of the parent node, the source node ID, destination ID (broadcast), number of considered channels, and the next synchronization period. Once the child nodes receive the SYNC frame, they are able to synchronize their internal clocks with the clock of their parent node (the PAN coordinator). The length of the SYNC frame is equal to the one employed in SCP-MAC and is sufficient to the child nodes to receive it on time, even if serious clocks drifts persist in the child nodes. The first synchronization procedure finishes with the transmission of another beacon frame (by the PAN coordinator), to indicate to the child nodes that the synchronization procedure has finished and they could go to sleep mode until the next wake-up to send data frames. This synchronization procedure is equal to the one employed in the beacon-mode from the IEEE 802.15.4 standard.

7.3 Fundamentals of the Protocol

Since the MC-SCP protocol is a scheduled-based protocol, nodes should first synchronize in order to send and receive frames with the correct timing. To access the medium and send messages, nodes deterministically choose a channel/frequency from a maximum of N_{ch} channels (i.e., N_{ch}=16 channels in IEEE 802.15.4) by applying a LCG. In each channel the transmissions do not interfere with the concurrent transmissions in other channels. SCP-like double slot selection in the enhanced two phase contention window mechanism allows the node to contend for the medium and transmit its frames. The node delivers frames to the sink node in a deterministic TDMA manner, based on a pseudo-random number generator that defines the choice of the channel for the node as well as the corresponding time for the delivery to the sink node.

7.3.1 Enhanced-Two Phase Contention Window Mechanism

During the nodes' contention period and data frame exchange through the slot channels, an enhanced-two phase contention window mechanism, similar to the one from in SCP-MAC [YSH06b], is employed. In these slot channels the contention based mechanism is preceded by the switching channel time, in order to facilitate that the child nodes communicate within the chosen slot channel. SCP [YSH06b] uses two contention windows, CW_1 and CW_2. CW_1 is divided into CW_1^{max} time slots of equal duration, as shown in Figure 7.4.

When a node wants to transmit a data frame, it randomly chooses a time slot, $\varphi_{slot}^{CW_1}$, in CW_1, with a uniform distribution, whilst sensing the shared medium for the duration of that time slot. The choice of the slot (from the ones available in the interval) follows a uniform distribution:

$$\varphi_{slot}^{CW_1} \in [1; CW_1^{max}] \tag{7.5}$$

In our case, the first contention window is not going to be divided into slots of equal duration. Instead, the CW_1 time interval ($\Delta\sigma$) is defined, in which a node that wants to transmit a data frame will randomly choose a time value within $\Delta\sigma$, with a resolution of $\rho_{CW_1} = 32$ μs. The maximum value defined for $\Delta\sigma$ is two times the time required by a sensor node to perform the carrier sense of the channel, as follows:

$$\Delta\sigma_{max} = 2 \cdot t_{CS} \tag{7.6}$$

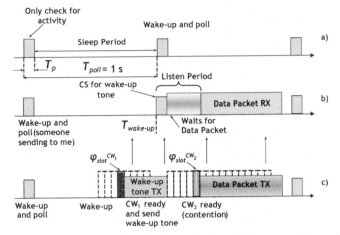

Figure 7.4   SCP two-phase contention window mechanism: a) Sensor node has neither data to send nor to receive; b) Sensor node is ready to receive data; c) Sensor node has data to be sent.

Table 7.1   Duration of CS, sensitivity channel switch time for CC2420 and AT86RF231 radio transceivers.

Radio transceiver	t_{CS} [μs]	S_{min} [dBm]	t_{switch} [μs]
CC2420	1920	-94	200
AT86RF231	528	-101	11

where the duration of the CS, t_{CS}, sensitivity, S_{min}, and channel switch time, t_{switch}, of the CC2420 and AT86RF231 radio transceivers are presented in Table 7.1.

The value chosen for the time to initiate the CS (of the medium) must guarantee that the CS is concluded before the end of the CW_1. The resolution parameter defines the minimum time distance between two consecutive randomly chosen values for the CW_1 (i.e., it is the duration of a slot). However, the conducted performance evaluation tests considered the uniform slot selection.

If no channel activity is detected (idle channel), a node sends a wake-up tone. The wake-up tone is sent after the node performs the CS, verifies that the medium is free and has a minimum duration of 62 Bytes (2 ms @ 250 kbps). This length, in Bytes (corresponding to a certain time duration) is extended for the duration of the remaining time slots (T_{subs_slots}) in CW_1, as presented in Figure 7.5.

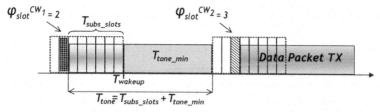

Figure 7.5   Wake-up tone duration as a function of the CW_1 chosen slot.

Like in SCP [YSH06b], the wake-up tone in the MC-SCP protocol announces to other potential senders that a node is preparing to send data. The node enters in CW_2 immediately after the wake-up tone transmission. The second contention window is divided into CW_2^{max} time slots of equal duration. Only the nodes which have data to send utilize the two-phase contention window mechanism, while the remaining ones poll the channel. In both contention windows, a collision occurs if more than one node chooses the same slot (i.e., the wake-up tones collide, or simultaneous transmissions start in the same slot). Only nodes that succeed in CW_1 enter CW_2. When more than one node succeeds in CW_1 (at least two nodes transmit the wake-up tones at the same lowest order slot), "effective" data frame collision happens in CW_2 if the nodes, which were successful in CW_1, randomly choose the same time slot in CW_2. This simultaneous choice of a time slot corresponds to an effective data frame collision event. If a node succeeds in sending the wake-up tone during CW_1, it randomly chooses a slot $\varphi_{slot}^{CW_2}$ from CW_2 by using a uniform distribution:

$$\varphi_{slot}^{CW_2} \in [1; CW_2^{max}] \tag{7.7}$$

The node then starts carrier sensing in CW_2 (with a duration of about 1920 μs [YSH06b]), as shown in Figure 7.4. If the channel is sensed idle, it starts transmitting its frame. In SCP, after sending/receiving the data frame, the sender/receiver node checks whether there are any new data frames pending to be sent. If so, the node repeats the steps described above before going into sleep mode. However, in MC-SCP the winning node does not perform all the procedures of the two-phase contention mechanism if it has more frames in the queue to be sent during the slot channel. If the node has more than one frame in the queue, the "more bit" and the "remaining number of frames" must be added to the control information of the first data frame sent to the parent node. If the parent node receives the data frame with the "more bit" enabled it will enable the ER phase, in order to receive the remaining frames.

The child nodes are the ones that request the ER phase, based on a decision mechanism that is further explained in Section 7.3.6. The maximum allowed frames that can be received by the parent node is limited to the remaining time of the slot channel and must take into account the ACK or NACK frame sent by the parent node. It allows for reducing the queuing delay at the parent node, especially when traffic bursts are considered. If the parent node receives the first data frame with the information that the child node has more frames ready to be sent in the queue and misses receiving the last data frame of the sequence, it has a time out timer, $t_{guard} = 400 \mu$ s. This timeout timer allows for the parent node to be aware of transmission errors in the frame sequence sent by the child node. As soon as the ER phase ends, the parent node sends an ACK or a NACK frame, depending on if it receives more than one frame or node from the child node, respectively. This ACK/NACK frame has a particular structure that allows for retransmissions and synchronization of the child nodes. The frame structure is going to be further discussed in Section 7.3.2. By considering two contention stages (CW_1 and CW_2), the collision probability decreases when compared to the single contention window case with equivalent length (sum of the lengths of CW_1 and CW_2). This is because only the nodes succeeding in CW_1 enter into CW_2. This choice decreases the number of nodes effectively accessing the medium during the second contention stage. After the first phase of the contention procedure, only the successful nodes from CW_1 (the first one(s) that transmit the wake-up tone) contend in CW_2. With less contending nodes, the effective collision probability is considerably reduced. Each slot channel can accommodate the transmission of multiple data frames from the child node that wins the contention, as shown in Figure 7.3. Each slot channel has a fixed length, but the NC and ER phases duration varies according to the number of data frames involved and the choice of slots in CW_1 and CW_2. Since the number of slot channels depends on the number of nodes initially deployed, the nodes may maintain the same frame time, t_F, and use one of the 15 channels, leading to lower energy consumption.

7.3.2 Frame Structure

In order to establish communication between child and parent nodes, different frames are needed. In the context of MAC protocols, data and control frames are involved in the communications. The content of the data frame transmitted during the NC or ER phase are as follows:

- **Source ID**: it represents the node ID of the sender;

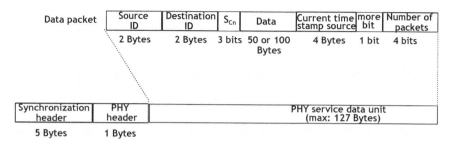

Figure 7.6   Structure for the data frame in the MC-SCP-MAC protocol.

- **Destination ID**: it represents the node ID of the parent node (receiver);
- **Slot channel** (S_c): it represents the chosen channel;
- **Data**: it represents the sensor data sent to the parent node;
- **Current time stamp of the source**: it represents the time instant when the child node sends the data frame, and is used for synchronization by new nodes that try to join the WSN;
- **More bit**: a sensor node receiving a data frame with this bit set will continue to listen the current slot channel after finishing receiving the first data frame;
- **Number of remaining frames**: it represents the remaining number of am frames that are in the sender's queue (if the "more bit" is enabled, this field is used by the parent node to know how long it must remain in the receiving mode, to receive the remaining data frames);

Figure 7.6 shows the corresponding fields for the data frames, along with their length. The control frames that may be exchanged during the MC-SCP are the following:

- **SYNC frame**: it contains synchronization information for the child nodes (with a fixed length);
- **Wake-up tone**: it contains the source ID, destination ID, current time stamp of source, and a field with random data (dummy bits) only used to occupy the frame (with variable size, and minimum length of 62 Bytes);
- **Beacon frame**: it consists of a frame that is used only to signal the beginning and end of the synchronization phase, and has a duration of 0.7 ms [BMS$^+$08];
- **ACK frame**: consists of a frame that contains the source ID, destination ID, missed frames from sequence and current time of the parent (with fixed length);

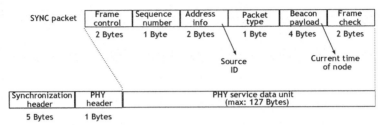

Figure 7.7   Structure for the SYNC frame.

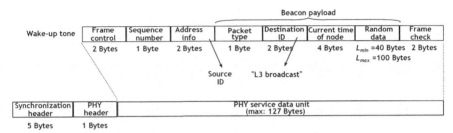

Figure 7.8   Frame structure for the wake-up tone in the MC-SCP-MAC protocol.

- **NACK frame**: it consists of a frame similar to the ACK one, corresponding to the cases when the parent node does not received any data frame (with a fixed length).

Some modifications are needed in some control frames, such as the beacon, ACK, and NACK frames compared to the IEEE 802.15.4 standard. This was decided to enable that the MC-SCP-MAC protocol presents higher agility with multi-channel.

The content of the SYNC frame is similar to the beacon frame in IEEE 802.15.4. However, here the SYNC frame has a length of 18 Bytes.

To comply with the IEEE 802.15.4 standard, the SYNC frame format has to be modified without removing any field of the frame needed. The required modifications are the addition of the field "frame type", which allows for the child nodes to check that the received frame is the SYNC one, and the consideration of 4 Bytes in the "beacon payload" field, to store the current time of the parent node, for synchronization purposes (internal clocks) in the child nodes. The wake-up tone is a frame whose fields are similar to the IEEE 802.15.4 beacon frame ones. However, it has a variable length (and a minimum length of 62 Bytes), since it depends on the remaining time/slots of the CW_1 in the NC phase. The content of the wake-up tone transmitted during the NC phase is shown in Figure 7.8. Compared to the IEEE 802.15.4,

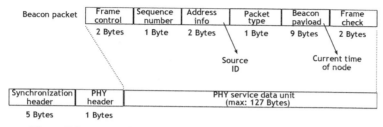

Figure 7.9 Beacon frame structure for the MC-SCP-MAC protocol.

the only difference is in the beacon payload fields. The first field is the "frame type", whose "wake-up" type is denoted by "2". The second field is the "destination ID", which, in this case, must be from the broadcast type, since the tone is used to warn potential contenders that other node has already won the medium in the CW_1. Since this wake-up tone is from the broadcast type, a field named as "current time of the child node" is added to the frame structure, in order to enable that other child nodes that are in the same slot channel and in the discovery-addition phase are able to obtain the first clock synchronization. Moreover, the parent node can keep track of the internal clocks of the child nodes and warn the child nodes when the clock drift is higher than the maximum allowable deviation, $\rho_{dev} = 120 \ \mu s$. The "random data" field presents a variable length and is used to increase or decrease the wake-up tone, depending on the remaining slots of CW_1. Considering the default fields lengths, the "random data" field length varies from 40 Bytes up to 100 Bytes.

The content of the beacon frame is the one from the IEEE 802.15.4 standard, as presented in Figure 7.9. Since the beacon frame has to present a duration of 0.7 ms, the required length for the field "beacon payload" is equal to 9 Bytes. With a total of 22 Bytes, the duration of the beacon frame is around 0.7 ms.

The ACK frame structure from the IEEE 802.15.4 standard, requires some modifications to facilitate that the MC-SCP protocol achieves a better performance. This modified ACK frame is named as MC-ACK frame. Its structure is simpler than the structure of the frames presented before. The contents of the IEEE 802.15.4 ACK frame and the necessary modifications to obtain the MC-ACK are presented in Figure 7.10. Compared with the standard structure of the IEEE 802.15.4 ACK frame, we verified the need to expand the structure, in order to include other important information, such as the "current time of the parent node", the missed frames sequence

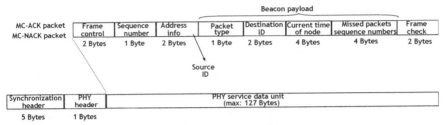

Figure 7.10 Structure for the MC-ACK and MC-NACK frames.

number, as well as the frame type, source ID and destination ID fields. The new ACK frame allows for suppressing the explicit synchronization frames, since all the child nodes that are in the same slot channel receive the MC-ACK frame and use it to synchronize their internal clocks. As shown in the SCP MAC protocol [YSH06b], the synchronization piggybacked in frames is truly efficient. If the parent node does not receive a data frame after sensing the wake-up tone in a slot channel, it sends a MC-NACK frame after the timeout timer expires. For the MC-NACK frame, the structure is similar to the MC-ACK frame presented in Figure 7.10. The "frame type" field is "5" for this type of frame.

Here, the overhearing of the MC-ACK and MC-NACK frames, by the child nodes that contend in the enhanced-two phase contention window but did not win the access to the medium, is considered to be beneficial. These child nodes receive the frame only to extract the synchronization information of the parent node.

Details on the influential range concept (that takes advantage of the overhearing in order to save energy) are addressed in [VBBC19].

7.3.3 Denial Channel List

In single-channel MAC protocols, the sensor nodes do not have the possibility to choose other channels when one of the channels is degraded due to severe interference. However, in multi-channel MAC protocols, as this degradation may also occur, the multi-channel MAC protocol can sense the channel conditions to mitigate this problem, in order to optimize the set of channels that present a higher reliability.

A similar mechanism is already considered by the EM-MAC [TSGJ11] protocol. Nevertheless, in our case the MC-SCP differs on the EM-MAC, since it is based on a two-phase contention window mechanism based on slots to resolve the access to medium, while the EM-MAC is based on the

CSMA scheme. Also this decision is performed at the application layer, which considers a cross-layer design in MC-MAC and allows for the WSN application to use this sensing information for cognitive radio purposes.

Each node maintains a denial channel list which contains the levels of "degradation" of each channel. This "degradation" level is a non-negative metric. When a node chooses a channel to transmit a frame, it switches to the chosen channel and performs a first CS within the CW_1. If the node detects an idle channel it decreases the degradation level of the channel in one unit (the metric cannot be lower than zero), and passes to the second contention window (CW_2). Otherwise, the node increases the channel degradation in one unit. The nodes that passed to the CW_2 will perform another CS to the channel. In case an idle channel is sensed, the degradation level of the channel is decreased in one unit. Otherwise, the same metric is increased in one unit. Other conditions imposed to channel's degradation level consider that a data frame collision increases the degradation metric by two units in all the nodes that won access to the CW_1 but lost the contention in the CW_2 while detecting a data frame collision. Besides, if a node sends a data frame and does not receives a MC-ACK it increases the channel's degradation level by two units.

Based on the channel degradation level stored in the denial channel list, a node in MC-SCP restricts the selection of channels it switches to. Since this denial channel list is locally maintained by each node, the node decides that a channel is considered severely degraded if the degradation level metric is higher than a defined threshold, D_{ch}. If so, the channel is added to the denial channel list and is avoided by the node during the data frame transmission. If the node chooses a degraded channel based on its pseudo-random generator, the node always chooses the next channel given by the pseudo-random generator and increases the sequence number from the sent of frames by one unit.

The maximum number of channels stored in the denial channel list is the total number of considered channels minus one. This limitation is required for the cases when all channels are considered degraded. However, the least degraded one is removed from the list. This forces the node to use a single-channel until the remaining channels reduce their degradation level metric and get removed from the denial channel list. If the number of degradation channel levels was not limited to the total minus one, the node can get isolated from the remaining nodes, mainly due to the restriction of not being allowed to use any channel to communicate. The denial channel list policies for the degradation level metric in the MC-SCP-MAC simulator are summarized in

Table 7.2. The degradation level metric allows for the nodes to consider the less degraded slot channels, which leads to a more efficient use of the medium to transmit the data frames.

Table 7.2 Reward and punishments for the degradation level metric from the denial channel list policies in MC-SCP-MAC.

Case scenario	Reward	Punishment
Performs CS in CW_1 and detects a free medium	+1	-
Performs CS in CW_1 and detects an occupied medium	-	−1
Performs CS in CW_1 passes to CW_2 and detects a free medium	+1	-
Performs CS in CW_1 passes to CW_2 and detects an occupied medium	-	−1
Performs CS in CW_1 passes to CW_2 and detects a free medium and wins contention in CW_2	+2	-
Performs CS in CW_1 passes to CW_2 and detects a free medium but loses contention in CW_2	-	−2
Performs CS in CW_1, detects an occupied medium, receives successfully the data frame	+2	
Performs CS in CW_1, detects an occupied medium, detects a data frame collision	-	−2
Node without data to TX, performs CS in CW_1, detects an occupied medium, receives successfully the data frame	+2	-
Node without data to TX, performs CS in CW_1, detects an occupied medium, detects a data frame collision	-	−2
Node receives a MC-ACK frame with success	$+1 \times n_{frames\_tx}$	-
Node receives a MC-NACK frame with success	-	$-1 \times n_{frames\_tx}$
Node TX data frames but does not receives MC-ACK or MC-NACK frames	-	−2
Timeout of the timer to receive MC-ACK or MC-NACK frames	-	−2
Timeout of the remaining timers in MC-SCP-MAC	-	−1

7.3.4 Frame Capture Effect

Existing MAC protocols assume a typical physical layer based on the IEEE 802.11 standard. In this context, when the radio transceiver simultaneously detects two frames the occurrence of a collision is assumed. The frames will interfere with each other. As a consequence, the receiver will not be able to decode either one of them. Although this is theoretically true, in the last years, the work from some researchers [RLLK08] has shown that this assumption is not absolutely true.

Under certain channel conditions and circumstances, the frame capture effect occurs when the strongest signal causes the other signals to be treated as noise (or interference) and is filtered out by the receiver. As a consequence, a frame is received even though a collision occurred due to concurrent transmissions [RLLK08]. This phenomenon (of being able to receive a frame even in the presence of another interference frame) is called FC effect. This effect neglects the assumption of *"collision as failure"* in which frame collision leads to frame corruption. This assumption is commonly accepted in simulations [WWJ⁺05, SKA04], theoretical analysis [KT75], and collision avoidance schemes [PCB00]. The FC effect is considered in our MAC protocol due to the beneficial revenues that it can offer, such as prevention of frame collisions, increase of the network throughput, and decrease of the delay. In theory, frame capture can provide additional gains through collision detection and the recovery of the strongest frame from collisions [WWJ⁺05]. The FC effect is going to be considered in the simulations of the MC-SCP, but it can be disabled to observe how the MAC protocol behaves when this effect is not considered. Besides the FC effect implemented in the simulator framework, a reliability decision algorithm has also been implemented, jointly with the FC effect. This decision algorithm allows for the physical layer to decide whether a frame is correctly decoded or not. It tries to mimic the real-world constraints in the simulator, in order to decide if it accepts the frame based on the channel conditions.

7.3.5 Node Topology and Envisaged Real-World Scenarios

The IEEE 802.15.4 standard supports three types of networks topologies: star, peer-to-peer and tree topologies. Depending on the application of the WSN, a different topology can be chosen. The star topology does not ensure a reasonable scalability of the network. Besides, the peer-to-peer topology does not guarantees low energy consumption, since nodes are always active. Since the tree topology supports a higher scalability and presents low energy

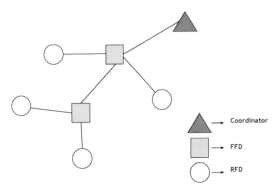

Figure 7.11   Considered topology: full tree topology scenario.

consumption, we focus on this topology as presented in Figure 7.11. The verification of the MC-SCP-MAC performance also aims at implementation on real-world scenarios. It is designed for dense deployment scenarios, such as tracking and localization, environmental monitoring or intrusion detection applications. Some works have been already done for the environmental monitoring, such as the one from the authors [TPS⁺05]. In turn, for the intrusion detection application the authors from [ADB⁺04] developed an application capable of detecting, classifying and tracking a target.

The MC-SCP MAC protocol considers the IEEE 802.15.4 standard. The sensor nodes are divided into two types: FFD and RFD. The FFD are capable of sensing events and route frames to other nodes. While the RFD has limited capabilities, since these devices can only communicate with FFD and perform the sensing of phenomena. Moreover, the FFD can support two types of roles:

- **PAN coordinator**: the PAN coordinator is the primary controller of the PAN. It initiates the network and often operates as a gateway to other networks. Each PAN must have exactly one PAN coordinator;
- **Coordinator**: is a FFD capable of relaying messages using data routing and network self-organization operations to achieve it.

In our case, we consider a coordinator that collects and processes information sent from the sensors (RFD and FFD). When an event is detected in a given area, every node senses it, creates a frame and sends the frame to the sink node (coordinator or parent node) throughout the FFD if multi-hop is considered. In multi-hop scenario the FFD forwards the frame to another FFD, until the frame reaches the coordinator.

The scenario presented in Figure 7.11 is considered as the full tree topology

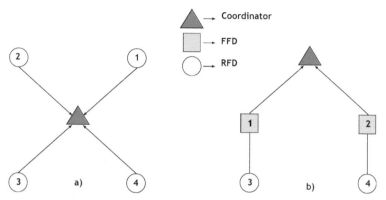

Figure 7.12 Considered single- and multi-hop scenarios.

scenario. In order to observe the behaviour of the MC-SCP MAC protocol in single- and multi-hop scenarios, simpler scenarios have to be considered for evaluation purposes. Figure 7.12 (a) presents a topology where all the RFD nodes (child nodes) are directly connected to the sink node (coordinator or parent node). In the case of single-channel, if C is the aggregated throughput (i.e., the capacity of the medium), then each of the RFD entities from Figure 7.12 (a) theoretically transmits a capacity of $C/4$ per RFD. In the multi-hop case, presented in Figure 7.12 (b) if a suitable scheduling mechanism is considered (no interference between nodes) an aggregated throughput per node of $C/4$ is achievable. If no scheduling mechanism is considered, all the transmissions will interfere with each other, leading to throughput decrease. The achievable throughput drops to $C/6$ per node, since RFD 1 and 2 have to forward RFD 3 and 4 frames besides their own frames. By considering non-interfering channels (multi-channel based), the interference can be eliminated and the RFD and FFD can achieve a maximum aggregated throughput per node of $C/4$.

These scenarios are useful for comparison with the MC-LMAC protocol [IVHJH11]. Since MC-SCP protocol is based on SCP-MAC protocol [YSH06b] it is worthwhile to compare both protocols in order to verify the improvements from using multi-channel based schemes. Other scenario that is going to be considered is the chain scenario (multi-hop) one, as it allows for observing the end-to-end delay of frames, as shown in Figure 7.13. The presented topologies are considered in the next sections to address the efficiency of the multi-channel MAC protocol, MC-SCP.

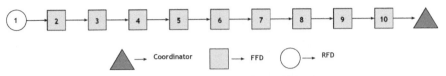

Figure 7.13   Chain scenario (multi-hop).

7.3.6 Extra Resolution Phase Decision Algorithm

Each slot channel has an ER phase, which can be used by the sensor nodes (child nodes) to transmit more frames that may exist in their queues after sending the first data frame. Since the channel can be degraded due to interference, or the nodes may move to other positions (mobility is considered), the node may transmit the data frames unnecessarily, since they will not be received by the parent node. Therefore, a decision algorithm is required to decide whether a child node sends all queued frames during the ER phase. The decision is based on the SNIR from the frame, the "degradation" level of the slot channel and the neighbouring nodes of the child node that intends to transmit the remaining data frames. The first component of the decision algorithm is based on the degradation level from the slot channel, already defined for the denial channel list. While the second component of this algorithm is based on the IR concept, on the last three accesses of the child node to the medium and the corresponding frame delivery results.

A node that has more frames in queue to transmit compares the degradation level of the current slot channel (stored in the denial channel list) with the threshold, D_{ch}. If the channel is considered has "good" to send frames, then it passes to the second phase of the EC phase decision algorithm. In the second phase, the node checks the last three calls of the influential range algorithm. If the node receives at least three positive feedbacks in the IR algorithm for that channel for the same neighbouring nodes, then the node can use the ER phase to transmit the remaining frames stored in queue. In case the IR algorithm feedback is not possible to be checked, the node will check in the last three channel transitions if it has successfully transmitted the frames to the parent node. If so, then the node can utilize the ER phase to transmit the frame in the queue.

7.4 State Transition Diagram and Description

Nodes are in the *startup* state (Figure 7.1) when they are recently deployed or rejoin the network, due to hardware reset or batteries replacement. During the *startup* state the node senses the medium for an incoming frame to synchronize with the network. To achieve this, the nodes switch to channel 11, which is the SYNC and control slot channel and wait for the parent node to send the SYNC frame. When the nodes are switched on, they wait for 10 s to stabilize and switch to the SYNC and control slot channel and start sensing the medium. It is assumed that the parent node is also switched on during the first 10 s, but it sends the first SYNC frame only after another 10 s. This allows for the child nodes stabilize properly the hardware related issues and be ready to receive the SYNC frame. If such frame is received, a node passes to the *synchronization* state and synchronizes with the parent node (who sends the SYNC frame). For the cases of child nodes with no direct link with the parent node, the child node has to send his frames to a FFD node and consequently the FFD node relays the frames to the parent node. To achieve this goal, during the *synchronization* state the nodes that receive the SYNC frame will have to retransmit this SYNC frame in the next attempt. This is essential for the nodes that do not have direct communication with the parent node and need to send their frames to the parent node. Therefore, the nodes that are going to retransmit the SYNC frames will utilize a slot choice based on a single contention window, since other nodes will utilize the same channel to retransmit SYNC frames during the *synchronization* state. After winning the access to the medium and retransmit the SYNC frames for the other nodes, it will be assumed by the nodes as their parent node.

Once the *synchronization* phase finishes, a node follows the schedule to deliver frames to the coordinator (single-hop) or the FFD (multi-hop). If a node needs to join to the network after the *synchronization* phase (with a maximum duration of 1 minute) it enables the first of the two mechanisms that allows for the synchronization. If the random channel slot selection algorithm allows the node to synchronize with a parent or FFD node, then it passes to the *synchronization* state. If the first mechanism fails, the node initiates the fast slot channel hopping mechanism (a secondary mechanism). Both these mechanism are fully described in Section 7.2.3. If the new node has success in one of these mechanisms, it enters in the *synchronization* phase. After the network synchronization, the child nodes wake-up periodically, to check the medium for any potential incoming frames. This check lasts for 1920 μs (time to perform the CS) and only occurs when the node does not have any frames in queue to be sent. The slot channel choice is given by the current pseudo-random generator value. If it does not detect any traffic during the CS, the node goes immediately to sleep mode, whilst scheduling the next wake-up. If the node has frames in the queue (it senses an event) it transits to the *slot channel choice* state. Here, the node chooses and switches to slot channel given by the pseudo-random generator of the parent node. Then, it passes to the medium access phase, in which it contends for the channel by employing the enhanced two-phase contention window mechanism. If the contention is successful in both contention windows the frame is transmitted and waits for the MC-ACK. If the contention fails, the node waits for the MC-ACK frame that may utilize to synchronize his internal clock.

Each generated frame will make the node to transit from the *medium access* state to the *slot channel choice* state, in order to the node gets the corresponding slot channel of the parent node and the time instant when the parent node is in the current slot channel. After a node collects the slot channel information, it enters into the *medium access* state, in which it performs the two-phase contention window mechanism resolution. If the node receives a warning of collision in a specific slot channel, it transits to the *denial channel list* state. After updating the degradation level of the slot channel in the Denial channel list, the node(s) passe(s) to the *slot channel* state. If a node detects or is warned by the parent node that the synchronization error is higher than the maximum allowable deviation time, ρ_{dev}, then the node transits to the *synchronization* state, to resynchronize its internal clock. The nodes enter in the *discovery and addition* state in two situations. First, when the nodes fail to receive a SYNC frame from a parent node, the node transit to the *discovery and addition* and tries to resynchronize later. Second, when nodes are turned on and try to join the WSN, the nodes try to receive a SYNC frame from a parent node.

We have proposed a possible FSM. The objective of this FSM is to give a detailed specification of why and how the simulator changes between the different states, while explaining the associated actions and events as shown in Figure 7.14. The corresponding transition events and actions are listed in Table 7.3.

This state machine is based on twenty five states, described as follows:

- **STARTUP**: The node is switched on;
- **WAIT FOR SYNC**: The node is waiting for SYNC frames. It switches to the SYNC and control slot channel ($S_C = 11$);
- **SYNCHRONIZATION**: All the nodes, during the startup, switch to SYNC and control slot channel and listen for the incoming SYNC frame sent by the coordinator. The synchronized nodes retransmit the SYNC frames to the neighbouring nodes by employing a SYNC contention window;

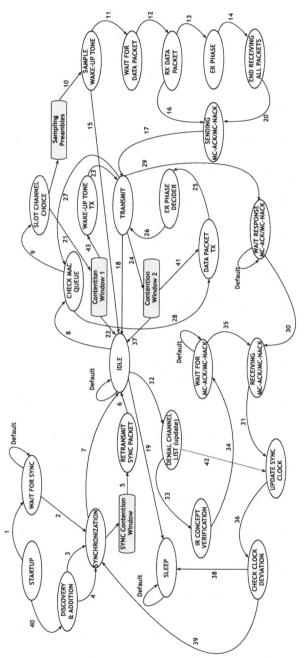

Figure 7.14 Detailed finite state machine for the MC-SCP-MAC protocol.

Table 7.3   MC-SCP events and actions.

Events	Actions
1	**State**: WAIT FOR SYNC **Schedule**: Switches to SYNC and control slot channel; Wait for the SYNC frame; SYNC_TIMEOUT_START;
2	**State**: SYNCHRONIZATION **Schedule**: Finish_SYNC; Remains in SYNC and control slot channel; Choose the slot in SYNC CW to retransmit SYNC frame;
3	**State**: SYNCHRONIZATION **Schedule**: New node joins the network; Random_channel_slot_selection mechanism enabled; Detect a frame from node that allows for synchronization; 1st mechanism time out;
4	**State**: SYNCHRONIZATION **Schedule**: New node joins the network; Random_channel_slot_selection mechanism fails; Fast_slot_channel_hopping mechanism enabled; Detect a frame from node that allows for synchronization; 2nd mechanism time out;
5	**State**: RETRANSMIT SYNC PACKET **Schedule**: SYNC_Contention_Window_Initiate; **Update**: Chooses a slot for SYNC CW and performs CS;
6	**State**: IDLE **Schedule**: SYNC_RETRANSMISSION_FINISHED; WAITS_FOR_PACKET_IN_MAC_QUEUE;
7	**State**: IDLE **Schedule**: If a new node that tries to join, transits directly to IDLE state;
8	**State**: CHECK_QUEUE **Schedule**: Verifies frames in queue;

9	**State**: SLOT CHANNEL CHOICE
	Schedule: SWITCH_TO_CHANNEL;
	if (number_frames_in_queue ≥ 1)
	    TX_DATA_PACKETS
	Update: Switches to same slot channel of parent node ;
	if (number_frames_in_queue $== 0$)
	    CHECK_MEDIUM_FOR_DATA_PACKETS
	Update: Switches to current slot channel of the node that initiates the sensing;

10	**State**: SAMPLE WAKE-UP TONE **Schedule**: CHECK_WAKE-UP_TONE; **Update**: Performs CS to the channel;

Events	Actions
11	**State**: WAIT FOR DATA PACKET **Schedule**: RX_DATA_PACKET; INITIATES_DATA_PACKET_TIMEOUT;

12	**State**: RX DATA PACKET
	Schedule: Check if "more bit" is enabled
	if (more bit == 1)
	  ER phase enabled;
	if (more bit == 0)
	  ER phase disabled;
	RECEIVE_PACKET;
	Update: Check piggybacked clock information of the child node by the parent node
	  for deviations;

13	**State**: ER PHASE **Schedule**: Finish receiving data frames; **Update**: Remaining frames from sender node;
14	**State**: END RECEIVING ALL PACKETS **Schedule**: Transmit MC-ACK/MC-NACK frame; **Update**: Frames in queue to be transmitted or retransmitted to coordinator;
15	**State**: IDLE **Schedule**: MEDIUM_FREE; Go to SLEEP state;
16	**State**: SENDING MC-ACK/MC-NACK **Schedule**: TX MC-ACK/MC-NACK with piggybacked SYNC information;
17	**State**: TRANSMIT **Schedule**: TX MC-ACK/MC-NACK frame;
18	**State**: IDLE **Schedule**: IDLE_TO_SLEEP; **Update**: Degradation channel level;
19	**State**: SLEEP **Schedule**: EVENT_SENSING; CHECK_QUEUE;
20	**State**: SENDING MC-ACK/MC-NACK **Schedule**: TRANSMIT MC-ACK/MC-NACK frames with piggybacked SYNC information;
21	**State**: WAKE-UP TONE TX **Schedule**: CONTENTION_WINDOW_1_INITIATE; **Update**: Chooses a slot/time for the CW_1; Performs CS to the channel;

Events	Actions
22	**State**: IDLE **Schedule**: MEDIUM_BUSY; Stops the transmission of the frame and goes to IDLE; Switches radio transceiver to RX mode;
23	**State**: TRANSMIT **Schedule**: TRANSMIT WAKE-UP TONE;
24	**State**: DATA PACKET TX **Schedule**: CONTENTION_WINDOW_2_INITIATE; **Update**: Chooses a slot in CW_2 to contend for the channel; Performs CS;
25	**State**: ER PHASE DECISION ALGORITHM **Schedule**: Verifies the three conditions to enable this phase: 1) Queue has more than one frame to be sent; 2) Slot channel does not has a high degradation level; 3) Mobility/reduced mobility is not present by means of IR algorithm metric; **Update**: if (all conditions == true) sets the "more bit" in data frame specify number of frames in queue of sender if (all conditions = false) resets the "more bit" in data frame ASSESS POSSIBILITY TO USE ER PHASE
26	**State**: TRANSMIT **Schedule**: Transmit N data frames;
27	**State**: CHECK MAC QUEUE **Schedule**: VERIFY MAC QUEUE FOR MORE PACKETS TO BE SENT;
28	**State**: DATA PACKET TX **Schedule**: TRANSMITS REMAINING DATA PACKETS; **Update**: Remaining number of frames to be sent in the data frame field;
29	**State**: WAIT RESPONSE MC-ACK/MC-NACK **Schedule**: MC-ACK/MC-NACK TIMEOUT; WAIT MC-ACK/MC-NACK;
30	**State**: RECEIVING MC-ACK/MC-NACK **Schedule**: UPDATES SYNC CLOCK; Degradation channel level;
31	**State**: UPDATE SYNC CLOCK **Schedule**: Extract piggybacked synchronization clock information; **Update**: Verifies clock deviation and synchronizes internal clock of child node;
32	**State**: DENIAL CHANNEL LIST **Schedule**: Influential range concept verification; **Update**: Degradation level of each slot channel;

Events	Actions
33	**State**: IR CONCEPT VERIFICATION **Schedule**: IR verification based on overheard data or MC-ACK/MC-NACK frames; **Update**: IR table;
34	**State**: WAIT FOR MC-ACK/MC-NACK **Schedule**: MC-ACK/MC-NACK TIMEOUT; WAIT MC-ACK/MC-NACK;
35	**State**: RECEIVING MC-ACK/MC-NACK **Schedule**:UPDATES SYNC CLOCK; **Update**: Degradation channel level metric;

36	**State**: CHECK CLOCK DEVIATION
	Schedule:Comparison of child node and parent node sync clocks;
	    if (deviation $>\rho_{dev}$)
	        Synchronization error
	        Node goes to the SYNCHRONIZATION state
	    if (deviation $\leq \rho_{dev}$)
	        Node goes to the SLEEP state

37	**State**: IDLE **Schedule**: MEDIUM_BUSY; Stops transmission of data frame; Waits for data frame or MC-ACK/MC-NACK frame; Applies the IR concept algorithm and synchronization procedure (if possible);
38	**State**: SLEEP **Schedule**: EVENT_SENSING; Checks MAC queue;
39	**State**: SYNCHRONIZATION **Schedule**: Retries to synchronize due to high internal clock deviations;
40	**State**: DISCOVERY & ADDITION **Schedule**: When node is added to the WSN after first synchronization. Random_channel_slot_selection mechanism is enabled;
41	**State**: DATA PACKET TX **Schedule**: Node successfully contends in CW_2; Check MAC queue; Schedules data frame transmission;
42	**State**: UPDATE SYNC CLOCK **Schedule**: Check for clock deviation of the sending node; This transition only appears in the coordinator node;
43	**State**: WAKE-UP TONE TX **Schedule**: Node successfully contends in CW_1; Transmit wake-up tone;

- **RETRANSMIT SYNC PACKET**: The node already synchronized will contend for the medium based on the chosen slot in the SYNC contention window; if it detects an idle channel it sends a SYNC frame for the neighbouring nodes (RFDs). This is useful when multi-hop is considered;
- **SLEEP**: The node "turns off" the radio and put Central processing unit (CPU) in low power consumption mode;
- **DISCOVERY & ADDITION**: The new nodes that want to join the WSN may employ two mechanisms that allow for the new nodes to synchronize with the remaining nodes of the WSN, including the coordinator (if it is in range);
- **IDLE**: The node is waiting for a task to perform;
- **CHECK MAC QUEUE**: Before choosing the slot channel (coordinator or parent node slot channel) the node verifies his MAC queue for any frames to be sent;
- **SLOT CHANNEL CHOICE**: After the node checks if it has frames in the MAC queue or not, the node changes to the receiving mode (in case it is an FFD and multi-hop is considered), or changes in transmission mode (in case it has frames to send), or go to sleep mode when the MAC queue has no frames and the node is not an FFD. It also checks the denial channel list to restrict the slot channel choices only to the best slot channels;
- **TRANSMIT**: The node transmits the frame, independently of its type;
- **SAMPLE WAKE-UP TONE**: The node samples for the presence of the wake-up tone;
- **WAIT FOR DATA PACKET**: When the node has no frames in the MAC queue (in case it is a FFD and multi-hop is considered), it waits for the reception of the data frame from the RFD node. If no wake-up is detected the node transits directly to IDLE state and goes to SLEEP state;
- **RX DATA PACKET**: The node receives the data frame and extracts the synchronization information, suppressing the transmission of explicit SYNC frames. Checks if the "more bit" included in the data frame is set;
- **ER PHASE**: In case the node receives a data frame with the "more bit" set, the receiving node extends its reception time in order to receive the remaining data frames from the sender node;
- **END RECEIVING ALL PACKETS**: The node enters this state after receiving all the data frames sent during the ER phase;

- **SENDING MC-ACK/MC-NACK**: After receiving all data frames the receiving node sends a MC-ACK / MC-NACK frame depending on if the frames have been successfully received or not, respectively;
- **DENIAL CHANNEL LIST**: The node that contends to transmit his data frames transit to this state if contention in CW_1/ CW_2 is lost, or in case it received a MC-NACK frame or a data frame. It updates the denial channel list based on these events;
- **IR CONCEPT VERIFICATION**: If the node loses contention in CW_1 or CW_2, it verifies if it is in the influential range of the node that has sent the overheard data frame. Based on this, the node in the next attempts checks if the frames (to be sent) stored in the queue contain the same information. If so, the node does not send the stored frame with the same information as the one overheard. This mechanism avoids redundant data frames transmissions;
- **WAIT FOR MC-ACK/MC-NACK**: After receiving the overheard data frame then node waits for the MC-ACK / MC-NACK frame in order to check his internal clock in comparison with the node that sends the overheard frame (parent node or coordinator node);
- **RECEIVING MC-ACK/MC-NACK**: The nodes receive the MC-ACK /MC-NACK frame;
- **UPDATE SYNC CLOCK**: After receiving the MC-ACK/MC-NACK frame the node extracts the synchronization time sent by the coordinator/parent node;
- **CHECK CLOCK DEVIATION**: The node verifies if the internal synchronization clock presents a high deviation error compared with the value of ρ_{dev};
- **WAKE-UP TONE TX**: The node transmits the wake-up tone after the node contends successfully in the CW_1;
- **DATA PACKET TX**: After successfully contending in both contention phases, the node is in conditions to transmit the data frame. If the ER phase decision algorithm allows to expand the delivery of more data before the receiving nodes go to SLEEP state. The sender node after sending the first data frame continues to sends the remaining data frames;
- **ER PHASE DECIDER**: The sender nodes assesses the possibility of using the ER phase to send more than one data frame before going to the SLEEP state. The decision is based on the denial channel list and IR concept information.

The different transitions between the states from the FSM are as follows:

o After the node is switched on (STARTUP) the node transits to WAIT FOR SYNC state (1);

o If the child node receives a SYNC frame from the coordinator it transits to the SYNCHRONIZATION state (2);

o After receiving the SYNC frame, the nodes will contend in the SYNC CW, in order to retransmit the SYNC frames to the remaining nodes that do not have a direct communication link with the coordinator (5);

o If the nodes successfully contends during the SYNC CW, it retransmits the SYNC frame for the neighbouring nodes. After this synchronization procedure the nodes pass to the IDLE state (6);

o If a new node wants to join the WSN after the main synchronization procedure the node transits from the STARTUP state to the DISCOV-ERY & ADDITION state (40);

o Here the node has two mechanisms that allow to synchronize with the WSN. The first mechanism is enabled, allowing the node to join the network. If the node is able to join the network it transits to the SYNCHRONIZATION state (3);

o If the first mechanism fails, the node enables the second mechanism (fast slot channel hopping mechanism). The node passes to the SYN-CHRONIZATION state if successfully synchronizes with the nodes already in the WSN (4);

o In the SLOT CHANNEL CHOICE state the node verifies if there are any frames in queue to be sent. If the node has frames in queue, the node transits to WAKE-UP TONE TX (21), in which the node contends for the medium by choosing a slot randomly within the CW_1;

o The node transits directly to the IDLE state if it loses the contention in CW_1 (22);

o Otherwise, the node initiates the transmission of the wake-up tone in the WAKE-UP TONE TX state(43);

o To transmit the tone, the node transits to the TRANSMIT state (23);

o After transmitting the wake-up tone, the node transits to the second contention window and chooses randomly a slot to contend for the channel (24);

o In case the node successfully contends for the channel than it transits to the DATA PACKET TX state (41) and consequently to the ER PHASE DECIDER state, in which the node decides whether it is necessary to use the ER phase or not (25);

o Once the decision is taken, the node enables the "more bit" if the ER phase is necessary and resets it if the ER phase is not necessary. In case it is necessary, the node passes to the TRANSMIT state (26);

o After transmitting the first data frame, if the ER phase is enabled, the node moves to the DATA PACKET TX state (25). This procedure is repeated until the MAC queue has no frames to be sent;

o The transitions (22), (37) and (18) to the IDLE state are followed by the transition to the DENIAL CHANNEL LIST state (32). In the cases of the nodes that lost the contention (in CW_1 or CW_2) the nodes will overhear the data frame(s) in order to compute the degradation level of the slot channel, while updating the Denial channel list. In the case of the node that finished transmitting the data frame(s) the Denial channel list is also updated;

o Then the node transits to the IR CONCEPT (33) and only the nodes that overheard the data frame(s) will verify the redundancy of their data frames, in terms of information. In the case of the nodes that end the transmission of the data frame(s), the node will only change to the IR CONCEPT state after receiving the MC-ACK/MC-NACK frame;

o The node transits to the WAIT FOR MC-ACK/MC-NACK state (34), while it waits for the MC-ACK or MC-NACK from the coordinator/parent;

o After initiating the reception of the MC-ACK/MC-NACK frame, the node waits until it finishes receiving completely the MC-ACK/MC-NACK frame. The node passes to the RECEIVING MC-ACK state (35);

o The node extracts the synchronization clock information included in the MC-ACK/MC-NACK frame from the piggybacked information and updates its internal clock information in the UPDATE SYNC CLOCK state (31);

o The clock deviation verification procedure is performed by the receiving node of the MC-ACK/MC-NACK frames in the CHECK CLOCK DEVIATION state (36);

o If the node does not detects a clock deviation higher than ρ_{dev}, the node transits directly to the SLEEP state (38);

o Otherwise, the node passes to the SYNCHRONIZATION state due to the high clock deviation error (39). Here the node tries to resynchronize again;

o The IR CONCEPT state does not applies to the coordinator node, since it only receives frames from the child nodes. After the coordinator receives the data frame it transits to the DENIAL CHANNEL LIST state in which it updates the list and transits to the UPDATE SYNC CLOCK state (42) for synchronization clock deviations checking.

7.5 Simulation Results for the MC-SCP-MAC Protocol

7.5.1 Collision Probabilities

In this section, the CW_1 and CW_2 collision probabilities for the MC-SCP-MAC and SCP-MAC are obtained as a function of the number of nodes whilst considering the contention window size as a parameter. Six different random seeds are considered in simulations and the 95 % confidence intervals are represented. In this set of results, we consider a deployment scenario of 50×50 m^2, a maximum transmitter power of $P_{tx} = 1$ mW for both MAC protocols, maximum contention window sizes of $CW_k^{max} = 8$, $k \in \{1, 2\}$, data frame size of $L_{data} = 50$ Bytes, as well as different values for the traffic generation rate. Saturated and unsaturated traffic load regimes are considered for the MC-SCP-MAC and SCP-MAC protocols. One assumes a number slot channels for data utilized in these tests of $N_{ch} = 15$ for the MC-SCP-MAC protocol and $N_{ch} = 1$ (single-channel) for SCP-MAC.

Figure 7.15 shows the comparison of the CW_1 collision probability for the MC-SCP-MAC and SCP-MAC protocols in the saturated and unsaturated regimes. Moreover, it addresses the verification of the validity of the CW_1 collision probability, under the saturated and unsaturated traffic regimes, as an extension to the protocol SCP-MAC. It is observable that the CW_1

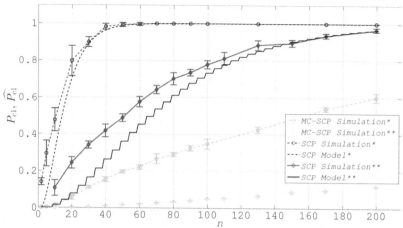

Figure 7.15 Comparison of the CW_1 collision probability between the MC-SCP-MAC and SCP-MAC protocols in saturated (*) and unsaturated (**) regimes, for $CW_1^{max} = 8$.

collision probability is much lower for the MC-SCP-MAC protocol than for SCP (in both regimes). This is an expected behaviour since MC-SCP-MAC envisages the use of multi-channel jointly with an enhanced two-phase contention window mechanism to cope with possible data frame collisions. These mechanisms allow for significantly reducing the CW_1 collision probability.

As an example, Figure 7.16 presents the expected number of nodes that pass from CW_1 to CW_2, in the saturated regime, $E[S]$, and unsaturated regime, $\widehat{E}[S]$. The Figure compares the MC-SCP-MAC and SCP-MAC protocols. The 95 % confidence intervals are also presented in the Figure. It is verified that the difference between the modelling and simulation results "is negligible" for all the range of n. Despite the smaller model accuracy for low number of nodes, as n increases, the curve for the simulation results converges to the curve of the numerical solution for the expected number of nodes that pass from CW_1 to CW_2. The expected number of nodes that pass from CW_1 to CW_2 in MC-SCP-MAC is much lower than for SCP, as expected. Since the number of nodes that pass from CW_1 to CW_2 decreases, the probability of a node to successfully deliver a frame is much higher than in the case of the single-channel SCP-MAC protocol.

Figure 7.17 shows a comparison between the simulation and analytical results for the collision probability in CW_2 as a function of n for the saturated and unsaturated traffic regimes. The values of the collision probability in CW_2 are much lower for the MC-SCP-MAC protocol than for the SCP. In addition, the curve for the MC-SCP-MAC protocol in the unsaturated case

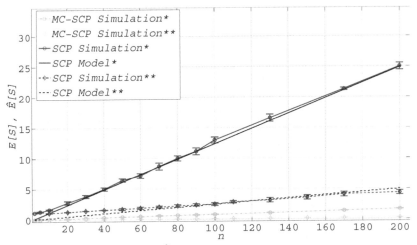

Figure 7.16   Comparison of $E[S]$, $\widehat{E}[S]$ between the MC-SCP-MAC and SCP-MAC protocols in the saturated (*) and unsaturated (**) regimes, for $CW_1^{max} = 8$.

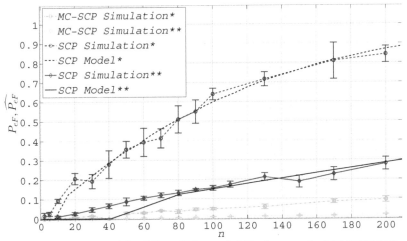

Figure 7.17   Comparison of the CW_2 collision probability between the MC-SCP-MAC and SCP-MAC protocols in the saturated (*) and unsaturated (**) regimes, for $CW_1^{max} = 8$.

is relatively close to zero. We can conclude from these tests that the use of multi-channel features, jointly with an enhanced two-phase contention window mechanism, allows for achieving a much lower collision probability for the data frames, which increases the overall performance of the network at the expense of a small increase of complexity in terms of the protocol algorithm.

Table 7.4 Typical values for CC1100, CC2420 and AT86RF231 radios (P_{tx}=1 mW).

Symbol	Meaning	CC1100	CC2420	AT86RF231
I_{tx}	Current in transmitting	16.9 mA	17.4 mA	11.6 mA
I_{rx}=I_{listen}	Current in receiving/listening	16.9 mA	18.8 mA	12.3 mA
I_{poll}	Current in polling	16.4 mA	18.8 mA	12.3 mA
I_{sleep}	Current in sleep	0.02 mA	0.000021 mA	0.00002 mA
r_B	Bit rate	250 kbps	250 kbps	250 kbps
S_{min}	Sensitivity	-111 dBm	-94 dBm	-101 dBm
f_b	Carrier Frequency	868 MHz	2.4 GHz	2.4 GHz

7.5.2 Energy Efficiency per Delivered Frame

This section addresses the evaluation of the energy consumption per frame successfully delivered when employing the MC-SCP-MAC and SCP-MAC protocols. This metric corresponds to the total energy consumed (Joule) divided by the total number of successfully delivered frames. In these tests, the energy consumed to receive and transmit as well as the energy consumed to forward the frames towards the PAN coordinator or sink node is analized. Also, the energy spent to transmit control messages, e.g., wake-up tones is also considered.

In these simulations we have considered a deployment scenario of 50×50 m^2, maximum transmitter power of $P_{tx} = 1$ mW for both MAC protocols, maximum contention window sizes of CW$_k^{max}$= 8, $k \in \{1, 2\}$, different values for the periodic traffic generation rate, data frame size of $L_{data} = 50$ Bytes, as well as the power consumption parameters of the different radio transceivers presented in Table 7.4.

Here, the saturated traffic regime is the only one considered for the MC-SCP-MAC and SCP-MAC protocols. One assumes a number slot channels for data utilized in these tests of N_{ch}= 15 for the MC-SCP-MAC protocol and N_{ch}= 1 (single-channel) for SCP-MAC. Figure 7.18 presents the comparison of the energy consumption per frame successfully delivered between MC-SCP and SCP protocols for the CC2420 and AT86RF231 transceivers.

The energy spent per delivered frame increases as the traffic generation interval increases due to the increase of the time that the node needs to dispatch all the frames. In comparison with the single-channel SCP-MAC protocol, the MC-SCP-MAC protocol presents the lowest values of energy spent per delivered frame for the AT86RF231 transceiver. This is due to the very high delivery rate that the use of multiple channels allows for,

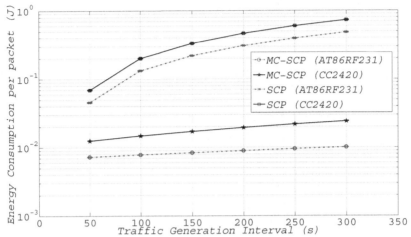

Figure 7.18 Comparison of the energy consumption per frame successfully delivered between MC-SCP-MAC and SCP-MAC protocols for the CC2420 and AT86RF231 transceivers.

combined with the ultra-low power consumption feature of the AT86RF231 transceiver. The same holds for the CC2420 transceiver, where the MC-SCP-MAC protocol also presents lower values of energy spent per delivered frame than the SCP-MAC protocol.

7.5.3 Collision Probability Performance

Figure 7.19 shows the CW_1 collision probability as a function of the number of slots channels in the saturated and unsaturated regimes, for maximum contention window sizes of $CW_k^{max} = 8$, $k \in \{1, 2\}$ for the MC-SCP-MAC and SCP-MAC protocols. For this set of results we consider a deployment scenario of 50×50 m^2, a maximum transmitter power $P_{tx} = 1$ mW for both MAC protocols, a periodic traffic generation rate of $\lambda = 1/2$ s^{-1} and data frame size of $L_{data} = 32$ Bytes. One assumes a number slot channels for data utilized in these tests of $N_{ch} \in \{3; 4; 7; 9; 15\}$ for the MC-SCP-MAC protocol and $N_{ch} = 1$ (single-channel) for SCP-MAC. The frame durations is $t_F = 1$ s. The time allocated for each slot channel is given by $\Delta t_{SC} = t_F/(N_{ch} + 1)$, for the MC-SCP-MAC. In the SCP-MAC protocol the duration of each frame is $t_F = 1$ s. Each node polls the medium after each second. The network size considered in these tests is $n = 99$ nodes, plus 1 PAN coordinator. The CW_1 collision probability for the SCP-MAC protocol does not vary as the number of slot channels increases, since it only

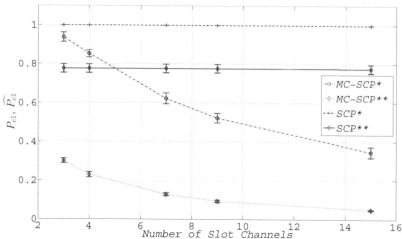

Figure 7.19 Comparison of the dependence of the CW_1 collision probability with the number of slots channels between the MC-SCP and SCP MAC protocols in saturated (*) and unsaturated (**) regimes, for CW_1^{max}=8.

considers one channel for data frames transmission. In Figure 7.19 for both MAC protocols the values of $\widehat{P}_{c1}$ under the unsaturated traffic regime hold the lowest values of the CW_1 collision probability. The 95 % confidence intervals are also presented in the Figure. However, for the MC-SCP-MAC protocol the CW_1 collision probability considerably decreases as the number of available slot channels increases. The increase of the slot channels allows for the nodes to select different slot channels; hence, decreasing the probability of a node to choose the same slot channel to transmit its data frames. The difference of the CW_1 collision probability between the MC-SCP and SCP MAC protocols is more noticeable in the unsaturated regime.

Figure 7.20 presents the dependence of the expected number of nodes that pass from CW_1 to CW_2 with the number of available data slot channels. In MC-SCP, the expected number of nodes that pass from CW_1 to CW_2 presents a behaviour similar to the one from the CW_1 collision probability. In the MC-SCP-MAC protocol and the saturated case, the expected number of nodes that pass from CW_1 to CW_2 considerably decreases as the number of slot channels increases. For the unsaturated case, this decreasing behaviour is smoother compared to the expected number of nodes that pass from CW_1 to CW_2 in the SCP protocol.

Figure 7.21 compares simulation and analytical results of the dependence of the collision probability in CW_2 with the number of data slot channels,

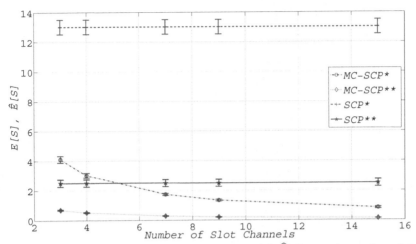

Figure 7.20 Comparison of the dependence of the $E[S], \widehat{E}[S]$ with the number of slots channels between the MC-SCP and SCP MAC protocols in saturated (*) and unsaturated (**) regimes, for $CW_1^{max} = 8$.

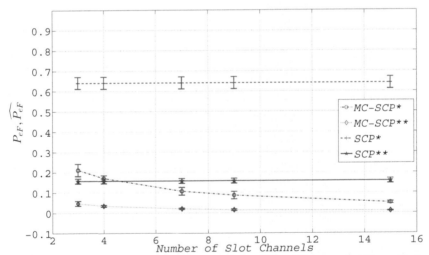

Figure 7.21 Comparison of the dependence of the CW_2 collision probability with the number of slots channels between the MC-SCP and SCP MAC protocols in saturated (*) and unsaturated (**) regimes, for $CW_k^{max} = 8$, $k \in \{1, 2\}$.

for two different traffic regimes. For the MC-SCP-MAC protocol the curves of the CW_2 collision probability decrease as the number of slot channels increases. The CW_2 collision probability of the MC-SCP-MAC protocol is always lower than for SCP. This is due to the low values of the CW_1 collision probability resulting from the multi-channel feature, which leads to a low expected number of nodes that pass from CW_1 to CW_2. It is worth noting that the unsaturated regime is the one that presents the lowest values for the CW_2 collision probability.

From this set of tests, the lessons learned are twofold:

- The increase of the number of slot channels along with an enhanced two-phase contention window mechanism allows for achieving a much lower collision probability for data frames;
- The difference of effective collision probability between the MC-SCP and SCP MAC protocols is more notorious in the saturated case, as the number of slot channels significantly increases.

7.5.4 Energy Efficiency with Multiple Slot Channels and Contention Window Sizes

This section analyzes the impact of the number of slot channels and different contention window sizes on the network performance. For this set of tests all the nodes initiate CBR streams towards the sink node and each node generates a frame every 2 s. In the MC-LMAC protocol, if the frame generation is quicker, buffer overflows start to occur. In the MC-SCP-MAC protocol this only happens when the traffic generation time is lower than the frame duration. Here, the frame duration is $t_F = 1$ s and the time allocated for each slot channel is given by $\Delta t_{SC} = t_F/(N_{ch} + 1)$ for the MC-SCP-MAC protocol. In turn, MC-LMAC considers a frame duration of $t_F = 1.6$ s. The MC-SCP-MAC considers $CW_k^{max} \in \{16\}$, $k \in \{1, 2\}$ for the maximum contention window sizes. For this set of results we consider a maximum transmitter power $P_{tx} = 0.8$ mW for both MAC protocols. The data frame size is $L_{data} = 32$ Bytes. The number of slot channels for data utilized in these tests is given by $N_{ch} \in \{3; 4; 7; 9; 15\}$ for MC-SCP whereas MC-LMAC considers $N_{ch} \in \{3; 4; 7; 9; 10\}$. In this set of tests we considered a network composed by $n = 99$ nodes. The terrain size considered in the MC-LMAC and MC-SPC-MAC protocols is 150×150 m^2 [IVHJH11]. However, the authors from MC-LMAC [IVHJH11] consider that, in the multi-hop scenario, all nodes are in the carrier sensing range of each other. Nodes in the 2nd tier cannot directly communicate with the sink node. This assumption does not considers the hidden terminals problem. Since the nodes in the MC-SCP-MAC are deployed randomly, the nodes may not be in the carrier sensing range of all neighbouring nodes. Therefore, as the MC-SCP-MAC

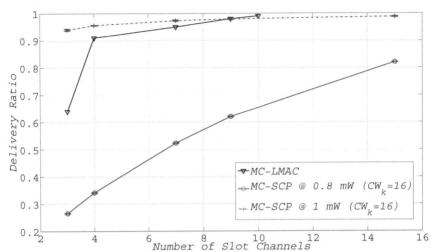

Figure 7.22   Comparison of the delivery ratio for different transmitter powers, for the MC-SCP-MAC and MC-LMAC protocols, $k \in \{1, 2\}$.

protocol envisages a more realistic scenario, it accounts for hidden terminals in the different deployments. The traffic generation can be considered as a periodic data collection, which is commonly used in WSN applications. In both MAC protocols retransmissions are disabled i.e., we considered only best effort delivery, such that the lost frames are not retransmitted. With sufficient channels and retransmissions enabled, MC-SCP-MAC achieves to delivers 99 % of the frames, on average.

Figure 7.22 presents the dependence of the delivery ratio on the number of slot channels for the MC-SCP-MAC and MC-LMAC protocols, for different transmitter powers. In these results, we consider maximum contention window sizes of $\mathrm{CW}_k^{max} = 16$, $k \in \{1, 2\}$. In both MC-SCP-MAC and MC-LMAC, as the number of slot channels increases the delivery ratio increases. Figure 7.22 shows that, for $P_{tx} = 0.8$ mW, the MC-LMAC protocol presents higher delivery ratios than MC-SCP-MAC. The low delivery ratio caused by the large number of hidden terminals leads to the failure in successful frames reception. For the MC-SCP-MAC, with the increase of the transmitter power the impact of the hidden terminals problem decreases, which leads to a significant increase of the delivery ratio.

The MC-SCP-MAC achieves delivery ratios higher than 90 % for all the number of slot channels presented in Figure 7.22.

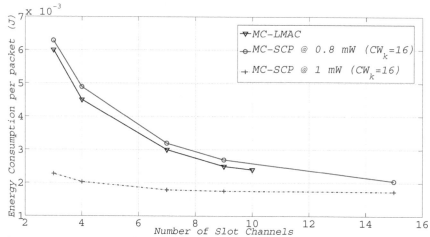

Figure 7.23 Comparison of the energy consumption per delivered frame for different transmitter powers, for the MC-SCP-MAC and MC-LMAC protocols, $k \in \{1, 2\}$.

Figure 7.23 shows the comparison of the energy consumption per delivered frame for different transmitter powers for the MC-SCP-MAC and MC-LMAC. The energy spent per delivered frame is high in the presence of three slot channels. Here, the MC-SCP-MAC protocol with higher transmitter power ($P_{tx} = 1$ mW) presents a much lower energy spent per delivered frame than the MC-LMAC protocol. This is explained by the high delivery ratio presented in Figure 7.22 as the number of slot channels increases. From this set of tests, we can conclude that, in denser scenarios, MC-SCP-MAC can handle high contention on the medium due to the use of the enhanced two-phase mechanism with the multi-channel feature. Moreover, the increase of the transmission power eliminates the majority of occurrences of hidden terminals, which allows for achieving higher delivery ratio and, consequently, lower energy consumed per delivered frame.

7.5.5 Influential Range Concept and Energy Performance Evaluation

Fixed Number of Channels

In this section, we analyze the impact, in terms of energy performance, of the influential range (IR) concept enabled in the MC-SCP-MAC. For this set of tests, all the nodes initiate CBR streams towards the sink node. We assume that each node generates a frame every 10 s and that the maximum contention window sizes are $CW_k^{max} = 16, k \in \{1,2\}$. Besides, we consider a maximum transmitter power $P_{tx} = 0.8$ mW, a deployment area of 50×50 m^2, data frame size $L_{data} = 32$ Bytes and a number of slot channels $N_{ch} = 15$. Here the frame duration is $t_F = 1$ s. The time allocated for each slot channel is $\Delta t_{SC} = 0.0625$ s. The objective of these tests is to analyze the gains achieved when the IR concept is enabled, for different values of the sensing range, $\Pi_{irmax} \in \{-90; -80; -70; -60\}$ dBm.

Variable Number of Slot Channels

In this section, we analyze the impact of the number of slot channels and different network sizes while enabling the influential range concept in MC-SCP-MAC. For this set of tests all the nodes initiate CBR streams towards the sink node. Each node generates a frame every 2 s. Here the duration of the frame, t_F, assumes two different values depending on the number of considered slot channels:

- For $N_{ch} = 8, t_F = 1.53$ s and $\Delta t_{SC} = 0.17$ s;
- For $N_{ch} = 15, t_F = 1$ s and $\Delta t_{SC} = 0.0625$ s.

The maximum contention window sizes are $CW_k^{max} = 80, k \in \{1,2\}$. These long values for the contention window sizes enables to test how long they can be. We assume a maximum transmitter power $P_{tx} = 0.8$ mW and data frame size of $L_{data} = 32$ Bytes. The considered terrain size is 50×50 m^2. Retransmissions are disabled in these tests.

Figure 7.24 shows the average energy consumption per node when the IR concept is enabled for different sizes of network. Each node generates and sends 600 frames to the PAN. From Figure 7.24, it is noticeable that the increase of the number of slot channels leads to higher energy consumption. This is due to the increase of the available slot channels to send data frames. As the value of Π_{irmax} increases the influential range of the node decreases, leading to the reduction of the overhearing and consequent decreasing of the discarded redundant data frames. Since discarding redundant data frames decreases as the values of the sensing range, Π_{irmax} increases, the node has to transmit more data frames which, in turn, increases the energy consumption per node. For $N_{ch} = 8$ slot channels, since a frame duration of $t_F = 1.53$ s is considered, the results presented in this section can be compared with the ones from MC-LMAC. This is possible because MC-LMAC considers a frame duration of $t_F = 1.6$ s.

The evaluation of the benefits achieved by applying the IR concept in these tests is presented in Figure 7.25, where the energy consumption is presented as a function of the values of Π_{irmax} for different sizes of the

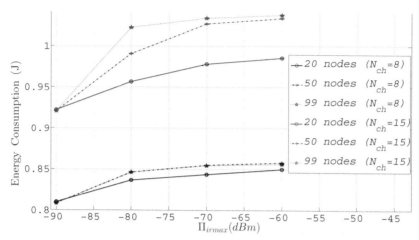

Figure 7.24 Average energy consumption per node when IR is enabled for different sizes of the network and number of data slot channels, i.e., $N_{ch} \in \{8, 15\}$ (600 frames).

network. For $\Pi_{irmax} = -90$ dBm the energy consumption saving is higher for $N_{ch} = 15$ slot channels. As explained before, the use of 15 slot channels leads to higher delivery ratio (compared to the case $N_{ch} = 8$), which corresponds to a larger number of redundant frames dropped due to the application of the IR concept. In turn, the drop of a larger number of redundant data frames increases the energy consumption savings. For higher values of the sensing range, as the values of Π_{irmax} increases the influential range between the nodes decreases, which causes the decrease of the energy consumption savings. While the energy saving can reach 11.42 % for $N_{ch} = 15$, it reaches up to 5.94 % for $N_{ch} = 8$ slot channels.

7.5.6 Impact of Traffic Periodic and Exponential Patterns in the Overall Performance

In Section 7.5.4 we analyzed the impact of the number of slot channels on the network performance in a deployment scenario of 150×150 m^2. Besides, the network performance metrics have been compared between MC-SCP and MC-LMAC. In this section, we analyze the different performance metrics, for the MC-SCP-MAC, MC-LMAC, CSMA and MMSN protocols while varying different parameters (e.g, number of sources nodes, number of slot channels, traffic generation intervals). The metrics include the aggregate throughput, delivery rate, energy consumption per node, energy consumption per frame and latency. We also assume that the nodes consider periodic and/or

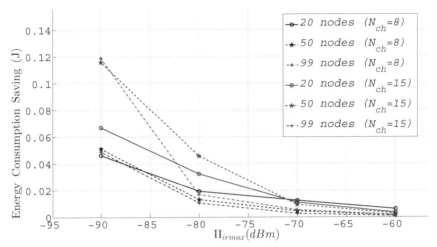

Figure 7.25 Energy consumption savings per node with IR enabled for different sizes of network and $N_{ch} \in \{8, 15\}$ (600 frames).

exponential traffic generation. Details on the considered hidden terminal scenarios, and overall vision on the impact of periodic traffic and exponential patterns is analyzed in [VBBC19], where the aggregate throughput is compared between MC-SCP-MAC, MC-LMAC, CSMA and protocols.

In [VBBC19] it is shown that with low density of nodes 150×150 m^2 (sparser scenario) in MC-SCP-MAC, under periodic and exponential traffic pattern, the aggregate throughput is much lower than the maximum achievable. However, in the 50×50 m^2 scenario the density of nodes is higher yielding an aggregate throughput close to the maximum, as shown in Figure 7.26.

These results show the best performance of the MC-SCP-MAC protocol in high density scenarios. However, the presence of hidden terminals may lead to degradation of the aggregate throughput as the number of source nodes increases and a method that decreases the impact of the hidden terminal problem may be needed. We can also conclude that the increase of the traffic generation frequency is well supported by MC-SCP-MAC when the hidden terminals is not an issue.

Figure 7.26 shows the variation of the delivery rate as a function of the number of source nodes for MC-SCP-MAC protocol for periodic and exponential traffic patterns, for $\lambda \in \{1/2; 1\}$ s^{-1}. It is shown that the delivery rate is always near the maximum, even for large number of source nodes.Figure 7.27 presents the dependence of the energy performance on

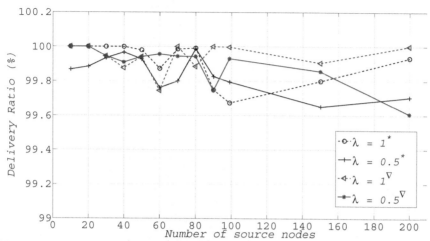

Figure 7.26 Delivery rate for the MC-SCP-MAC protocol with periodic (*) and exponential (∇) traffic patterns for different number of sources in the 50×50 m^2 deployment scenario.

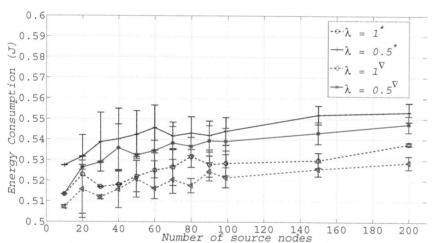

Figure 7.27 Energy consumption per node of MC-SCP-MAC with periodic (*) and exponential (∇) traffic generation for different number of sources in deployment scenario of 50×50 m^2.

the number of source nodes from MC-SCP-MAC for the traffic generation patterns considered within these tests.

The 95 % confidence intervals are also presented in this Figure. Here, the energy consumption of the PAN coordinator is not taken into account, since this node has extended capabilities and is not limited by any energy

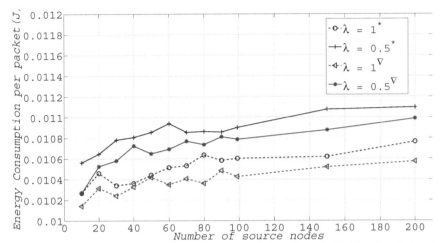

Figure 7.28   Energy consumption per frame successfully delivered of MC-SCP-MAC with periodic (*) and exponential (∇) traffic generation for different number of sources in deployment scenario of 50×50 m^2.

constraint. It is observable in Figure 7.27 that the reduction of the area from the deployment scenario leads to a slight increase of the energy consumption, whilst attaining higher aggregate throughput and delivery rate. These tests show that higher traffic generation rates lead to the lowest values for the energy consumptions.

Figure 7.27 evaluates the energy consumption per frame successfully delivered by the MC-SCP-MAC protocol, for periodic and exponential traffic patterns with $\lambda \in \{1/2; 1\}$ s^{-1} for a scenario with high density of nodes than the one in Section 7.5.6.

In Figure 7.28 it is noticeable that the curves for $\lambda = 1/2$ s^{-1} are the ones that present the highest energy consumption. This is explained by the highest energy consumption observed in Figure 7.27 jointly with the high delivery rate presented in Figure 7.26 that cause an effective decrease of the energy consumption per delivered frame. Therefore, we conclude that a deployment scenario with higher density of nodes allows for mitigating the hidden terminals problem while achieving less energy waste per successfully delivered frame. Figure 7.29 addresses the results obtained for the end-to-end latency. The 95 % confidence intervals are also presented in this Figure. The exponential traffic patterns is the one that presents the lowest values for the latency (as the number of source nodes increases). For higher density of nodes we conclude that the values for the latency present similar values up to $n= 30$.

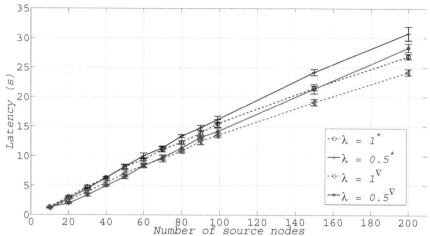

Figure 7.29 Latency of MC-SCP-MAC with periodic (*) and exponential (∇) traffic generation for different number of sources in deployment scenario of 50×50 m².

However, for a number of source nodes higher than 30, the latency is longer in the 50×50 m² deployment scenario. This is explained by the increase of the number of contention nodes and the capability of all nodes to sense each other which facilitates to detect nodes more often in the data frame transmission phase. Hence, less collisions are originated but the latency becomes longer.

7.5.7 Impact of the Density of Nodes

The results for the aggregate throughput, and delivery ratio, from MC-SCP-MAC are compared with the ones from MC-LMAC, and CSMA protocols. In order to test the scalability of the MC-SCP-MAC protocol we have varied the density of the node deployments from 50×50 m² to 75×75 m², 100×100 m², 125×125 m², 150×150 m² and 200×200 m². The authors from MC-LMAC [IVHJH11] claim that with random deployment beyond a side length (L) of 225 m, unconnected nodes appear. Although we have performed tests with larger deployments, unconnected nodes only start to appear after for side lengths larger than 400 m. However, for the sake of fair comparison with the other MAC protocols we only consider side lengths up to 200 m. For this set of tests nodes we assume periodic or exponential traffic patterns. Here, the frame duration is $t_F = 1.57$ s and the time allocated for each slot channel is $\Delta t_{SC} = 0.17$ s for MC-SCP-MAC, and $t_F = 1.6$ s for MC-LMAC and CSMA consider. Maximum contention window sizes are

$CW_k^{max} = 80$, $k \in \{1,2\}$ for the MC-SCP-MAC protocol. We consider a maximum transmitter power $P_{tx} = 0.8$ mW for all MAC protocols and data frame size of $L_{data} = 32$ Bytes. The number slot channels considered in these tests for data is $N_{ch} = 8$ for the MC-SCP-MAC and MC-LMAC protocols, while a single-channel ($N_{ch} = 1$) is considered in CSMA. The objective of these tests is to analyze the gains achieved when the IR is enabled for different values of $\Pi_{irmax} \in \{-90; -80; -70; -60\}$ dBm. In all the MAC protocols, the retransmissions are disabled for the sake of fairness in comparisons. Figure 7.30 shows the variation of the aggregate throughput with the side length, L, for the MC-SCP-MAC, MC-LMAC, and CSMA protocols. A **periodic** traffic with a generation rate of $\lambda = 1/2$ s^{-1} is considered. MC-SCP-MAC presents high values of the aggregate throughput if the scenario deployment is denser and IR is enabled. Beyond a side length of 100 m, the aggregate throughput from MC-SCP-MAC starts to decrease, while the aggregate throughput of MC-LMAC continues to increase. However, for a side length of 150 m, the aggregate throughput of MC-LMAC starts to decrease due to the sparsity of the scenario, while the aggregate throughput of MC-SCP-MAC starts to increase. If the IR is disabled, the aggregate throughput of the MC-SCP-MAC achieves its highest values. When the IR is enabled the values of aggregate throughput are always lower owing to of the drop of redundant frames. It reduces the aggregate throughput but it increases the node lifetime while decreasing the latency, as discussed in earlier sections. For denser scenarios (e.g., L= 50 m) the MC-SCP-MAC protocol achieves much higher aggregate throughput than the remaining MAC protocols.

Figure 7.31 shows the results for the delivery ratio as a function of the side length of the MC-SCP, MC-LMAC, and CSMA protocols, when considering a **periodic** traffic profile with a generation frequency of $\lambda = 1/2$ s^{-1}. MC-SCP-MAC presents higher values of delivery ratio when the scenario deployment is denser for both the situations, i.e., when IR is enabled and disabled. Compared with the MC-LMAC protocol, MC-SCP-MAC achieves higher delivery ratios for side lengths that vary from 50 to 100 m. For longer side lengths, the delivery ratio of MC-SCP-MAC decreases sharply, while the MC-LMAC is maintained constant around 92 %. In Figure 7.31, an interesting behaviour of MC-SCP-MAC is the increase of the delivery ratio for side lengths longer than 125 m. For denser scenarios i.e., L = 50 m the MC-SCP-MAC achieves much higher delivery ratios than the considered MAC protocols.

Figure 7.32 presents the end-to-end latency of MC-SCP-MAC as a function of the side length of the deployment scenario for **periodic** traffic profile,

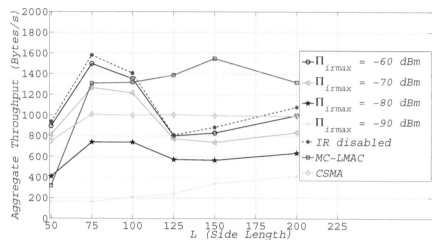

Figure 7.30 Aggregate throughput of MC-SCP-MAC, MC-LMAC and CSMA protocols with periodic traffic generation ($\lambda = 1/2$ s^{-1}) for different deployment scenario side lengths when IR enabled or disabled.

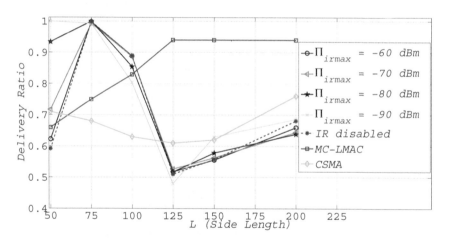

Figure 7.31 Delivery ratio of MC-SCP-MAC, MC-LMAC and CSMA as a function of the deployment scenarios side lengths with periodic traffic generation ($\lambda = 1/2$ s^{-1}) when IR enabled or disabled.

with generation rate of $\lambda = 1/2$ s^{-1}. The 95 % confidence intervals are also presented in this Figure. The MC-SCP-MAC presents shorter latency if IR is enabled for all the values of the sensing range, Π_{irmax}. As the value of sensing range, Π_{irmax}, increases, the latency also increases, for all the

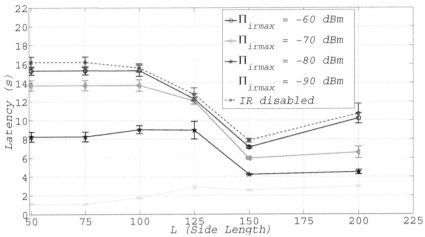

Figure 7.32   Latency of MC-SCP-MAC as a function of the deployment scenarios side lengths with periodic traffic generation ($\lambda = 1/2\ \mathrm{s}^{-1}$) when IR enabled or disabled.

considered values for the side length. The short values for the latency if IR is enabled is due to the reduction of the redundant data frames waiting to be transmitted in the queue. As shown in Figure 7.30, the use of IR may reduce the aggregate throughput but the latency is considerably reduced, too, as shown in Figure 7.32. The longest latency is verified when the IR is disabled, while the case, Π_{irmax}= -90 dBm always presents the shortest latency. For side lengths that vary between 50 and 100 m the values are different for all the considered values of sensing range, Π_{irmax}, when the IR is enabled and disabled. The latency decreases between side lengths of 100 and 150 m, except for sensing range of Π_{irmax}= -90 dBm. For side lengths longer than 150 m the values for the latency are relatively constant, except for IR disabled and IR enabled with sensing range Π_{irmax}= -60 dBm, as shown in Figure 7.32.

Here, we evaluate the energy consumption of MC-SCP-MAC as a function of the side length whilst considering periodic traffic profile with a generation rate of $\lambda = 1/2\ \mathrm{s}^{-1}$, considering the cases where IR is enabled and disabled. These results are presented in Figure 7.33, where the 95 % confidence intervals are also presented. As the value of the side length from the deployment scenario increases the energy consumption is pretty constant up to the value of 125 m for the side length. This is due to the high delivery ratio that MC-SCP-MAC presents in denser scenarios along with the consequent decrease of hidden terminals on the deployment. For side lengths

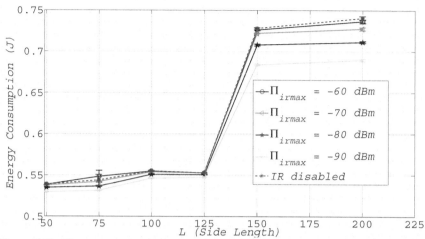

Figure 7.33 Energy consumption per node of MC-SCP-MAC as a function of the deployment scenarios side lengths with periodic traffic generation ($\lambda = 1/2$ s^{-1}) when IR enabled or disabled.

larger than 125 m the MC-SCP-MAC presents high energy consumption. This is explained by the increase of the presence of more hidden terminals, which cause the MC-SCP-MAC to decrease the delivery ratio and, consequently suffer from an increase of the energy consumption. MC-SCP-MAC presents the highest values for the energy consumption when the IR is disabled. However, we can observe in Figure 7.33 that, by enabling the IR, the energy consumption slightly decreases. This is due to the slight increase of delivery ratio verified when the IR is enabled.

It is now worthwhile to evaluate the energy saving gains that can be achieved with the employment of the IR concept, as shown in Figure 7.34. The respective percentages of saving gain are shown as a function of the side lengths. It considers the **periodic** traffic profile with a generation rate of $\lambda = 1/2$ s^{-1}. It is observable that the use of IR is only efficient for denser scenarios and for sensing values $\Pi_{irmax} \in \{-90; -80; -70\}$ dBm. For side lengths larger than 100 m, the saving gains are negligible.

We can conclude that the employment of IR allows for achieving shorter latency values, higher delivery ratios and considerable energy consumption savings, at the expense of reduced aggregate throughput. Moreover, the denser scenarios allow for MC-SCP-MAC to achieve better overall performance.

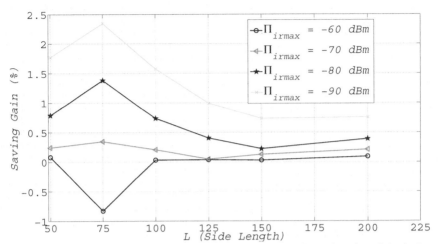

Figure 7.34   Saving gain percentage per node of MC-SCP-MAC as a function of the deployment scenarios side lengths when IR is enabled with periodic traffic generation ($\lambda = 1/2$ s^{-1}).

7.5.8 Performance Analysis in the Cluster Topology

This set of tests aim to assess the impact of enabling the influential range concept and the dependence on the available number of slot channels for data exchange in the MC-SCP-MAC. Different traffic patterns and data generation rates are considered. We consider a heterogeneous sensor network, where nodes are hierarchically different, and mainly focus on the operation inside a cluster. Since MC-SCP-MAC is more robust and efficient in denser scenarios (e.g., deployment scenario comprises sensor nodes in a forest or agricultural field), we consider that the performance evaluation in a cluster topology can be useful to extract insights. The main advantages of using heterogeneous sensor networks (i.e., clusters) are twofold:

- it has improved scalability than a flat network without hierarchies;
- the majority of nodes in the network, can be made very simple and inexpensive, which reduces the overall cost of the network.

Asymmetric communications is, however, a special and unique feature in clusters: The message sent by a cluster head can be received directly by all sensors in the cluster, whereas the message sent by a sensor node (Full Function Device (FFD) or Reduced Function Device (RFD)) may have to be relayed by other sensors, by multi-hop connections, to reach the cluster head. In heterogeneous sensor networks, the basic sensors can be randomly

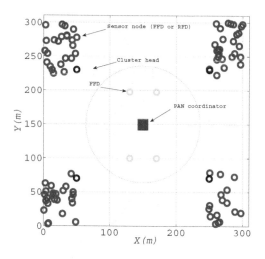

Figure 7.35 Network deployment in the cluster topology.

deployed as in homogeneous sensor networks. The cluster heads, on the other hand, should be more carefully deployed, to ensure that all basic sensors are covered, i.e., that each sensor node can sense and hear from at least one cluster head. Nevertheless, since the number of cluster heads is small, their best locations can be found within a reasonable amount of time. Besides, the cluster heads can even increase their transmitter power to cover remote sensors. For this set of tests we consider a deployment scenario of 300×300 m^2. Four clusters have been deployed composed by 22 FFD's or RFD's in each cluster, four cluster heads, 4 FFD's and one PAN coordinator, illustrated in Figure 7.35.

MC-SCP-MAC considers maximum contention window sizes of CW$_k^{max}$ = 80, $k \in \{1, 2\}$. For this set of results we consider for the maximum transmitter power P_{tx} = 0.8 mW, data frame size L_{data} = 32 Bytes, a variable number of slot channels $N_{ch} \in \{4; 7; 15\}$ and that each node generates 600 frames. Here, the frame duration is t_F = 1s for all the considered number of slot channels, while the values for the time allocated for each slot channel are Δt_{SC} = 0.0625 s, Δt_{SC} = 0.125 s and Δt_{SC} = 0.2 s, for N_{ch}= 4, N_{ch}= 7 and N_{ch}= 15, respectively. The objective of these tests is the analysis of the gains achieved when the IR is enabled for different values of $\Pi_{irmax} \in \{-90; -80; -70; -60\}$ dBm while varying the number of slot channels, as shown in Figure 7.36 (λ = 1 s^{-1}).

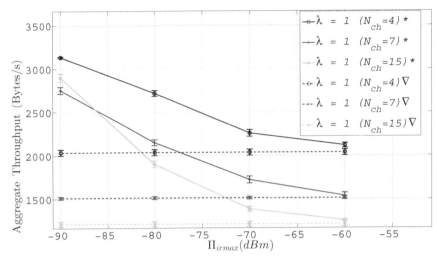

Figure 7.36   Aggregate throughput as a function of Π_{irmax} for the MC-SCP-MAC protocol when IR enabled (*) or disabled (∇) while varying the number of slot channels, for periodic traffic generation ($\lambda = 1\,s^{-1}$).

Figure 7.36 presents the aggregate throughput as function of the Π_{irmax} for MC-SCP-MAC when IR is enabled or disabled with the number of slot channels as a parameter for a traffic generation rate of $\lambda = 1\,s^{-1}$ and **periodic** profile. The 95 % confidence intervals are also presented. This analysis enables to understand how the MC-SCP-MAC behaves when the data generation rate is increased. The highest values for the aggregate throughput occur when $\Pi_{irmax} = -90$ dBm, due to the increased sensing range (that enables to overhear more frames). When IR is disabled, the achieved aggregated throughput is much lower than when the IR is enabled, as expected. As the value of the sensing range, Π_{irmax}, increases the influential range decreases, which, in turn, causes the aggregate throughput to decrease. The number of available slot channels also play an important role in the aggregate throughput for higher traffic generation rates since, in Figure 7.36, the highest values of the aggregate throughput occur when there are 4 slot channels available. From these results, we are able to conclude that MC-SCP-MAC supports higher traffic generation rates in the cluster topology.

Figure 7.37 shows the results for the delivery ratio as a function of the sensing range, Π_{irmax}, for the MC-SCP protocol considering the IR feature (enabled or disabled) with the number of slot channels as a parameter for a traffic generation rate $\lambda = 1/2\,s^{-1}$ and a periodic profile. The 95 %

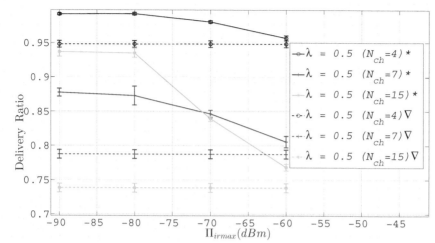

Figure 7.37 Delivery ratio as a function of Π_{irmax} for the MC-SCP-MAC protocol when IR is enabled (*) and disabled (∇) while varying the number of slot channels for periodic traffic generation ($\lambda = 1/2 \text{ s}^{-1}$).

confidence intervals are also presented. When the IR is enabled, the high values for the delivery ratio verified in Figure 7.37 indicate a high aggregate throughput. The highest values of delivery ratio occur when $\Pi_{irmax} = -90$ dBm, since it corresponds to the situation where more redundant data frames are dropped, due to the considered reduced value of IR sensing range, Π_{irmax}. As the value of the sensing range, Π_{irmax}, increases the influential range decreases which, in turn, causes the delivery ratio to decrease. As so, when the number of available slot channels increases, the aggregate throughput decreases and, therefore, the delivery ratio decreases.

We intend to evaluate the energy consumption per node of MC-SCP-MAC when IR is enabled or disabled, for different number of slot channels, whilst considering a **periodic** traffic generation profile with $\lambda = 1/2 \text{ s}^{-1}$. Here the computation of the energy consumption does not take into account the energy spent by the PAN coordinator. However, it considers the energy consumed by the cluster heads and FFD nodes, placed between the cluster heads and the PAN coordinator. Since these nodes have to process and forward more frames than the ones located within the cluster, their energy consumption is higher than for the remaining ones inside the cluster. Therefore, the energy consumption presented in the following figures takes into account this difference. By analyzing the results from Figure 7.38 it is noticeable that the use of IR allows for achieving a lower energy consumption per node since there are

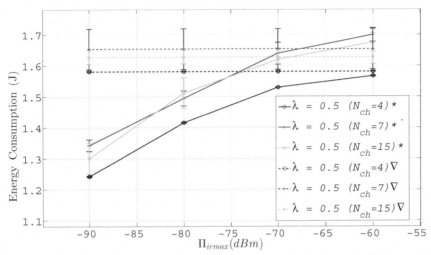

Figure 7.38 Energy consumption per node as a function of Π_{irmax} for MC-SCP-MAC protocol when IR is enabled (*) and disabled (∇) while varying the number of slot channels for periodic traffic generation ($\lambda = 1/2\ \mathrm{s}^{-1}$).

less frames to be transmitted by the sensor nodes. The lowest values of energy consumption occur when $\Pi_{irmax} = -90$ dBm. As the value of the sensing range, Π_{irmax}, increases the energy consumption also increases, up to the value of energy consumption that corresponds to disabled IR. It is worthwhile to note that, in Figure 7.38, the energy consumption presents lower values of energy consumption per node for a reduced number of slot channels. This is justified by the higher values for the delivery ratio from Figure 7.37. For the case of 4 slot channels a slot channel duration of 0.2 s allows for the nodes to transmit more frames in shorter time. Decreasing the active time to transmit all the frames consequently causes the reduction of the energy consumption per node.

From Figure 7.38 it is noticeable that increasing the data generation frequency reduces the energy consumption when IR is enabled and disabled. This is due to the quick generation and transmission of the frame from the sensor nodes to the PAN coordinator, which leads to a significant reduction of active time needed to transmit and receive all the frames. This reduction of the active time causes the decrease of the energy consumption per node.

It is worthwhile to evaluate the end-to-end latency of MC-SCP-MAC when IR is enabled and disabled, for different values of the number of slot channels, whilst considering a periodic traffic generation profile with

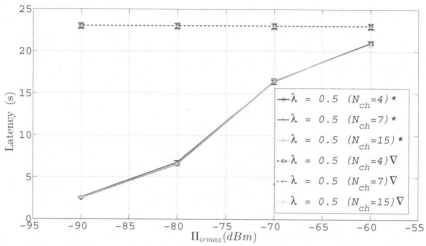

Figure 7.39 Latency for MC-SCP-MAC as a function of Π_{irmax} when IR is enabled (*) and disabled (∇) while varying the number of slot channels for periodic traffic generation ($\lambda = 1/2 \, \mathrm{s}^{-1}$).

$\lambda = 1/2 \, \mathrm{s}^{-1}$. The computation of the end-to-end latency considers in average 3 hops until the frame reaches the PAN coordinator. It is expected that the values for the latency when the IR is not enabled are going to be quite high. However, we expect that enabling IR helps to reduce the latency since it will reduce the redundant data frames and therefore avoiding high traffic loads.

By observing the results from Figure 7.39 it is evident that the use of the IR allows for obtaining a shorter latency since there are less frames in queue (to be transmitted by the sensor nodes). For a value of the sensing range of $\Pi_{irmax} = -90$ dBm, the MC-SCP can achieve values for the latency around 2.8 s. It corresponds to a latency of around 0.933 s per hop, almost a frame time duration (per hop). As the values of the sensing range, Π_{irmax}, increases the latency also increases. This increase of the latency is due to the increase of the number of frames waiting in queue to be transmitted. When the IR is disabled, the value of the latency is around 23 s for all the considered number of slot channels. The same holds if the IR is enabled since for all the considered number of slot channels, the latency is similar and increases as the value of the sensing range increases.

In Figure 7.39, this increase of the latency is due to the increase of the number of frames in the queue that are awaiting to be transmitted. As the value of the sensing range, Π_{irmax}, increases the latency also increases.

When the IR is disabled, for all the considered values for the number of slot channels, the latency is around the 24 s. The same holds if the IR is enabled since for all the considered number of slot channels, the latency is similar and increases as the value of the sensing range increases.

7.5.9 Fairness Index Evaluation for the Throughput

In this section we evaluate the throughput fairness index of the MC-SCP-MAC protocol for different network sizes and number of slot channels. The throughput fairness index follows the following definition [JDB99]:

$$\vartheta_{fi} = \frac{\left(\sum_{i=1}^{n} x_i\right)^2}{\left(n \sum_{i=1}^{n} (x_i)^2\right)} \tag{7.8}$$

where x_i represents the throughput of the node i. The closer the fairness index is to one, the fairer the system is. Fairness problems usually result from location dependent contention which is very common in multi-hop networks. The fairness problem is especially conspicuous when very few nodes are competing in a region. However, when n is larger, the fairness problem is less severe, as it is much more difficult for a node to win exclusive access to the shared channel from all the other competing nodes at the same time.

The fairness problems appear in MAC protocols that give more opportunities to win the shared medium to some nodes. One of cases is the Bynary Exponent Backoff (BEB) protocol. In slotted contention based protocols, like MC-SCP-MAC, if the system is not fair enough, a solution to provide better fairness is the increase of the maximum contention window size value when successful transmission happens. It causes an increase in fairness because it gives opportunities to others that cannot get access to the channel. This way provides fairness improvement.

In this simulation tests, we consider 10, 30 and 99 contending nodes and vary the simulation duration, in order to observe the short and long term fairness.

As the simulation duration gets longer, the fairness index indicates the long term fairness. Note that the duration is not the program running time, rather the time considered for the simulation. We consider a deployment scenario of 50×50 m^2, in which the transmitter power P$_{tx}$= 0.8 mW, the IR is disabled. It is considered a periodic traffic generation profile with $\lambda = 1/2$ s^{-1}. Each node can transmit a maximum of 3000 frames. However, depending on the simulation time considered, the node can transmit more or

Table 7.5 Correspondence between N_{ch} and Δt_{SC} for the throughput fairness index for the MC-SCP-MAC protocol.

t_F [s]	N_{ch}	Δt_{SC} [s]
	3	0.25
	7	0.125
1	9	0.1
	15	0.0625
1.53	8	0.17

less frames. For each network size we consider four simulation scenarios: a) scenario A considers a simulation time $t_{simulation}$= 100 s (short-term fairness); b) scenario B considers $t_{simulation}$= 400 s ; c) scenario C considers $t_{simulation}$= 1000 s and d) scenario D considers $t_{simulation}$= 3000 s (long-term fairness). The throughput fairness index is evaluated considering different number of slot channels. We consider the correspondence between N_{ch} and Δt_{SC} from Table 7.5.

Figure 7.40 shows the solution of the throughput fairness index of MC-SCP-MAC with the simulation duration for n= 10 (n here is not the number of frames, n_{frames}, which, in Chapter 6, is represented by n in the view charts). We averaged 6 simulation runs, with different seeds for each point. The simulation results show that, for a low number of nodes, the short and long-term fairness index stays high for all the considered number of slot channels. Here, all nodes access the medium with the same probability.

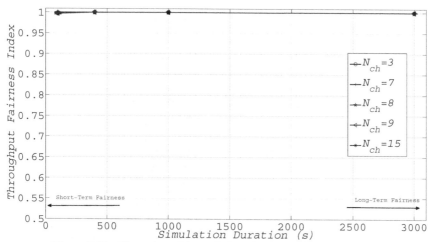

Figure 7.40 Throughput fairness index of MC-SCP-MAC for n= 10.

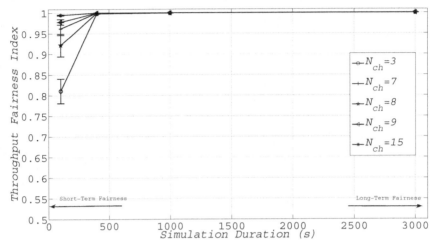

Figure 7.41   Throughput fairness index of MC-SCP-MAC for n= 30.

Figure 7.41 shows the simulation results of the throughput fairness index for the MC-SCP-MAC for n= 30. The results shows that the increase of the number of slot channels improves the short-term fairness of MC-SCP-MAC. However, the long-term fairness stays high for all the considered number of slot channels.

Figure 7.42 shows the results of the throughput fairness index for n= 99 nodes. This scenario considers a dense deployment scenario. The obtained results show that, as the number of slot channels increases, the short-term fairness improves. By comparing these results with the ones for n= 30, the highest network sizes lead to a less fair behaviour for the short-term fairness among the nodes. The throughput long-term fairness index is high for all the network sizes.

7.6 Enhancements to be implemented in MC-SCP-MAC

In order to evaluate different performance metrics after performing all sort of tests with MC-SCP-MAC, some enhancements can be proposed and implemented in order to achieve better results.

The FFDs nodes possess two wake-up states. One wake-up state aims to sense the medium for any node that wants to send a data frame to the node. The other wake-up state is the one in which the node wakes-up to send the data frames stored in queue to its parent node (it can be the PAN coordinator or other FFD node).

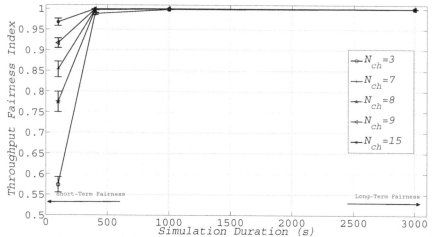

Figure 7.42   Throughput fairness index of MC-SCP-MAC for n= 99.

To achieve better results in terms of latency, a solution is to suppress the first wake-up state from the RFDs nodes when, after some attempts, the node verifies that no neighbouring node wants to send frames to it. For example, by establishing a limit of 20 wake-ups without any frame reception in the first wake-up state, the disabling of this wake-up state can be triggered. As so, only one wake-up state can be used to switch to the same slot channel of its parent node, in order to send data frames to it. Other solution that aims to decrease the latency time is to decrease the frame size. However, decreasing the frame size may cause the decrease of the delivery ratio (to the sink). A fair tradeoff between the frame size, delivery ratio and latency needs therefore to be envisaged. It must not impair network performance or jeopardize network integrity.

Other improvement that can be proposed to the MC-SCP-MAC protocol is the inclusion of a second wake-up state to FFDs nodes that have the PAN coordinator as its parent node. In the current version of the MC-SCP-MAC, the nodes that have direct connection with the PAN only consider one wake-up state in which they perform the sensing of the channel for incoming data frames (when the queue is empty) from neighbouring nodes, and transmit data frames to the PAN coordinator when the node has frames in queue to be transmitted. Since these border nodes consider only one wake-up state, in denser deployments this may limit the throughput performance and lead to the increase of the latency.

A solution to mitigate this limitation is the addition of a second wake-up state. Until now, these 1-hop nodes consider that the slot channel in which they wake-up and switch is given by their Linear Congruential Generator (LCG). The slot channel choice given by the LCG is used to wake-up and sense the medium, or to wake-up and send data frame to the PAN. The addition of a second wake-up state imposes special attention in the choice of the slot channel by these 1-hop nodes (to send the data frames to the PAN). Since the PAN coordinator is continuously hopping the slot channels, after a certain duration of time, the best solution for the slot channel choice of the 1-hop nodes (for the new wake-up state) is the addition of one unit to the slot channel choice, given by their LCG. If the LCG slot channel choice corresponds to the last slot channel available, the slot channel choice for the second wake-up state is the first lowest slot channel available for data frames transmission. The addition of this second wake-up state can significantly increase the delivery ratio and aggregate throughput, as well as decreasing the latency, due to the higher delivery ratio.

7.7 Summary and Conclusions

This work addressed the proposal of the Multi-Channel-Scheduled Channel Polling (MC-SCP-MAC) protocol, a new MAC protocol based on the SCP-MAC protocol that envisages multi-channel features. The new protocol states, namely startup, synchronization, slot channel choice, discovery and addition to the network and medium access algorithm, frames structure, energy optimization techniques are described, as well as the transition from one state to the other. These techniques comprise the influential range concept, the proposal of the denial channel list technique which considers the degradation metric of each slot channel and other cognitive- based capabilities. In addition, the study of the capture effect in this new protocol is covered, as well as the extra resolution phase algorithm useful to mitigate frame losses. To better understand the development and implementation of the protocol a state transition diagram is proposed and described for MC-SCP-MAC, jointly with the associated table of the events and transitions between states.

The MC-SCP-MAC performance has been extensively investigated. For the first set of tests, the goal has been to evaluate the collision probability of MC-SCP-MAC and compare the results with the ones from SCP. It is observable that the curves for the MC-SCP-MAC protocol achieve much lower values for the CW_1 and CW_2 collision probabilities (in both regimes) for the MC-SCP-MAC are much lower than SCP-MAC.The improvement is due to the use of multi-channel jointly with an enhanced two-phase contention window mechanism to cope with the possible data frame collisions in MC-SCP-MAC. It causes an increase of the overall network performance at the expense of a small increase in the of complexity of the protocol algorithm.

In almost all the tests performed the simulation results consider mostly the CC2420 radio transceiver. However, one set of experiments addresses the comparison between the energy consumption per frame (successfully delivered) between MC-SCP-MAC and SCP-MAC, for the CC2420 and AT86RF231 transceivers. With the AT86RF231 transceiver the MC-SCP-MAC protocol presents the lowest values for the energy spent per delivered frame, in comparison with the single-channel SCP-MAC protocol. The lowest values for the energy spent per delivered frame are justified by the very high delivery rate, owing to the use of multiple channels, combined with the ultra-low power consumption features from the AT86RF231 transceiver (with low energy consumption, shorter duration to perform CS, shorter duration to perform channel switching). The same holds for the CC2420 transceiver, in which the MC-SCP-MAC protocol also presents lower values for the energy spent per delivered frame. Since MC-SCP-MAC is a multi-channel based protocol, we have shown how the collision probability varies with the number of slot channels, $N_{ch} \in \{3; 4; 7; 9; 15\}$. Here, SCP considers a single-channel for data ($N_{ch} = 1$). The network size considered in these tests is $n = 99$ nodes plus 1 PAN coordinator. The increase of the number of slot channels allows for nodes selecting different slot channels. Therefore, it decreases the probability of a node choosing the same slot channel to transmit their data frames.

The difference of the CW_1 collision probability between the MC-SCP and SCP protocols is more noticeable in the unsaturated regime. The curves for the CW_2 collision probability for the MC-SCP-MAC protocol decreases as the number of slot channels increases. In comparison with the SCP protocol, the CW_2 collision probability of the MC-SCP-MAC protocol is always lower. This is due to the low values of CW_1 collision probability that result from the multi-channel features, and lead to a low number of expected number of nodes that pass from CW_1 to CW_2. There are two main conclusions arising from these tests: the number of slot channels increases along with the proposed enhanced two-phase contention window mechanism, allowing for achieving a much lower collision probability of the data frames.

The difference of the effective collision probability between the MC-SCP and SCP protocols is more notorious in the saturated case. Other tests have been performed to evaluate the energy efficiency with multiple slot channels and contention window sizes. The comparison of the energy efficiency is performed for the MC-LMAC and MC-SCP-MAC whilst considering equal deployment area and transmitter power.

This work, also analyzes the adaptation of the IR concept to a multi-channel based protocol along with the impact in terms of energy performance of enabling of the IR concept in the MC-SCP-MAC. The objective of these tests is to analyze of the gains achieved when the IR concept is enabled, for different values of sensing range, $\Pi_{irmax} \in \{-90; -80; -70; -60\}$ dBm. By enabling the IR, the node that overhears a data frame and verifies that it has redundant data stored in the queue to send to the sink, it discards these redundant frames. The adapted IR concept implies improvements, providing significant reduction of the redundancy of data frames and, consequently, the reduction of the energy consumption per node. As the value of Π_{irmax} increases the influential range of each node decreases which leads to a decrease of the overhearing and the consequent diminishing in the dropping of redundant data frames.

An additional performance metric has also been defined: the energy consumption saving. This metric translates the energy that a node can save when applying the IR concept (for different values of Π_{irmax}). The energy consumption saving is larger if the nodes have to send 600 frames to the PAN (long-term evaluation), compared with the case when nodes have to send 50 frames (short-term evaluation). Moreover, larger network sizes present the largest energy consumption savings when $\Pi_{irmax} =$ -90 dBm. For 99 nodes, there is a gain of 21.42 % while for 50 and 20 nodes the gain is 11.70 % and 7.07 %, respectively. These savings decrease as the value of Π_{irmax} increases. This is due to the reduction of the influential range area which, in turn, reduces the overhearing of data frames by the node (as there is a decrease of the level of data frame redundancy), leading to the increase of the energy consumption per node, i.e., there is a reduction of the energy consumption savings.

Additional tests have been conducted with variable number of slot channels to assess the benefits gained from the use of the IR concept in a multi-channel MAC protocol. We conclude that if the number of slot channels used by the MC-SCP-MAC increases the energy consumption also increases. However, the employment of IR allows for reducing the energy consumption for low values of Π_{irmax}. As the value of sensing range, Π_{irmax}, increases the number of redundant frames increases. Therefore, more frames are going to be exchanged, leading to an increase of the energy consumption. The energy savings decrease as the value of the sensing range, Π_{irmax}, increases. In turn, these gains are higher as the number of available slot channels increases.

In WSNs, the most usual traffic pattern is the periodic one. In all the tests performed the periodic traffic pattern is always considered in the simulations. However, the impact of different traffic patterns in the MC-SCP-MAC performance is also addressed. MC-SCP-MAC supports longer traffic generation intervals than other multi-channel based MAC protocols (e.g., MC-LMAC). The results for sparser scenarios have shown that the aggregate throughput does not attain a desirable value. However, tests have been conducted considering a denser scenario for $\lambda = 1$ s^{-1} (under periodic and exponential traffic generation). An aggregate throughput was achieved close to the maximum. With these results we can conclude that the MC-SCP protocol presents a better performance in high density scenarios but the presence of hidden terminals may lead to a degradation of the aggregate throughput as the number of source nodes increases. A method that enables to decrease the impact of the hidden terminals problem is therefore imperative.

Another conclusion is that the increase of the traffic generation frequency is well supported by the MC-SCP-MAC in the scenarios that do not present the hidden terminals problem. In terms of latency, as the number of source nodes increases the values of the latency increase. However, in scenarios with larger density of nodes the values for the latency are longer than in scenarios with sparser nodes density. Since MC-SCP-MAC is more robust and efficient in denser scenarios (and the envisaged scenario for this protocol is a deployment of sensor nodes in a forest or an agricultural field), we have considered that the performance evaluation in a cluster topology is useful to extract insights for further improvements in the protocol. The objective has been to assess the impact of the IR concept enabling and the number of slot channels available for data exchange in MC-SCP-MAC, whilst considering different traffic patterns and data generation rates. In the cluster topology, the increase of the number of slot channels leads to a decrease of the aggregate throughput when the IR is disabled. This is due to the high contention between nodes that have a higher number of frames in queue to be transmitted and are trying to win the medium in order to transmit their frames. Enabling the IR concept allows for achieving the highest values of aggregate throughput when $\Pi_{irmax} \in \{-90; -80\}$ dBm. Higher values for the the sensing range, Π_{irmax}, allows for nodes to overhear higher number of frames and therefore increasing the number of discarded frames. Energy consumption per node increases as the number of slot channels increases, whereas the enabling of the IR concept allows for achieving lower values of energy consumption per node. As the sensing range, Π_{irmax}, increases the energy consumption also increases (up to the value of the energy consumption corresponding to the case when the IR is not enabled). The use of the IR allows for obtaining shorter latency, since there are less frames to be transmitted by the sensor nodes in the queue. For a sensing range values of Π_{irmax}= -90 dBm, the MC-SCP achieves values for the latency around 2.8 s. It corresponds to a latency around 0.933 s per hop, almost a frame time duration per hop. As the value for the sensing range, Π_{irmax}, increases, the latency also increases. This increase for the latency is due to increase of the number of frames waiting in queue to be transmitted. If the IR feature is disabled, the latency value is around 23 s for all the considered number of slot channels. The same holds for the case when the IR is enabled since for all the considered number of slot channels, the latency is similar and increases as the value of the sensing range increases.

In order to test the scalability of the MC-SCP-MAC protocol, we have varied the density of nodes in a homogeneous network. The deployment scenario has been varied between 50×50 m^2 and 200×200 m^2. The objective of these tests has been to analyze the gains achieved when the IR is enabled for different values of $\Pi_{irmax} \in \{-90; -80; -70; -60\}$ dBm. Interesting results are presented for the performance metrics such as, the aggregate throughput, delivery ratio, latency, energy consumption and energy consumption savings when the MC-SCP-MAC is deployed in scenarios with side lengths larger than 125 m while varying the values for the sensing range, Π_{irmax}.

The throughput fairness index of MC-SCP-MAC has also been evaluated, for different network sizes and number of slot channels. This metric allows for verifying if the MC-SCP-MAC is fair in terms of opportunities to use the channel by all the nodes (in the network in order to transmit their frames). Fairness aspects are especially conspicuous when very few nodes are competing in a region. Short and long term fairness has been analyzed as a function of the number of slot channels for different sizes of networks. Results for network sizes n= 10 and 30 nodes shows that the increase of the number of slot channels improves the short-term fairness of MC-SCP-MAC. The long-term fairness stays high for all the considered values of the number of slot channels. Results for the throughput fairness index for n=99 nodes consider a dense deployment scenario, and show that, as the number of slot channels increases, the short-term fairness improves. By comparing these results with the ones for n= 30, the highest network sizes lead to a less fair behaviour among the nodes for the short-term fairness. High values of throughput fairness index are presented for the long-term fairness index for all the network sizes. Finally, it is a worth mentioning that the throughput fairness index is very important if the envisaged scenario of the MC-SCP-MAC considers QoS policies.

8

Conclusions and Suggestions for Future Research

8.1 Conclusions

The interdisciplinary research on Wireless Sensor Networks (WSNs) is stimulating the development of totally new WSN services and applications to be supported by sensor nodes in the context of smart environments. In five years time, a massive scale deployment of WSNs applied to a wide range of applications and will certainly change the way people interact, live or even work within the surrounding ambient [BVL14].

The book started by deeply addressing the PHY layer and MAC sublayer of Wireless Sensor Networks, covering a comprehensive taxonomy of MAC sub-layerprotocols, followed by further insights into the IEEE 802.15.4 standard. Then, conceptual aspects of the Scheduled Channel Polling MAC protocol were described, followed by its performance evaluation in relevant scenarios, including aspects of its energy efficiency, and the consideration of IEEE 802.15.4.

In Chapter 6, the enhancement of the IEEE 802.15.4 MAC layer has been proposed by employing RTS/CTS combined with the frame concatenation feature for WSNs. The use of the RTS/CTS mechanism improves channel efficiency by decreasing the deferral time before transmitting a data frame. The proposed solution has shown that, even for the case with retransmissions, if the number of TX frames is lower than 5 (i.e., the number of aggregated frames), IEEE 802.15.4 employing RTS/CTS combined with concatenation achieves higher values for the throughput in comparison to IEEE 802.15.4 with no RTS/CTS even for shorter frame sizes. The advantage comes from not including the backoff phase into the retransmission process like the IEEE 802.15.4 basic access mode (i.e., $BE = 0$). By comparing the analytical and simulation results, we conclude that there is a perfect match. This actually verifies the accuracy of our proposed retransmission model. Performance results for the minimum average delay, D_{min}, as a function of the number of

TX frames and by assuming a fixed payload size of 3 bytes (i.e., L_{DATA}=3 bytes) show that, by using IEEE 802.15.4 with RTS/CTS with frame concatenation, for 5, 7 and 10 aggregated frames, D_{min}, decreases by 8% 14% and 18%, respectively. For more than 28 aggregated frames, D_{min} decreases approximately 30%.

Moreover, we observe that by considering IEEE 802.15.4 in the basic access mode, the maximum average throughput, S_{max}, does not depend on the number of TX frames, and achieves the maximum value of 5.2 kb/s. For IEEE 802.15.4 with RTS/CTS combined with the frame concatenation feature, the maximum achievable throughput is 6.3 kb/s. In IEEE 802.15.4 employing RTS/CTS with frame concatenation, results for S_{max} as a function of the number of TX frames show that, for 5, 7 and 10 aggregated frames, S_{max} increases by 8%, 14% and 18%, respectively. For a number of aggregated frames higher than 28, S_{max} increases approximately 30%.

Chapter 6 also presents the SBACK-MAC, a new innovative MAC protocol that uses a BACK mechanism to achieve channel efficiency for WSNs. The use of a BACK mechanism improves channel efficiency by aggregating several ACK into one special frame, the *BACK Response*. Two innovative solutions were proposed to improve the IEEE 802.15.4 performance. The first one considers the SBACK-MAC protocol in the presence of *BACK Request* (concatenation mechanism), while the second considers the SBACK-MAC in the absence of *BACK Request* (piggyback mechanism). The results showed that, for the shortest payload sizes (i.e., $L_{DATA} = 3$), it is possible to improve the network performance by using the SBACK-MAC with and with no *BACK Request* by using a NAV extra time.

A set of results arising from our research were published in [NBC20]. By analyzing the results as a function of the payload size, for short frame sizes, S_{max}, is increased by 17% for IEEE 802.15.4 with RTS/CTS and SBACK-MAC with *BACK Request* and 25% for the SBACK-MAC with no *BACK Request*. By considering longer frame sizes, S_{max}, is increased by 8.8% for IEEE 802.15.4 with RTS/CTS, 8.6% for SBACK-MAC with *BACK Request* and 25% for SBACK-MAC with no *BACK Request*. A similar behaviour occurred for D_{min}, which instead of increasing (as S_{max}) in fact decreased.

From the results as a function of the number of TX frames (aggregated). For 10 aggregated frames, S_{max}, is increased by 20% for IEEE 802.15.4 employing RTS/CTS, 21% for SBACK-MAC with *BACK Request* and 30% for the SBACK-MAC with no *BACK Request*. By considering 80 aggregated frames, S_{max} is increased by 30% for IEEE 802.15.4 with RTS/CTS, 43% for

SBACK-MAC with *BACK Request* and 45% for the SBACK-MAC with no *BACK Request*. Again, a similar behaviour occurred for D_{min}, which instead of increasing (as S_{max}) in fact decreased.

This work has given novel contributions on the following main issues:

- Evaluation of the IEEE 802.15.4 MAC layer performance by using the RTS/CTS combined with frame concatenation through numerical and simulation results;
- Proposal of an innovative MAC protocol that uses a BACK mechanism to achieve channel efficiency, the SBACK-MAC protocol. The SBACK-MAC allows the aggregation of several ACK responses in one special frame. Two different solutions are addressed. The first one considers the SBACK-MAC protocol in the presence of *BACK Request* (concatenation) while the second one considers the SBACK-MAC in the absence of *BACK Request* (piggyback);
- Mathematical derivation of the maximum average throughput and the minimum average delay for the proposed mechanisms, either under ideal conditions (a channel environment with no transmission errors) or non ideal conditions (a channel environment with transmission errors), for IEEE 802.15.4 with and with no RTS/CTS and SBACK-MAC with and with no *BACK Request*.

In chapter 7 the Multi-Channel-Scheduled Channel Polling (MC-SCP-MAC) protocol has been proposed, a new MAC protocol based on the SCP-MAC protocol that envisages multi-channel features. In addition, the study of the capture effect in this new protocol is covered, as well as the extra resolution phase algorithm useful to mitigate frame losses. The MC-SCP-MAC performance has been evaluate in terms of collision probability. It is observable that the curves for the MC-SCP-MAC protocol achieve much lower values for the CW_1 and CW_2 collision probabilities (in both regimes) for the MC-SCP-MAC are much lower than SCP-MAC.The improvement is due to the use of multi-channel jointly with an enhanced two-phase contention window mechanism to cope with the possible data frame collisions in MC-SCP-MAC. It causes an increase of the overall network performance at the expense of a small increase in the of complexity of the protocol algorithm.

In almost all the tests performed the simulation results consider mostly the CC2420 radio transceiver. However, one set of experiments addresses the comparison between the energy consumption per frame (successfully delivered) between MC-SCP-MAC and SCP-MAC, for the CC2420 and AT86RF231 transceivers. With the AT86RF231 transceiver the MC-SCP-MAC protocol presents the lowest values for the energy spent per delivered frame, in comparison with the single-channel SCP-MAC protocol. The lowest values for the energy spent per delivered frame are justified by the very high delivery rate, owing to the use of multiple channels, combined with the ultra-low power consumption features from the AT86RF231 transceiver (with low energy consumption, shorter duration to perform CS, shorter duration to perform channel switching). The same holds for the CC2420 transceiver, in which the MC-SCP-MAC protocol also presents lower values for the energy spent per delivered frame. Since MC-SCP-MAC is a multi-channel based protocol, we have shown how the collision probability varies with the number of slot channels, $N_{ch} \in \{3; 4; 7; 9; 15\}$. Here, SCP considers a single-channel for data ($N_{ch} = 1$). The network size considered in these tests is $n = 99$ nodes plus 1 PAN coordinator. The increase of the number of slot channels allows for nodes selecting different slot channels. Therefore, it decreases the probability of a node choosing the same slot channel to transmit their data frames.

The difference of the CW_1 collision probability between the MC-SCP and SCP protocols is more noticeable in the unsaturated regime. The curves for the CW_2 collision probability for the MC-SCP-MAC protocol decreases as the number of slot channels increases. In comparison with the SCP protocol, the CW_2 collision probability of the MC-SCP-MAC protocol is always lower. This is due to the low values of CW_1 collision probability that result from the multi-channel features, and lead to a low number of expected number of nodes that pass from CW_1 to CW_2. There are two main conclusions arising from these tests: the number of slot channels increases along with the proposed enhanced two-phase contention window mechanism, allowing for achieving a much lower collision probability of the data frames.

The difference of the effective collision probability between the MC-SCP and SCP protocols is more notorious in the saturated case. Other tests have been performed to evaluate the energy efficiency with multiple slot channels and contention window sizes. The comparison of the energy efficiency is performed for the MC-LMAC and MC-SCP-MAC whilst considering equal deployment area and transmitter power.

This work, also analyzes the adaptation of the IR concept to a multi-channel based protocol along with the impact in terms of energy performance of enabling of the IR concept in the MC-SCP-MAC. The objective of these tests is to analyze of the gains achieved when the IR concept is enabled, for different values of sensing range, $\Pi_{irmax} \in \{-90; -80; -70; -60\}$ dBm. By enabling the IR, the node that overhears a data frame and verifies that it has redundant data stored in the queue to send to the sink, it discards these redundant frames. The adapted IR concept implies improvements, providing significant reduction of the redundancy of data frames and, consequently, the reduction of the energy consumption per node. As the value of Π_{irmax} increases the influential range of each node decreases which leads to a decrease of the overhearing and the consequent diminishing in the dropping of redundant data frames.

An additional performance metric has also been defined: the energy consumption saving. This metric translates the energy that a node can save when applying the IR concept (for different values of Π_{irmax}). The energy consumption saving is larger if the nodes have to send 600 frames to the PAN (long-term evaluation), compared with the case when nodes have to send 50 frames (short-term evaluation). Moreover, larger network sizes present the largest energy consumption savings when Π_{irmax} = -90 dBm. For 99 nodes, there is a gain of 21.42% while for 50 and 20 nodes the gain is 11.70% and 7.07%, respectively. These savings decrease as the value of Π_{irmax} increases. This is due to the reduction of the influential range area which, in turn, reduces the overhearing of data frames by the node (as there is a decrease of the level of data frame redundancy), leading to the increase of the energy consumption per node, i.e., there is a reduction of the energy consumption savings.

Additional tests have been conducted with variable number of slot chan-nels to assess the benefits gained from the use of the IR concept in a multi-channel MAC protocol. We conclude that if the number of slot channels used by the MC-SCP-MAC increases the energy consumption also increases. However, the employment of IR allows for reducing the energy consumption for low values of Π_{irmax}. As the value of sensing range, Π_{irmax}, increases the number of redundant frames increases. Therefore, more frames are going to be exchanged, leading to an increase of the energy consumption. The energy savings decrease as the value of the sensing range, Π_{irmax}, increases. In turn, these gains are higher as the number of available slot channels increases.

In WSNs, the most usual traffic pattern is the periodic one. In all the tests performed the periodic traffic pattern is always considered in the simulations. However, the impact of different traffic patterns in the MC-SCP-MAC performance is also addressed. MC-SCP-MAC supports longer traffic generation intervals than other multi-channel based MAC protocols (e.g., MC-LMAC). The results for sparser scenarios have shown that the aggregate throughput does not attain a desirable value. However, tests have been conducted considering a denser scenario for λ= 1 s^{-1} (under periodic and exponential traffic generation). An aggregate throughput was achieved close to the maximum. With these results we can conclude that the MC-SCP protocol presents a better performance in high density scenarios but the presence of hidden terminals may lead to a degradation of the aggregate throughput as the number of source nodes increases. A method that enables to decrease the impact of the hidden terminals problem is therefore imperative.

Another conclusion is that the increase of the traffic generation frequency is well supported by the MC-SCP-MAC in the scenarios that do not present the hidden terminals problem. In terms of latency, as the number of source nodes increases the values of the latency increase. However, in scenarios with larger density of nodes the values for the latency are longer than in scenarios with sparser nodes density. Since MC-SCP-MAC is more robust and efficient in denser scenarios (and the envisaged scenario for this protocol is a deployment of sensor nodes in a forest or an agricultural field), we have considered that the performance evaluation in a cluster topology is useful to extract insights for further improvements in the protocol. The objective has been to assess the impact of the IR concept enabling and the number of slot channels available for data exchange in MC-SCP-MAC, whilst considering different traffic patterns and data generation rates. In the cluster topology, the increase of the number of slot channels leads to a decrease of the aggregate throughput when the IR is disabled. This is due to the high contention between nodes that have a higher number of frames in queue to be transmitted and are trying to win the medium in order to transmit their frames. Enabling the IR concept allows for achieving the highest values of aggregate throughput when $\Pi_{irmax} \in \{-90; -80\}$ dBm. Higher values for the the sensing range, Π_{irmax}, allows for nodes to overhear higher number of frames and therefore increasing the number of discarded frames. Energy consumption per node increases as the number of slot channels increases, whereas the enabling of the IR concept allows for achieving lower values of energy consumption per node. As the sensing range, Π_{irmax}, increases the energy consumption also increases (up to the value of the energy consumption corresponding to the case when the IR is not enabled). The use of the IR allows for obtaining shorter latency, since there are less frames to be transmitted by the sensor nodes in the queue. For a sensing range values of $\Pi_{irmax} = $ -90 dBm, the MC-SCP achieves values for the latency around 2.8 s. It corresponds to a latency around 0.933 s per hop, almost a frame time duration per hop. As the value for the sensing range, Π_{irmax}, increases, the latency also increases. This increase for the latency is due to increase of the number of frames waiting in queue to be transmitted. If the IR feature is disabled, the latency value is around 23 s for all the considered number of slot channels. The same holds for the case when the IR is enabled since for all the considered number of slot channels, the latency is similar and increases as the value of the sensing range increases.

In order to test the scalability of the MC-SCP-MAC protocol, we have varied the density of nodes in a homogeneous network. The deployment scenario has been varied between 50×50 m^2 and 200×200 m^2. The objective of these tests has been to analyze the gains achieved when the IR is enabled for different values of $\Pi_{irmax} \in \{-90; -80; -70; -60\}$ dBm. Interesting results are presented for the performance metrics such as, the aggregate throughput, delivery ratio, latency, energy consumption and energy consumption savings when the MC-SCP-MAC is deployed in scenarios with side lengths larger than 125 m while varying the values for the sensing range, Π_{irmax}.

The throughput fairness index of MC-SCP-MAC has also been evaluated, for different network sizes and number of slot channels. This metric allows for verifying if the MC-SCP-MAC is fair in terms of opportunities to use the channel by all the nodes (in the network in order to transmit their frames). Fairness aspects are especially conspicuous when very few nodes are competing in a region. Short and long term fairness has been analyzed as a function of the number of slot channels for different sizes of networks . Results for network sizes $n = 10$ and 30 nodes shows that the increase of the number of slot channels improves the short-term fairness of MC-SCP-MAC. The long-term fairness stays high for all the considered values of the number of slot channels. Results for the throughput fairness index for $n = 99$ nodes consider a dense deployment scenario, and show that, as the number of slot channels increases, the short-term fairness improves. By comparing these results with the ones for $n = 30$, the highest network sizes lead to a less fair behaviour among the nodes for the short-term fairness. High values of throughput fairness index are presented for the long-term fairness index for all the network sizes. Finally, it is a worth mentioning that the throughput fairness index is very important if the envisaged scenario of the MC-SCP-MAC considers QoS policies.

8.2 Suggestions for Future Work

As future research, we propose the creation of a webpage enabling for access the data in real time in the cloud for the real time applications developed in the scope of this work.

Another suggestion for further research is to implement the SBACK-MAC with no *BACK Request* in the OMNeT++ simulator, since in the scope of this work we only derive the analytical model for the maximum average throughput and minimum average delay. In addition, we plan to compare the IEEE 802.15.4 with and with no RTS/CTS with SBACK-MAC with and with no *BACK Request* by considering the new optional CSS PHY layer at the 2.4 GHz frequency band, that enables to achieve a maximum data rate of 1 Mb/s against the results obtained in this work where the maximum data rate is 250 kb/s.

Finally, the performance of the innovative MAC sub-layer protocols proposed in this research work will be verified through experimental measurements by using WSN hardware platforms (e.g., Waspmote platform), while varying the number of transmitted frames, the payload and the number of nodes.

For the MC-SCP-MAC protocol Another we intend to investigate the collision probability model by considering limited retransmissions and other queueing models, such as the M/M/1/K Markov chain model, which assumes finite buffers. In addition, we plan to use the model in the routing layer of wireless sensor networks, in a real testbed, in order to use the model's output to decide whether to transmit a frame, depending on the number of neighbour nodes. It can also be suggested is to analytically compare the advantages/disadvantages of using one or two contention windows in slotted contention based MAC protocols. The idea is to apply the two phase contention window mechanism of the SCP-MAC protocol with the specfications of the IEEE 802.11 Distributed Coordination Function (DCF) access method. The objective is to derive the achieved throughput based on the different slot probabilities andfind the optimal values of the contention windows sizes that maximize the throughput for a certain network size. In addition, a throughput comparison between the IEEE 802.11 DCF and the modified IEEE 802.11 DCF access methods is planned.

Deployment in MC-SCP-MAC can be performed by taking into account if the IDs from the neighbouring nodes are sequential or random.

A

IEEE 802.15.4 PHY Layer

The main purpose of the PHY layer is to establish an interface with the physical medium where the communications are done. The PHY layer is the lowest layer in the WSN protocol stack and it is responsible by the control (enable and disable) of the radio transceiver, energy detection, link quality, CCA, channel selection and the transmission and reception of message frames that are exchanged in physical medium [GCB03].

A.1 IEEE 802.15.4 Country Regulations

The governments or a telecommunications regulatory institute of a country regulates and administer the radio-frequency spectrum. In almost all cases, there are some frequency bands allocations just for unlicensed operation, in order to (the manufacturer can) ensure operation within some pre-established limits in output power, duty cycle, modulation and other parameters. In the case of Europe, Japan, Canada and the United States of America, the regulation consists in unlicensed bands but they are type-approved DSSS services. The IEEE 802.15.4 standard is written so that the conforming devices can be manufactured to operate in any of the three bands. Two of these bands are limited to specific geographical regions, but one band is available for nearly worldwide service. In Europe, the European Telecommunications Standards Institute (ETSI) has published recommendations about the IEEE 802.15.4 standard and, within the European service area, there is one common band with operation allowed between 868.0 and 868.6 MHz, which supports a single channel of low data rate service with less than 1 % transmission duty cycle [GCB03]. In United States of America, the national regulatory agency is the Federal Communications Commission (FCC) and just like in Europe there is only one band available for service. This band is called the 915 MHz and it covers the range between 902 and 928 MHz. This band enables the use of ten channels of low data rate service. Worldwide, in order to obtain economies of scale in product design, it is desirable to employ a single band that is available on nearly all world. To decide which band should be employed, the studies concluded that the band has to be unlicensed, have sufficient width (to enable the use of many channels) and be high enough in the spectrum, so that efficient antennas could be applied. The selected band is the 2.4 GHz ISM band, which extends from 2400 to 2483.5 MHz. This band presents a wavelength of 12.25 cm and allows for employing efficient antennas. A 2.4 GHz transceiver may also support the 868/915 MHz bands, but it is not required by IEEE 802.15.4.

A.2 IEEE 802.15.4 Frequency Bands

In the IEEE 802.15.4 version which was released in September 2006, three bands for unlicensed operation were defined:

- **868-868.6 MHz (868 MHz band)**: This unlicensed band is available in most European countries for a 20 kbps DSSS service. IEEE 802.15.4 standard also refers to this band as the "868 MHz" band;
- **902-928 MHz (915 MHz band)**: Some portions of this unlicensed band are available in North America, Australia, New Zealand, and some countries in South America for 40 kbps DSSS service. IEEE 802.15.4 standard also refers to this band as the "915 MHz" band;
- **2400-2483.5 MHz (2.4 GHz band)**: This last unlicensed band is available in most countries worldwide for the faster 250 kbps DSSS service. This band is referred as the 2.4 GHz band.

A.3 IEEE 802.15.4 Data Rates

Due to the physical characteristics of each band (and the regulations that are applied depending on the country), the IEEE 802.15.4 specifies different data rates and modulations for the three bands used by the two PHY layers.

The IEEE 802.15.4 working group specifies a total of 27 half-duplex channels across the three frequency bands mentioned before and they are described as follows:

- The 868 MHz band has a frequency between 868.0 MHz and 868.6 MHz and it is used in Europe. The modulation format used is the BPSK, with a DSSS at a chip-rate 300 kchip/s. In terms of channels a single channel is available with data rate of 20 kb/s and devices shall be capable of achieving a sensitivity of -92 dBm or better. A pseudo-random sequence of 15 chips is transmitted in a 50 μs symbol period.

Table A.1 IEEE 802.15.4 data rates and frequencies of operation.

Frequency (MHz)	number of channels	DSSS spreading parameters		Bit rate (Kb/s)	Symbol rate (Ksymbol/s)	Spreading method
		Modulation	Chip rate (Kchip/s)			
868 – 868.6	1	BPSK	300	20	20	Binary DSSS
902 – 928	10	BPSK	600	40	40	Binary DSSS
868 – 868.6	1	ASK	400	250	12.5	20−bit PSSS
902 – 928	10	ASK	1600	250	50	5−bit PSSS
868 – 868.6	1	O−QPSK	400	100	25	16−array Orthogonal
902 – 928	10	O−QPSK	1000	250	62.5	16−array Orthogonal
2400 – 2483.5	16	O−QPSK	2000	250	62.5	16−array Orthogonal

- The 915 MHz band presents a range between 902 MHz and 928 MHz and is used in the North American and Pacific area, adopting a binary phase-shift keying BPSK modulation format, with DSSS at a chip-rate of 600 kchip/s. In terms of channels it has ten channels available with data rate of 40 kb/s and devices shall be capable of achieving a sensitivity of -92 dBm or better. A pseudo-random sequence of 15 chips is transmitted in a 25 μs symbol period.
- The unlicensed 2.4 GHz ISM band, with a range from 2400 MHz to 2483.5 MHz and since it is used worldwide, adopts a offset quadrature shift keying (O-QPSK) modulation format, with DSSS at 2 Mchip/s. In terms of channels it has sixteen channels with data rate of 250 kb/s and devices shall be capable of achieving a sensitivity of -85 dBm or better.

Table A.1 provides details regarding how the three frequency bands mentioned previously are used in the IEEE 802.15.4 standard, in terms of number of channels, DSSS spreading parameters, bit rate, symbol rate and spreading method. IEEE 802.15.4 requires that if a transceiver supports the 868 MHz band, it must support 915 MHz band as well, and vice versa as mentioned before. Therefore, these two frequency bands are always bundled together as the 868/915 MHz frequency bands of operation. Note that one 'symbol' is equivalent to four 'bits'.

A.4 IEEE 802.15.4 Network Topologies

The IEEE 802.15.4 network, independently of the type of topology, is always created by a PAN coordinator. Therefore, a PAN coordinator controls the network and performs the following minimum tasks:

- Allocate a unique address (16 bit or 64 bit) to each device in the network;
- Initiate, terminate and route the messages throughout the network;
- Select a unique PAN identifier for the network, in order to allow the devices within a network to use the 16 bit short addressing method and still communicate with other devices of other neighbouring networks;

A.4.1 IEEE 802.15.4 Star Topology

In the star topology all devices establish a communication link with a single central controller, called the PAN coordinator. In this type of topology the PAN coordinator may be main powered while the other devices of the network will most likely be battery powered. When an FFD is activated for the first time, it may establish its own network and become the PAN coordinator. To establish its own network the PAN coordinator select a unique PAN identifier that is not used by any other network in the surrounding radios range (range around the device in which its radio can establish a communication link with other radios). This allows each star network to operate independently.

A.4.2 IEEE 802.15.4 Peer-to-peer Topology

In the peer-to-peer topology there is also one PAN coordinator and in contrast to the star topology each device can communicate directly with any other device if the devices are placed close enough together to establish a successful communication link. A peer-to-peer network can be ad hoc, self-organizing and self-healing. Any FFD in a peer-to-peer network can play the role of the PAN coordinator. The method of how to decide which device will be the PAN coordinator is done by picking the first FFD device that starts communicating as the PAN coordinator. In a peer-to-peer network, all the devices that are capable of relaying messages are FFDs, because an RFD is not capable of relay messages. On the other hand, an RFD will communicate with only one particular device (a coordinator or a router) in the network. A peer-to-peer network can be defined in to different shapes just by defining restrictions on the devices that can communicate among them. If no restriction exists, then the peer-to-peer network topology is called a *mesh topology*. It also allows multiple hops to route messages from any device to any other device in the network. It can provide reliability by multipath routing.

A.4.3 IEEE 802.15.4 Cluster Tree Topology

The cluster tree topology is a special case of a peer-to-peer network in which most devices are FFDs and an RFD may connect to a cluster-tree network as a leave node at the end of a branch. This last network topology type is not part of the IEEE 802.15.4 standard, but it is described in the ZigBee Alliance specifications. In this type of network the PAN coordinator establishes the initial network and any FFD can act as a coordinator and provide synchronization services to other devices and coordinators. The PAN coordinator forms the first cluster by establishing itself as the Cluster Head (CLH) with a Cluster Identifier (CID) of zero, choosing an unused PAN identifier, while broadcasting beacon frames to neighbouring devices. A candidate device receiving a beacon frame may request to join the network at the CLH. If the PAN coordinator allows the device to join, it will add this new device as a child device in its neighbour list. The newly joined device will add the CLH as its parent in its neighbour list and begin transmitting periodic beacons such that other candidate devices may then join the network at that device. Other feature of the cluster tree topology is the use of *multihopping*. For example, the PAN coordinator needs to send a message an RFD located in the edge of the branch, but there is a barrier between them that is hard for the signal to penetrate. The tree topology helps by relaying the message around the barrier and reach device. The advantage of using this type of topology is the increase of the network coverage area at the cost of increased message latency.

A.5 IEEE 802.15.4 PHY Specifications

As described before the physical layer of the IEEE 802.15.4 is responsible by the control of several parameters. The parameters controlled by the IEEE 802.15.4 are described below.

A.5.1 Receiver Energy Detection

The receiver EnD value is an estimation of the received signal power within the bandwidth of an IEEE 802.15.4 channel and there is no attempt to decode or identify the type of signal that is currently occupying the channel. In other words, when performing an EnD it does not reveal whether this signal is an IEEE 802.15.4 standard compliant or not. The EnD time should be equal to 8 symbol periods. For example, if the required receiver sensitivity is -85 dBm, the EnD procedure must be able to detect and measure the energy of signals as low as -75 dBm. The minimum EnD value (0) indicates received power less than 10 dB above the specified receiver sensitivity. The range of received power spanned by the EnD values shall be at least 40 dB and the mapping from the received power in dB to EnD values shall be linear with an accuracy of +/- 6 dB.

A.5.2 Link Quality Indication (LQI)

The RSS is a measure of the total energy of the received signal. Other way to evaluate the signal quality is using the SNR measure, which is the ratio of the desired signal energy to the total in-band noise energy. When considering a signal with high SNR it means that the signal has high-quality. The LQI measurement is performed for each received frame and LQI must have at least eight unique levels. This LQI measurement is generated in the PHY layer and is available to the NWK and APL layers. The minimum and maximum LQI values should be associated with the lowest and highest quality IEEE 802.15.4 signals detectable by the receiver and LQ values should be uniformly distributed between these two limits.

A.5.3 Carrier Sense (CS)

The CS technique is quite similar to the EnD technique and it is used to perform verification if whether a frequency channel is available to use. When a wireless device has the intention to transmit a message, it first switches to receive mode in order to detect the type of any possible signal that might be present in the desired frequency channel. While in EnD the signal detected in the channel is not decoded, in the CS technique the signal is demodulated so in order to perform verification whether the signal modulation and spreading are compliant with the parameters of the PHY used by the receiver device. After this verification and if the signal is compliant with the IEEE 802.15.4 PHY, the wireless device will decide if the channel is busy or not, regarding to the signal energy level.

A.5.4 Clear Channel Assessment (CCA)

When performing the CCA, the results of EnD or CS can be used to decide whether a frequency channel should be considered available or busy. The CCA period must be eight symbols.

Three CCA modes are contemplated in the IEEE 802.15.4 compliant PHY and can operate in any one of them:

- **CCA mode 1**: when considering this mode, the EnD result is the only one that is taken into account. Therefore if the energy level is above the

EnD threshold, the channel is considered busy. This EnD threshold can be set by the manufacturer of the radio transceiver;

- **CCA mode 2**: when considering this mode, the CS result is the only one that is taken into account and the channel is considered busy only if the occupying signal is compliant with the PHY layer of the device that is performing the CCA;
- **CCA mode 3**: this mode is the result of a logical combination (AND/OR) of mode 1 and mode 2. Therefore, there are two solutions to detect if the channel is busy:
    - If the detected energy level is above the threshold and a compliant carrier is sensed then the channel is busy;
    - If the detected energy level is above the threshold or a compliant carrier is sensed then the channel is considered busy.

A.5.5 Channel Selection

The IEEE 802.15.4 standard initial release has a total of 27 channels and the implementation of multiple operating frequencies bands could not be supported. For each channel page it could have a maximum of 27 channels. A channel page is a concept introduced by the IEEE 802.15.4 in 2006 to distinguish between supported PHYs, because in the previous releases of the standard the frequency channels were identified by channel numbers and there were no optional PHYs. The distribution of channels selection in different PHYs could be resumed by the Table A.2.

A.6 IEEE 802.15.4 PHY Frame Structure

The structure of the PPDU encapsulates all data structures from the higher levels of the protocol stack.

- **PPDU synchronization header:** It enables the receiver to synchronize and lock into the bit stream. The PPDU header consists of two fields, a preamble and a Start-of-Frame Delimiter (SFD). The preamble field is used by the receiver to obtain chip and symbol synchronization. The bits in the preamble field in all PHYs, except for the ASK PHYs, are binary zeros. The SFD allows the receiver to establish the beginning of the frame in the stream of bits. Remind that one "octet" is equal to one "byte" and one "symbol" is equal to four "bits". For the lengths and durations of the preambles in all PHY options it is resumed in Table A.3 and for the length of the SFD it is resumed in Table A.4.

Table A.2   Channel assignments.

Channel page	Channel number	Frequency band	Modulation
0	0	868 MHz	BPSK
	1-10	915 MHz	BPSK
	11-26	2.4 GHz	O-QPSK
1	0	868 MHz	ASK
	1-10	915 MHz	ASK
	11-26	Reserved	-
2	0	868 MHz	O-QPSK
	1-10	915 MHz	O-QPSK
	11-26	Reserved	-
3-31	Reserved	Reserved	-

Table A.3   Preamble field lengths and durations.

PHY option	Modulation	Length		Duration [μs]
868 MHz	BPSK	4 octets	32 symbol	1600
915 MHz	BPSK	4 octets	32 symbol	800
868 MHz	ASK	5 octets	2 symbol	160
915 MHz	ASK	3.75 octets	6 symbol	120
868 MHz	O-QPSK	4 octets	8 symbol	320
915 MHz	O-QPSK	4 octets	8 symbol	128
2.4 GHz	O-QPSK	4 octets	8 symbol	128

- **PPDU PHY header**: The PHY header is a single 8 bit field with the MSB reserved and the remaining low order bits used to specify the total number of octets in the PHY payload. The PHY payload length can be any value from 0 to 127 bytes. Frames lengths of 0 to 4 and 6 to 8 bytes are reserved. Frames of length 5 bytes are MPDU acknowledgement

Table A.4 SFD field lengths.

PHY option	Modulation	Length	
868 MHz	BPSK	1 octet	8 symbol
915 MHz	BPSK	1 octet	8 symbol
868 MHz	ASK	2.5 octets	1 symbol
915 MHz	ASK	0.625 octet	1 symbol
868 MHz	O-QPSK	1 octet	2 symbol
915 MHz	O-QPSK	1 octet	2 symbol
2.4 GHz	O-QPSK	1 octet	2 symbol

frames and frames with 9 or more bytes are MPDU payloads for the MAC protocol layer service.

- **PPDU PHY payload**: The PHY payload is composed of only one field called the Physical Layer Service Data Unit (PSDU). The PSDU is variable length and carries the data payload of the PPDU. All frames carry an MPDU payload for the MAC layer.

B

IEEE 802.15.4 Standard MAC Sub-Layer

B.1 SuperFrame Structure

B.1.1 Timing Parameters

The timing parameters in beacon enabled operating mode that correspond to the superframe structure discussed in Section 2.6.3 are presented in Table B.1.

B.1.2 InterFrame Spacing

When transmitting data from one device to another, the device must wait during a short time interval between its successive transmitted frames in order to allow the receiving device to process the received frame before the next frame arrives. This feature is known as *Interframe Spacing* (IFS). The length of IFS depends on the transmitted frame size [Far08]. If the MPDU has sizes less than or equal to *aMaxSIFSFramesSize* (default value of 18 octects) it is considered as short frames and if a MPDU size exceeds *aMaxSIFSFramesSize* octets than it is considered as a long frame. The wait time after a short frame is named as Short IFS (SIFS) and the minimum value of SIFS is equal to *macMinSIFSPeriod*. The same way as in the short

Table B.1 Timing parameters in beacon enabled operating mode.

Time period	MAC attribute	Duration (symbols)
Unit backoff period	$aUnitBackoffPeriod$	20
Basic superframe slot	$aBaseSlotDuration$	$3 \cdot aUnitBackoffPeriod = 60$
Superframe slot		$aBaseSlotDuration \cdot 2^{SO}$
Superframe duration	SD	$aBaseSuperframeDuration \cdot 2^{SO}$
Beacon interval	BI	$aBaseSuperframeDuration \cdot 2^{BO}$

313

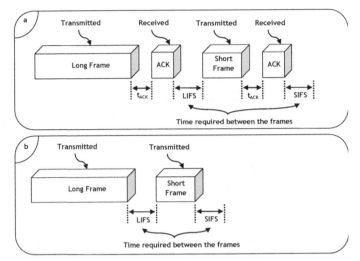

Figure B.1 The IFS in (a) Acknowledged transmission and (b) unacknowledged transmission.

frame, the long frame is followed by a Long IFS (LIFS) and the minimum length is given by *macMinLIFSPeriod*. The values of *macMinSIFSPeriod* and *macMinLIFSPeriod* are 12 and 40 symbols, respectively.

B.2 IEEE 802.15.4 MAC Frames

This section presents the details on the four IEEE 802.15.4 MAC frames structures addressed in Section 2.6.4.

B.2.1 Beacon Frames

When a beacon-enabled network is considered, a FFD device may transmit beacon frames. The entire MAC frame is used as a payload in a PHY frame. The content of the PHY payload is mentioned as the PSDU. In a beacon frame, the address field contains the source PAN ID and the source device address.

The MAC frame consists of three sections: the MAC Header (MHR), the MAC payload and the MAC Footer (MFR). The frame control field in the MHR has the information that defines the frame type, addressing fields and other control flags. The sequence number specifies the Beacon Sequence

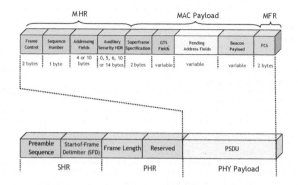

Figure B.2 MAC beacon frame structure and the PHY frame structure.

Number (BSN). The addressing field provides the source and destination addresses.

The MAC payload of a beacon frame is divided into four fields:

- *SuperFrame specification field*: contains the parameters that specify the superframe structure (if exists);
- *Pending address specification field*: contains the number and type of addresses listed in the Address List Field;
- *Address list field*: contains a list of devices addresses with data available at the PAN coordinator;
- *Beacon payload field*: optional field that will contain broadcast data to the devices that make part of its network within its range of coverage.

The receiver uses the Frame Check Sequence (FCS) field to check for any possible error in the received frame. The format of the beacon frame is shown in Figure B.2.

The subfields from the beacon frame are explained with more detailed in Figure B.3. The Beacon Order (BO) subfield determines the transmission intervals between the beacons.

The superframe is divided into 16 equal and contiguous time slots and the final CAP slot subfield defines the last time slot of the CAP. If there are any remaining time slots after the CAP, they are used as GTSs.

If the BLE subfield is set to one, it means that the beaconing device will turn off its receiver after a certain period of time to conserve its energy. When the beacon frame is transmitted by the PAN coordinator, the PAN coordinator subfield is set to one, helping to distinguish between a frame received from the PAN coordinator and any other coordinator in the same network.

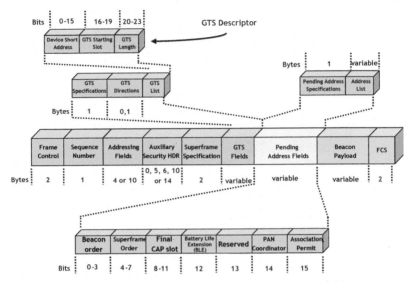

Figure B.3   Detailed description for the beacon frame fields.

The association permit subfield is the last bit in the superframe specification and if is set to zero, it means that the coordinator does not accept association requests at this moment.

The main objectives of the beacon frames are the establishment of GTSs. The GTS subfield defines whether the PAN coordinator currently accepts GTS requests and determines the number of GTSs listed in the GTS list field. A way to define direction of the GTSs is using the GTS direction subfield to set the GTSs to receive-only or transmit-only modes. The GTS list are stored the list of all GTSs that are currently maintained (GTS descriptors). The GTS descriptor contains the short address of the device that will use the GTS, the GTS starting slot and the GTS length. Finally, the pending address field contains the addresses of all devices that have data pending at the coordinator. Each device checks for its address in this field and if there is a match with the node address, then the device will contact the coordinator and request the data to be transmitted to it.

B.2.2 Data Frames

The MAC data frame, presented in Figure B.4, is used by the MAC sublayer to transmit data. The address field contains the PAN ID and the device ID of the source and/or destination, as specified in the MCPS-DATA primitive. The data payload field is provided by the NWK layer. The data contained in the

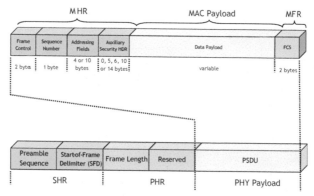

Figure B.4 The MAC data frame structure and the PHY frame structure.

MAC payload is referred to as the MAC Service Data Unit (MSDU). All the fields in this frame are similar to the beacon frame except for the superframe, GTS and pending address fields that are not present in the MAC data frame. The MAC data frame is referred to as the MAC Protocol Data Unit (MPDU) and becomes the PHY payload.

B.2.3 Acknowledgement Frames

The MAC acknowledgement frame, presented in Figure B.5, is the simplest MAC frame format and there is no MAC payload present in this frame [Far08]. These frames are sent by the MAC sublayer to confirm a successful frame reception to the device that has sent the message. The acknowledgement was designed by the IEEE 802.15.4 design group to be very short in order to minimize network traffic. The acknowledgement frame does not contains the address field in the MAC header and there is no MAC payload. When a network device receives an acknowledgement frame, it first verify if it was expecting an acknowledgement frame and then it checks if the frame sequence number matches the one it is expecting. Otherwise the acknowledgement frame is discarded. The acknowledgement frame is generated only if the received message is requesting acknowledgement and the FCS is evaluated as good or not by the receiving device.

B.2.4 MAC Command Frames

The MAC command frame is generated by the MAC sublayer and is responsible by all the MAC control transfers for each MAC command type presented in Table B.2.

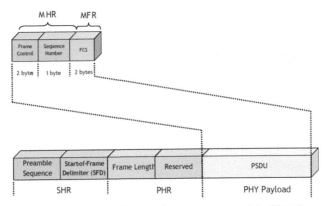

Figure B.5 The MAC acknowledgement frame structure and the PHY frame structure.

Table B.2 MAC commands frame types.

Command identifier	Comamand type	RFD	
		Transmit	Receive
1	Association Request	√	-
2	Association Response	-	√
3	Disassociation Notification	√	√
4	Data Request	√	-
5	PAN ID Conflict Notification	√	-
6	Orphan Notification	√	-
7	Beacon Request	-	-
8	Coordinator Realignment	-	√
9	GTS Request	-	-
10-255	Reserved	-	-

The MAC commands such as requesting association or disassociation with a network are transmitted using the MAC command frame, Figure B.6. The MAC payload has two fields, the MAC command type and the MAC command payload. The MAC command payload contains information specific to the type of command in use. The entire MAC command frame is placed in the PHY payload as a PSDU [Far08].

B.2.5 Slotted and Unslotted CSMA-CA Algorithm Phases

The slotted CSMA-CA algorithm can be described as follows [Far08, GCB03]:

- **Phase 1** - Initially sets the number of backoffs counter and the contention window variable (NB= 0 ; CW= 2). The backoff exponent is used

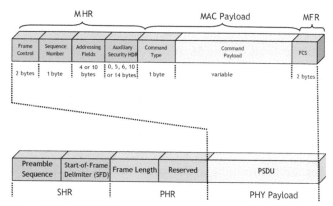

Figure B.6 IEEE 802.15.4 MAC command frame [GCB03, Far08].

to determine the number of backoff periods a device should wait before attempting to access the channel. If the device operates on battery power, the attribute *macBattLifeExt* is set to true, BE is set to 2 or to the constant *macMinBE*, depending on the one that is smaller. Otherwise, it is set to *macMinBE*, the default value of which is 3.

- **Phase 2** - Then the CSMA-CA mechanism locates the backoff period boundary and and a random number in the range 0 to $2^{BE} - 1$ is generated. The algorithm then counts down for this number of backoff periods. This period is defined as the Random Backoff Countdown (RBC) and during this period the channel activity is not accessed and the backoff counter is not stopped if detects activity in the channel. In order to disable the CA procedure at the first iteration, *BE* must be set to 0, and thus the waiting delay is null;

- **Phase 3** - Once the backoff counter reaches zero, the algorithm first checks if the remaining time within the CAP area of the current super-frame is sufficient to accommodate the necessary number of CCA checks, the frame transmission and the acknowledgement (alignment with time slots). Then if this is the case, the algorithm performs the CCA at the BP boundary to assess channel activity and if the channel is busy then the algorithm goes to phase 4, otherwise the algorithm goes to phase 5;

- **Phase 4** - If the CCA result is an idle channel the frame may be transmitted. Otherwise, if the result of the CCA is a busy channel, the node concludes that there is an ongoing transmission by another device and the current transmission attempt is aborted. The CSMA-CA mechanism is then restarted, the NB and the BE are incremented by one unit and the CCA counter is reset to two. The BE must not exceed *macMaxBE* (by default equal to 5). If the number of unsuccessful backoff cycles NB exceeds the limit of *macMaxCSMABackoffs* (default value equal to 5), the algorithm finishes with channel access failure status which is reported to higher protocol layers;

- **Phase 5** - If the channel is idle, the CW variable is decremented by one unit. The CCA operation is repeated until CW $\neq$ 0 (phase 3). With this step the algorithm ensures the performing of two CCA operations to prevent potential collisions of acknowledgement frames, when two or more modes are transmitting at the same time. When channel is again sensed as idle (CW= 0), the node attempts to transmit.

The unslotted CSMA-CA algorithm is used in a non-beacon network and can be described as follows [Far08, GCB03]:

- **Phase 1** - The unslotted CSMA-CA algorithm initially sets the number of backoffs and the backoff exponent is equal to *macMinBE* (NB=0, BE= *macMinBE*). The CW variable is not used, since the non-slotted CSMA-CA has no need to iterate the CCA procedure after detecting an idle channel;
- **Phase 2, 3, and 4** - These steps are quite similar to the one in slotted CSMA-CA, but there are some differences namely the fact that there is no synchronization to the backoff period boundary. Therefore, the random backoff countdown begins immediately after the arrival of the data frame from the upper layers of the protocol stack and then the CCA is performed immediately after the expiration of the random backoff delay generated in the phase 2. Since there is no superframe in unslotted CSMA-CA, the device performs the CCA check, followed by the frame transmission and acknowledgement (if requested), as soon as the random backoff countdown is finished;
- **Phase 5** - Unlike the slotted CSMA-CA, the unslotted version starts immediately the transmission of the frame after it detects an idle channel.

B.3 MAC Protocols Taxonomy

B.3.1 Unscheduled MAC protocols

- **PicoRadio**
  PicoRadio [RAdS+00] protocol solves the access conflict by using multiple channels, by using a scheme based on dynamic channel assignment,

where each node is assigned with a unique channel. The Code Division Multiple Access (CDMA) scheme employed also uses a wake-up radio to wake-up neighbours nodes. Here the node listen to a Common Control Channel (CCC) and broadcasts a Channel Assignment Frame (ChAP) to the neighbours, in order to notify them about the channel which is going to be used. These ChAP exchanges between nodes will allow to update the channel assignment table of each node (records the channel usage in the network). During the channel setup period the nodes that wake-up will listen the channel for a certain time, in order to obtain information about the channels used by the one-hop and two-hop neighbours. After gathering this information the node chooses an unused channel and broadcasts its channel and its neighbours channel throughout the network. If two nodes choose the same channel, the first one to detect it chooses another unused channel. In PicoRadio, two channel assignment techniques can be used: sender based and receiver based. In sender based technique the sender transmits frames in its transmission channel; whilst in receiver based the receiver wakes-up to receive the frames. The wake-up radio used in this protocol is very simple which could be very sensitive, leading to the appearance of false orders to wake-up caused by noise, activating the data radio unnecessarily. Moreover, the wake-up radio is capable of carrying information, which means that the sender can send a wake-up tone addressed to a certain node.

- **Practical Multi-Channel MAC**
Practical Multi-Channel [LHA08] protocol utilizes multiple channels efficiently. The channel assignment is based on a control theory approach that dynamically allocates channels to nodes, and groups of nodes that are sharing the channel. This protocol assumes that the nodes do not require time synchronization. The channel assignment algorithm employed specifies that all nodes start on the same channel. Then, as nodes exchange status frames, the frame loss ratio and degrees of estimated communication success probability is measured, allowing the nodes to know the amount of interference experienced on their channel. These nodes then choose the next available channel and others ones from the same cluster will choose the same one. When a node needs to send a message to another on a different channel it changes momentarily to the destination's node channel. The employed mechanism keeps separated the routing and application layer, allowing minimizing the cost of cross-channel communication, while maximizing the traffic loads in channels. The status frames are exchanged periodically, leading to a

decrease of the energy efficiency. Moreover, as nodes detect channel interference/congestion they change to another available channel which deals with the narrowband long lasting interference.

- **Dynamic MAC (DSMAC)**

  DSMAC [LQW04] protocol adjusts dynamically the node's duty cycle, depending on the current utilization efficiency and average latency of the sensor. All nodes assume the same duty cycle at the beginning. A node will double the original duty cycle by decreasing the sleep time period length when it needs to decrease the latency or when the traffic exchanged is higher. When a node changes its duty cycle, it informs about the additional active schedule the other nodes by sending the SYNC frame that has the SYNC initiator's duty cycle. Each node maintains a synchronization table for its neighbouring nodes. Once neighbouring nodes receive the SYNC frame, they will decide if the duty cycle is increased in order to be same as the one sent by the SYNC frame sender node. The duty cycle can return to initial one, has the traffic conditions or latency becomes more stable.

- **Transmitted Initiated CyclEd Receiver (TICER)**

  TICER [LRW04] protocol utilizes the cycled receiver scheme, where nodes are powered on and off periodically and beacons are transmitted on the same and single data channel. For unicast transmissions, when the node has no data frame to transmit, it wakes-up to check the channel and goes back to sleep. When it has a data frame to transmit it wakes-up and checks the channel for certain duration. If it does not sense any ongoing transmissions, it starts transmitting RTS frames. When the receiver node wakes-up it detects the RTS frames and answers with CTS frames. The sender node receives the CTS frames, which in turn starts to transmit the data frame. The exchange is finished by an Acknowledgement (ACK) frame transmitted by the receiver node. This protocol will lead to an idle listening increase at the receiver node side.

- **Receiver Initiated CyclEd Receiver (RICER)**

  RICER [LRW04] protocol shares, the same principles of TICER, where a sensor node changes communication initiation from the transmitter side to the receiver side. When a sensor node has no data frame to send it wakes-up and transmits a beacon just to announce that it is awaken. After it sends the beacon, the node senses the channel for a response. If no response is detected, the node goes back to sleep. Otherwise, if a response is received, it transmits the data frame. In order to send the data frame the sender node stays awake and monitors the channel, while waiting for a wake-up beacon. After it receives the beacon the data frame is transmitted. The exchange also ends with an ACK frame. Higher energy savings are achieved for unicast and anycast communications, but for broadcast and multicast communications cannot be used because the communication must start in the receiver. It is basically a preamble sampling technique, but here the transmitter has an inverted role, which keeps receiving instead of sending a preamble. For low load of traffic this periodic beacon report can lead to an overhead.

- **SyncWUF**

  SyncWUF [SS07] protocol combines the simple signalling technique with an innovative Wake-Up Frame (WUF) technique. The simple signalling technique is just a signal, while the WUF contains the information. The main idea of SyncWUF is that the sender node stores the receivers' schedules, while adapting the Wake-Up Signal (WUS). If a receiver node schedule is out-of-date the WUS length is going to be long. Multiple Short Wake-Up Frames (SWUFs) are used by the WUS in order to reduce the extra waiting time of the WUF scheme. These SWUFs contain the destination MAC address and the current SWUF state in the whole WUF. Based on the information contained in the SWUFs the receiver nodes can decide when to turn on the transceiver in order to receive the data frames, reducing the waiting time. However, if the sender node does not exchange frames during the active period of the receiver node, it must wait until the next active period, leading to an increase of the transmission delay.

- **Energy-Efficient Reliable MAC (E2RMAC)**

  E2RMAC [JaBA07] follows some of RMAC principle, in order to improve the energy-efficiency by reducing the idle listening and over-hearing. It achieves this by using an additional wake-up radio. This wake-up radio is used every time the node wants to transmit a frame, where it sends a tone on its low power radio. By transmitting this wake-up tone the neighbours will know when they will sleep. After sending a frame, the node goes to sleep mode for a duration equal to the time needed by the receiver to forward the frame (similar to adaptive listening). E2RMAC employs immediate data forwarding to save on ACK transmissions (same as in RMAC).

- **Sparse Topology and Energy Management (STEM)**

  STEM [JBA07] tries to conserve energy by using two separate channels: i) wake-up channel and ii) data channel. The wake-up channel is used to allow a time organization between the transmitter and the receiver, whilst the data channel is only used to data exchange when the transmitter and the receiver encounters. Therefore, to guarantee the encounter between the transmitter and the receiver, nodes follow a preamble sampling approach. Two versions of STEM have been proposed: i) STEM Tone (STEM-T) and ii) STEM Beacon (STEM-B). STEM-T employs a tone on a separate channel to wake-up neighbouring nodes, while in STEM-B the traffic generating node sends a series of beacons, which carry the MAC address of the transmitter and of the receiver. In the STEM-B version the beacons are sent through the wake-up channel in order to the sleeping nodes turn on their radios to receive the messages. The wake-up channel is formed by synchronized time slots. Comparing both versions of STEM the STEM-T will guarantee the lower latency, but with requires more overhead than STEM-B since the receivers must be idle, listening all of the time for the tones. However, STEM-T may require a separate radio transceiver for the wake-up channel.

B.3.2 Scheduled MAC Protocols

- **Timeout-MAC (T-MAC) protocol**
  T-MAC [vDL03] tries to improve some flaws of the S-MAC. It tries to eliminate idle energy by adapting the length of the active period of the frame. Instead of transmitting a message during a predetermined period (as in S-MAC), messages are sent in burst at the beginning of the frame. Activation events, (such as firing of the frame timer or any radio activity) will wake-up the node to handle the type of activation event. In Figure B.7, during the first active period there is a sensor node involved in a message transmission and the second active period has a SYNC transmission. The time-out period is used when a node does not detect any activity within the time-out interval, it can assume with sure that no neigbhour wants to communicate with it and goes to sleep. If there is a node that wants to communicate with it a new time-out timer is initiated. To improve the message latency the Future Request To Send (FRTS) control message is proposed. The FRTS frame informs the next hop that it has a future message transfer. T-MAC considers the buffer size of the sensor nodes when calculating the contention period. If the buffer is full, the sensor nodes will have higher priority and control the channel, avoiding the buffer overflow.
- **Mobility Aware MAC (MS-MAC) protocol**
  The MS-MAC [PJ04] is similar to S-MAC to conserve energy when nodes are static. MS-MAC employs a new mechanism to handle

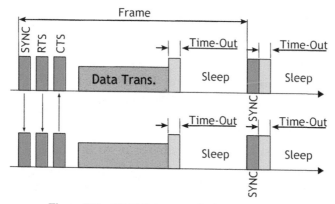

Figure B.7   T-MAC data transferring mechanism.

mobility. This mechanism is based on the RSSI as an indication of mobility and, when necessary, triggers the mobility handling mechanism. The mechanism relies on the RSSI of periodical SYNC messages from its neighbours. When the signal changes, the receiving node presumes that the neighbour or itself are moving and the level of change of the RSSI signal can predict the mobile node's speed. This protocol allows for the creation of cluster, formed by a mobile node and its neighbours (forms an active zone). The mobile nodes can cross from one cluster to other ones, but during this crossing the node runs the synchronization periods more times resulting in higher energy consumption with lower time needed to create new connections. This mobility aware mechanism of MS-MAC allows nodes to work efficiently in both stationary and mobile scenarios. Considering a stationary scenario or when nodes move only inside a single cluster, all nodes work efficiently.

- **EMACS**

  EMACS [DHNH03] is a TDMA based protocol, which divides the time slot into three sub periods, (communication request, Traffic control and Data sub periods. The traffic control sub period is used to transmit the periodic control information of the nodes. The sensor node must transmit this information within his time slot, while the neighbouring nodes listen for the control frame of the neighbours. All the data transmissions occur within the data sub period. The communication request sub period can be used by a node that wants to transmit in a time slot it does not own, by transmitting a request in this sub period. Three types of modes are possible in EMACS. The active nodes have full use of the sub periods, since they own a slot and transmit a control message within each slot they own. The passive nodes do not own a time slot and can only transmit a message after requesting a slot it does not own. The dormant nodes do not participate in the sensor network until it decides to become an active or a passive node. This allows conserving energy by activating only the number of nodes needed to perform the application functionality.

- **Lightweight MAC (LMAC)**
  LMAC [VHH04] is also a TDMA based protocol. Unlike traditional TDMA-based systems, the time slots in LMAC protocol are not divided among the networking nodes by a central entity. Instead a distributed algorithm is used. Concerning its time slot, the node will always transmit a message formed by two parts: control message and a data unit. The node is collision free, because a time slot can only be controlled by a single node. The control message carries the ID of the time slot controller, the distance of the node to the gateway in hops, contains the destination address and the length of the data unit. Moreover, the control data is used to maintain synchronization. If the control message is not addressed to some nodes, the nodes will switch off their transceivers and only wake-up at the next time slot. In this protocol, the node can only transmit a single message per frame.

- **Node Activation Multiple Access (NAMA)**
  NAMA [BGLA01] is a protocol where the access to the channel is controlled by assigning priorities to sensor nodes in each slot. It employs a TDMA scheme with time divided into blocks of Sb sections. Each Sb section has Ps parts and the parts contain Tp time slots, as presented in Figure B.8. Then a node selects a single part, to contend with other nodes that have chosen the same part. The last section of each block is reserved for signalling purposes. The priority is computed by each node with its neighbours and this is used to determine who will access to current time slot within the sensor node's chosen part. The sensor with the highest priority will be one that will transmit. Other protocols derived from NAMA, such as Link Activation Multiple Access (LAMA) and Pairwise-link Activation Multiple Access (PAMA). LAMA activates links to destinations sensor nodes based on the DSSS code assigned to the receiver and the priority assigned to the transmitter. PAMA activates links between sensor nodes by assigning priorities to the links, while varying the codes and priorities of links based on the current time slot. The aforementioned protocols all require a sensor node to compute the

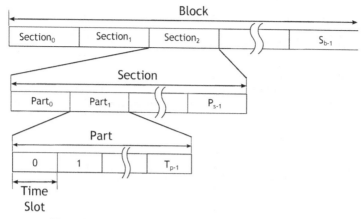

Figure B.8 Time division structure of NAMA protocol.

priorities of each neighbouring node for each time slot, leading to an increase of energy consumption.

- **Fast Path Algorithm (FPA)**
  FPA [LKR04] guarantees the relaying of frames by waking up the nodes for an additional time. A node uses its hop-distance from the sender to estimate when its neighbour can send a frame to it. When the node wakes-up at the estimated time only to receive and forward the frame to its downstream neighbour. These additional times are set from the information piggybacked in the first data frame sent.

- **Flexible MAC (FlexiMAC)**
  FlexiMAC [LDCO06] is able to handle the network dynamic and node mobility. A contention period is used in order to nodes exchange frames to build a data gathering tree rooted at the sink. The slot distribution is based on a Depth First Search (DFS) and the slot numbering starts with number 2 because slot number 1 is reserved for the network dynamics. The reserved slot has the name Fault Tolerant Slot (FTS). The slot assignment algorithm follows the tree. Two types of slots exist: the data gathering slots and the multifunctional ones. Data-gathering slots are used to uplink traffic from nodes to the sink, while multifunctional slots are used for the downlink traffic and synchronization. This data gathering tree has the disadvantages of the need for tree reconstruction when a link fails and only parent-child and child-parent communications are optimal. The period assigned to handle the network dynamics may lead to higher energy consumptions.

- **Low Energy Adaptive Clustering Hierarchy (LEACH)**
  LEACH [HHT02] tries to solve the energy dissipation in sensor networks. Nodes are selected randomly as cluster heads, in order to the high energy dissipation in communicating with the base station is divided to all sensor nodes. LEACH consists in two phases: the setup phase and the steady phase. To reduce the overhead the duration of the steady phase is longer in the duration of the setup phase. In the setup phase a sensor node chooses a random number between 0 and 1. If the chosen number is less than a certain threshold, the sensor node is selected as the cluster head. After that the cluster heads inform all the surrounding nodes that he is the cluster head. Once this advertisement is performed the other nodes choose the cluster that they want to belong and the cluster heads assign the time on which the sensor nodes can send data back to the cluster heads based on a TDMA scheme. In the steady phase, the nodes can begin sensing and transmitting data to cluster heads. Data aggregation is another feature of the cluster heads. This process of selecting cluster heads is repeated after a certain period of time. Some disadvantages are pointed, such as the need of a complex radio transceiver with DSSS and power scaling, which increases the energy consumption. Another disadvantage is the time it takes to redo the cluster. Finally, LEACH assumes that each node communicates directly with the cluster head, which can consume larger amounts of energy or restrict the sensor nodes to a certain geographical area.

B.3.3 Hybrid MAC Protocols

- **Wireless MAC (WiseMAC) protocol**
  WiseMAC [EHD04a] is a preamble sampling based protocol which employs TDMA/CSMA techniques and where all sensor nodes have two communication channels. The TDMA technique is used to access the data channel, while the CSMA technique is used to access the control channel. The use of preamble sampling jointly with a non-persistent CSMA decreases the idle listening. The preamble sampling technique involves the transmission of a preamble before each data frame, to inform the potential receiving node. The size of the preamble is initially set to be equal to the sampling period. However, the potential receiver may not be ready at the end of preamble to receive the data frame, causing energy waste. WiseMAC offers a method that dynamically determines the length of the preamble, with the intention

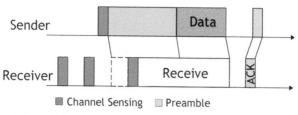

Figure B.9   Data frame transfer in WiseMAC protocol.

to reduce the power consumption due to a fixed length preamble. This method uses the knowledge of the sleep schedule of the transmitter node's direct neighbours. Based on neighbours' sleep schedule table, WiseMAC schedules transmissions so that the destination node's sampling time corresponds to the middle of the preamble sent. Figure B.9 describes the WiseMAC data frame transmission. Main disadvantage of WiseMAC is that decentralized sleep-listen scheduling leads to different sleep and wake-up times for each neighbour of a node. Moreover, the hidden terminal problem occurs with WiseMAC, because it is based on a non-persistent CSMA.

- **Power Efficient and Delay Aware Medium Access Control for Sensor Network (PEDAMACS)**
  PEDAMACS [EV06] gathers information concerning traffic and topology during the setup phase in the sink. Then, based on the collected information the sink calculates a global scheduling and sends it to the entire network. In PEDAMACS is assumed that the hardware of the sink is more powerful than the remaining nodes, in order to reach all nodes when it transmits. It also follows a TDMA based scheme for the uplink feature. For the topology phase, a CSMA based scheme is used to send information to the sink. In this phase the sink node sends topology-learning frames, in order to build the spanning tree and the sink has the knowledge of the entire topology. One of the disadvantages of PEDAMACS is that only supports convergecast traffic type. Moreover, the assumption that all nodes all reachable by the sink is not always satisfied. Nodes that did not receive the schedule information sent by the sink have to wait for the next topology learning phase and inform the sink by piggybacking this information in the reply frames to the sink.
- **Zebra MAC (ZMAC)**
  ZMAC [RWA+08] combines both CSMA and TDMA schemes in order to adapt to the level of contention in the network. If the network is

under low contention the ZMAC switches to CSMA scheme, and when is under high contention it behaves like TDMA based scheme. ZMAC employs a CSMA as baseline MAC scheme, but for contention resolution it employs a TDMA schedule where the time slot assignment is performed at the time of deployment (higher overhead at the beginning). ZMAC uses a scalable channel scheduling algorithm named as DRAND. After slot assignment, each node reuses periodically its assigned slot to transmit. In ZMAC the time slot has an owner and the others are called the non-owners of that slot. In ZMAC a time slot can have a maximum of two owners, but it has to be separated two hops at least. Before transmitting each node performs a carrier sense and transmits the frame if the channel is free. ZMAC reduces the collisions since the owners of time slots have higher priority than the non-owners to transmit. By mixing CSMA with TDMA, ZMAC becomes less susceptible to timing failures, slot assignment failures and topology changes than a standard TDMA scheme. When the TDMA and CSMA mixing fails, it always relies on the CSMA by default. A disadvantage is energy consumption when slot assignment and synchronization occur frequently in the network.

- **X-MAC**

  X-MAC [BYAH06] is an adaptive energy-efficient MAC layer protocol for duty cycled WSNs. Other duty cycle MAC protocols employ an extended preamble and preamble sampling. This long preamble introduces excess latency at each hop, while leading to suboptimal energy consumption and excessive energy consumption at non-target nodes. X-MAC proposes solutions to alleviate these problems by employing a shortened preamble approach that retains the advantages of low power listening. To solve the overhearing problem X-MAC embeds in the preamble address information of the target so that non-target receivers can quickly go back to sleep. To reduce the excessive latency and energy waste a strobed preamble is used by splitting the full-length preamble into frames with a gap between consecutive frames, giving an advantage of not always requiring the extended preamble. In Figure B.10 is presented a comparison between the LPL and X-MAC mechanisms. With this strobed preamble the listening nodes do not need to wait for the entire preamble. An adaptive algorithm for automatically adjusting the duty cycle of receivers to the offered traffic load is also employed in X-MAC. However, when light traffic loads are considered there is an increase of the idle listening at the receivers, leading to energy wastes.

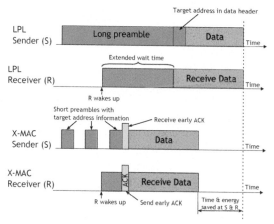

Figure B.10 Comparison of the timelines between LPL's and X-MAC's approaches.

- **Y-MAC**

  Y-MAC [KSC08] is a multi-channel MAC protocol for high deployment WSNs. This protocol employs a scheduled access, but with some modifications. The timeslots are assigned to receivers rather than to senders. Potential senders contend in the beginning of each timeslot for the same receiver. When there is the need to transmit multiple frames, the sender and receiver choose a new channel according to a pre-established sequence. The main disadvantage of Y-MAC is the increase of the contention time when there is a high number of nodes around the sink node, which increase the data rate.

B.3.4 QoS MAC protocols

- **Q-MAC**

  Q-MAC [LEQ05] is referred as a QoS-aware MAC protocol. A multi-hop scenario is assumed where nodes generate frames with different priorities. The main objectives of Q-MAC are to minimize the energy consumption, while providing QoS policies. To achieve these objectives Q-MAC is composed of intra-node and inter-node QoS scheduling mechanisms. The intra-node QoS scheduling mechanism classifies the sent frames according to their priorities, while the inter-node scheduling scheme controls the channel access while minimizing the energy consumption. After a frame is scheduled for transmission, the inter-node scheduling mechanism MACAW, is executed to obtain Loosely

Prioritized Random Access (LPRA) among sensor nodes. The contention time is randomly generated with a contention window size CW, where the size is defined by each node's transmission urgency, frame criticality, residual energy, number of hops and queue load. In case there is a collision the CW size is doubled and the frame is retransmitted. When this size reaches a maximum value the frame is dropped.

- **CoCo**
  CoCo [LSR04] protocol is a colouring-based real-time communication scheduling designed for multi-hop scenarios that use IEEE 802.11 MAC protocols and where all exchanged frames are of the unicast type. It is assumed that node locations are available at all time, and central node with CoCo running is responsible of communication scheduling. The main objective of CoCo is the schedule of real-time communication, while avoiding collisions and minimizing the overall frame transmission time. CoCo protocol is based on weighted directed graphs and can be compared to the assignment of a colour to each edge of the graph. Here, each colour represents a set of simultaneous communication during disjoint time periods. It employs a colouring selection heuristic algorithm to solve the colouring problem of the graph. CoCo tries to schedule a set of communication events, while it envisages the minimum communication time possible in real-time WSNs. One drawback of CoCo is the central node computational limitation that does not allows for the applicability of CoCo in large-scale WSNs deployments.

B.3.5 Multiple based MAC protocols

- **Reinforcement Learning based MAC (RL-MAC) protocol**
  RL-MAC [LE06] tries to increase while saving energy, by optimizing the active and sleep periods of the frame. Nodes actively infer the state of other nodes, using reinforcement learning based control mechanism. It employs a Markov Decision Process (MDP) to model the process of active time reservation. However, there is the disadvantage of relying on a constant traffic load over a long period of time.
- **Gateway MAC (G-MAC)**
  G-MAC [BMFD06] defines a node acting as a gateway for a certain time. After some time this role is rotated by other nodes, in order to balance the load among them. The G-MAC frame is composed by three sub periods: the collection period, the traffic indication period and the distribution period. In the collection period nodes contend for the channel to send

frames. In the traffic indication period all nodes wake-up and listen to the channel in order to receive the Gateway Traffic Indication Information (GTIM). This GTIM also maintain synchronization among nodes. G-MAC allows time critical frames to be sent during the collection period. However, this induces to have larger overhead in the gateway node even if the roles are rotating by the nodes this will induce a large overhead and consequently an energy consumption increase. Another disadvantage is the need of the nodes to communicate directly with the gateway, requiring more gateways in the network.

B.4 Comparison of the WSN MAC Protocols

Tables 10.3 to 10.14 present a comparison of the characteristics of the WSN MAC protocols.

Table B.3   Comparison of unscheduled MAC protocols.

MAC characteristics	Picoradio	PMC MAC	DSMAC	E2RMAC	BMAC	EA-ALPL	STEM	CSMA-MPS	TICER	WOR
Action against energy waste	-	-	-	Il & Oh	Il&Co	Il&Co	-	Co	Co	Co
Type of scheduling	-	-	-	-	-	-	-	-	-	-
Roles for the nodes	Fixed	Fixed	Fixed	Fixed	Fixed	Fixed	Fixed	Fixed	Rt	Fixed
Node mobility	Static	Static	Static	Static	Static	Static	Static	Static	Static	Static
Node traffic behaviour	Fixed	Fixed	Ad	Fixed	Fi	Ad	Fi	Fi	Fi	Fi
Type of Traffic	Sensor	Sensor	Sensor	Sensor	Sensor	Sensor	Sensor	Sensor	Sensor	Sensor
Node Traffic exchange		Uni	Uni	Uni	Uni	Uni	Uni	Uni	Uni	Uni
Traffic load	Me	Me	Hi&Me&Lo	Lo	Me	Hi&Me&Lo	Hi&Me& Lo	Hi	Hi	Hi
Slot assignment	-	-	-	On	-	-	-	-	-	-
Channel Assignment	Sd & Rc	Sd & Rc	-	-	-	-	-	-	-	-
Frequency allocation	Multi	Multi	-	-	-	-	Multi	-	-	-
Frame subinterval Time reduction	-	-	Sleep	-	-	-	-	-	-	-
Duty cycle tunning	-	-	App	-	-	Tf	Pre	-	-	-
Preamble Length Reduction	-	-	-	-	-	-	-	Pack & Pigg	Pack & Pigg	Pack & Pigg
Channel allocation	Dual	Multi	-	Dual	Single	Single	Dual	Single	Single	Single
Preamble reception filtering	-	None	-	-	-	-	-	-	-	-
Communication modes	-	-	-	-	-	-	Be	Be	Be	Be
Multiplex techniques	CDMA	-	-	CSMA	CSMA	CSMA	CSMA	CSMA	CSMA	CSMA
Wake-up techniques	Radio	-	Schd	Radio	Schd	Schd	Radio	Schd	Schd	Schd
Contention window	-	-	Sg & Fi	Sg & Fi	-	-	-	-	-	-
Clustering	-	Yes	No	No	No	No	No	No	No	No
Synchronization	No	No	Yes	Yes	Yes	Yes	Yes	Yes	No	Yes
Channel Technique	-	-	-	-	CCA	CCA	PrS	PrS	PrS	DtS

Table B.4   Unscheduled MAC protocols comparison (cont.).

MAC characteristics	SpeckMAC-D	MX-MAC	DPS-MAC	RICER	SyncWUF	PAMAS	CC-MAC	SIFT	Alert	Traw-MAC
Action against energy waste	Co	Co	Il&Co	Co	Co	Il&Co&Oh	Il&Co	Il&Co		
Type of scheduling	-	-	-	-	-	-	-	-	-	-
Roles for the nodes	Fixed	Fixed	Fixed	Rt	Fixed	Fixed	Fixed	Fixed	Fixed	Fixed
Node mobility	Static	Static	Static	Static	Adapt	Static	Static	Static	Static	Adapt
Node traffic behaviour	Fi	Fi	Fi	Fixed	Fixed	Fixed	Fi	Fi	Fi	Fixed
Type of Traffic	Sensor	Sensor	Sensor	Sensor	Sensor	Multi	Sensor	Sensor	Sensor	Sensor
Node Traffic exchange	Uni	Uni	Uni	Uni	Uni	Uni	Var & Uni	Uni	Uni	Uni
Traffic load	Hi	Hi	Hi	Hi	Hi&Me&Lo	Hi	Hi&Me&Lo	Me	Me	Hi&Me&Lo
Slot assignment	-	-	-	-	-	-	-	-	Rc&Ngb	On
Channel Assignment	-	-	-	-	-	Sd & Rc	-	-	Sd	-
Frequency allocation	-	-	-	-	-	-	-	-	Multi	-
Frame subinterval Time reduction	-	-	-	-	-	-	-	-	-	Sleep
Duty cycle tunning	-	-	-	-	Dt	-	-	-	-	Tf
Preamble Length Reduction	Pack & Pigg	Pack & Pigg	Samp	Pack & Pigg	Pigg	-	-	-	-	DpR
Channel allocation	Single	Single	Single	Single	Single	Dual	Single	Single	Multi	Single
Preamble reception filtering	-	-	Be	Be	-	-	-	-	-	Micro
Communication modes	Be	Be	Be	Be	-	-	-	-	-	-
Multiplex techniques	CSMA	CSMA	CSMA	-	CSMA	CSMA	CSMA	CSMA	Both	CSMA
Wake-up techniques	Schd	Schd	Schd	Schd	Schd & tone	Radio	Schd	Schd	Schd	Schd
Contention window	-	-	-	-	-	-	-	Sg & Fi	Sg & Fi	-

Table B.5   Unscheduled MAC protocols comparison (cont.).

MAC characteristics	SBRP	ARBP	RARBP
Action against energy waste	Co	Co	Co
Type of scheduling	-	-	-
Roles for the nodes	Fixed	Fixed	Fixed
Node mobility	Static	Static	Static
Node traffic behaviour	Fixed	Fixed	Fixed
Type of Traffic	Sensor	Sensor	Sensor
Node Traffic exchange	Uni	Uni	Uni
Traffic load	Hi&Me&Lo	Hi&Me&Lo	Hi&Me&Lo
Slot assignment	-	On	On
Channel Assignment	-	-	-
Frequency allocation	Single	Single	Single
Frame subinterval Time reduction	-	-	-
Duty cycle tunning	-	App	Range
Preamble Length Reduction	-	-	-
Channel allocation	Single	Single	Single
Preamble reception filtering	-	-	-
Communication modes	-	-	-
Multiplex techniques	CSMA	CSMA	CSMA
Wake-up techniques	Schd	Schd	Schd
Contention window	Sg & Fi	Sg & Fi	Sg & Fi
Clustering	-	-	-
Synchronization	Yes	Yes	Yes
Channel Technique	-	-	-

Table B.6 Scheduled MAC protocols comparison.

MAC characteristics	ARISHA	BitMAC	EMACs	PMAC1	MMAC	FlexiMAC	PMAC2	O-MAC	LMAC	AI-MAC
Action against energy waste	Co	Co	Co	Co	Co&Ho	Co	Co&Il	Ho	Co	Co
Type of scheduling	Cent	Cent	Local	Local	Local	Dist	Dist	Local	Dist	Dist
Roles for the nodes	Fixed	Fixed	Fixed	Adapt	Fixed	Fixed	Fixed	Fixed	Fixed	Fixed
Node mobility	Adapt	Static	Static	Static	Adapt	Static	Static	Static	Static	Static
Node traffic behaviour	Fixed	Fixed	Fixed	Fixed	Adapt	Fixed	Adapt	Fixed	Fixed	Fixed
Type of Traffic	Sensor	Sensor	Sensor	Sensor	Sensor	Sensor	Sensor	Sensor	Sensor	Sensor
Node Traffic exchange	Uni	Uni	Uni	Uni	Var	Uni	Uni	Uni	Uni	Uni
Traffic load	Me	Me	Lo	Lo	Me	Me	Hi&Me&Lo	Me	Hi&Me&Lo	Hi&Me&Lo
Slot assignment	Rc	Sd & Rc	Dist	Dist	Mob	Rc&Sd	Rc&Sd	Rc	Rc&Sd	Rc&Sd
Channel Assignment	-	-	-	-	-	-	-	-	-	-
Frequency allocation	-	-	-	-	-	-	-	-	-	-
Frame subinterval Time reduction	-	-	-	-	-	-	-	-	-	-
Duty cycle tunning	-	Tf	Pre	Pre	Mob	-	Tf	-	-	-
Preamble Length Reduction	-	-	-	-	-	-	-	-	-	-
Channel allocation	Single	Multi	Single	Single	Single	Single	Single	Single	Single	Single
Preamble reception filtering	-	-	-	-	-	-	-	-	-	-
Communication modes	-	Be	-	-	-	-	-	-	-	-
Multiplex techniques	TDMA	TDMA	TDMA	TDMA	TDMA	TDMA	TDMA	TDMA	TDMA	TDMA
Wake-up techniques	Schd	Schd	Schd	Schd	Schd	Schd	Schd	Schd	Schd	Schd
Contention window	Sg & Fi	-	-	-	Sg & Fi	Sg & Fi	Sg & Fi	Sg & Fi	Sg & Fi	Sg & Fi
Clustering	Yes	Yes	No	No	Yes	No	No	No	No	No
Synchronization	Yes	Yes	Yes	Yes	Yes	Yes	Yes	Yes	Yes	Yes
Channel Technique	-	PrS	-	-	-	-	-	PrS	PrS	PrS

Table B.7 Scheduled MAC protocols comparison (cont.).

MAC characteristics	SS-TDMA	SMAC	TMAC	E2MAC	SWMAC	nanoMAC	FPA	DMAC	QMAC	MS-MAC
Action against energy waste	Co	Co&Il&Ho&Ov	Il		Co	Co&Il		Oh&Il	Co&Il	
Type of scheduling	Dist	Dist	Dist	Dist	Dist	Local		Stag	Dist	Dist
Roles for the nodes	Fixed	Fixed	Fixed	Fixed	Fixed	Fixed	Fixed	Fixed	Fixed	Fixed
Node mobility	Static	Static	Static	Static	Static	Static	Static	Static	Static	Adapt
Node traffic behaviour	Adapt	Fixed	Fixed	Fixed	Fixed	Fixed	Fixed	Adapt	Adapt	Fixed
Type of Traffic	Sensor	Sensor	Sensor	Sensor	Sensor	Sensor	Sensor	Conv	Sensor	Sensor
Node Traffic exchange	Uni	Uni	Uni	Uni	Uni	Uni	Uni	Var	Uni	Uni
Traffic load	Hi&Me&Lo	Hi&Me&Lo	Hi&Me&Lo	Hi&Me&Lo	Hi	Hi&Me&Lo	Hi&Me&Lo	Hi&Me&Lo	Hi&Me&Lo	Hi&Me&Lo
Slot assignment	Rc&Sd	Rc&Sd	Rc&Sd	Rc&Sd	Dist	-	-	Rc&Sd	Rc&Sd	-
Channel Assignment	-	-	-	-	-	-	-	-	-	-
Frequency allocation	-	-	-	-	-	-	-	-	-	-
Frame subinterval Time reduction	-	-	Sleep & Act	Sleep & Act	Act	Sleep	Sleep	-	Act& Sleep	Act
Duty cycle tunning	-	Pre	Tf	Tf	-	-	-	Tf	Tf&Dt	Mob
Preamble Length Reduction	-	-	-	Pack	-	-	-	-	-	-
Channel allocation	Single	Single	Single	Single	Single	Single	Single	Single	Single	Single
Preamble reception filtering	-	-	-	-		-	Pigg	-	-	-
Communication modes	-	-	-	-		-	-	-	-	-
Multiplex techniques	TDMA	CSMA	CSMA	CSMA	CSMA	CSMA	CSMA	CSMA	CSMA	CSMA
Wake-up techniques	Schd	Schd	Schd	Schd	Schd	Schd	Schd	Schd	Schd	Schd
Contention window	Sg & Fi	Sg & Fi	Sg & Fi	Sg & Fi	Sg & Fi	-	-	Sg & Fi	Sg & Fi	-
Clustering	No	Yes	Yes	Yes	No	No	No	Yes	Yes	Yes
Synchronization	Yes	Yes	Yes	Yes	Yes	Yes	Yes	Yes	Yes	Yes
Channel Technique	PrS	Cs	Cs	Cs	Cs	Cs		DaP	-	-

Table B.8   Scheduled MAC protocols comparison (cont.).

MAC characteristics	GSA	Opti. MAC	ADCA	LEACH	NAMA	LAMA	PAMA	En. TDMA	Preamble sampling ALOHA
Action against energy waste	Co&Il&Ho&Ov	Il&Ho&Ov	Co&Ov	Co	Co	Co	Co	Co	Co
Type of scheduling	Glob	Dist	Stag	Cent	Dist	Dist	Dist	Cent	Dist
Roles for the nodes	Fixed	Fixed	Fixed	Fixed	Fixed	Fixed	Fixed	Fixed	Fixed
Node mobility	Static	Static	Static	Static	Static	Static	Static	Adapt	Static
Node traffic behaviour	Fixed	Adapt	Adapt	Fixed	Fixed	Fixed	Fixed	Adapt	Fixed
Type of Traffic	Sensor	Sensor	Sensor	Sensor	Sensor	Sensor	Sensor	Multi	Sensor
Node Traffic exchange	Uni	Uni	Uni	Uni	Var	Uni	Uni	Var	Cte
Traffic load	Hi&Me&Lo	Hi&Lo	-	-	Hi&Me&Lo	Hi&Me&Lo	Hi&Me&Lo	Hi&Me&Lo	Hi&Lo
Slot assignment	Rc&Sd	Rc&Sd	Rc&Sd	-	Sd	Sd	Sd	Rc	Sd & Rc
Channel Assignment	-	-	-	-	-	-	-	Rc	-
Frequency allocation	-	-	-	-	-	-	-	-	-
Frame subinterval									
Time reduction	Sched	Sched	-	-	-	-	-	-	-
Duty cycle tunning	Pre	Tf	Dt&Tf	-	-	-	-	Tf	-
Preamble Length Reduction	-	-	-	-	-	-	-	-	-
Channel allocation	Single	Single	Single	Single	Single	Single	Single	Multi	Dual
Preamble reception filtering	-	-	-	-	-	-	-	-	-
Communication modes	-	-	-	-	-	-	-	-	-
Multiplex techniques	CSMA	CSMA	CSMA	TDMA	TDMA	TDMA	TDMA	TDMA	CSMA/TDMA
Wake-up techniques	Schd	Schd	Schd	Schd	-	-	-	Schd	Schd
Contention window	Sg & Fi	Sg & Fi	Sg & Fi	-	Sg & Fi	Sg & Fi	Sg & Fi	-	Sg & Fi
Clustering	Yes	No	No	Yes	No	No	No	No	No
Synchronization	Yes	Yes	No	Yes	Yes	Yes	Yes	Yes	Yes
Channel Technique	Cs	-	-	-	NoB	LiB	Lib	-	PrS

Table B.9 Hybrid MAC protocols comparison.

MAC characteristics	PEDAMACS	TRAMA	FLAMA	μMAC	PACT	BMA	f-MAC	MMSN	Y-MAC
Action against energy waste		Co	Co	Co&Ov			Co	Co	
Type of scheduling	Cent	Local	Local	Local	Local	Local	-	-	Cent
Roles for the nodes	Fixed	Fixed	Fixed	Fixed	Rt	Rt	Fixed	Fixed	Fixed
Node mobility	Adapt	Static	Static	Static	Static	Static	Static	Static	Static
Node traffic behaviour	Fixed	Adapt	Adapt	Adapt	Adapt	Adapt	Fixed	Fixed	Adapt
Type of Traffic	Conv	Both	Both	Sensor	Sensor	Sensor	Sensor	Sensor	Sensor
Node Traffic exchange	Var	Var	Var	Var	Var	Var	Var	Uni	Var
Traffic load	Hi&Me&Lo	Hi&Me&Lo	Hi&Me&Lo	Hi&Lo	Hi&Lo	Hi&Lo	Hi&Lo	Hi&Me&Lo	Hi&Me&Lo
Slot assignment	-	-	-	Rc&Nghb	NoB	NoB	-	Rc	Rc
Channel Assignment	-	-	-	-	-	-	-	-	Rc
Frequency allocation	-	-	-	-	-	-	Multi	Multi	Multi
Frame subinterval Time reduction	-	-	-	-	-	-	-	-	-
Duty cycle tunning	-	Tf	Tf	Tf&App	Tf	Tf	Pre	-	-
Preamble Length Reduction	Pigg	-	-	-	-	-	-	-	-
Channel allocation	Single	Single	Single	Single	Single	Single	Multi	Multi	Multi
Preamble reception filtering	-	-	-	-	-	-	-	-	-
Communication modes	-	-	-	Be	-	-	-	Be	-
Multiplex techniques	CSMA/TDMA	CSMA/TDMA	CSMA/TDMA	CSMA/TDMA	TDMA	TDMA	-	CSMA/FDMA	TDMA/FDMA
Wake-up techniques	Schd	Schd	Schd	Schd	Schd	Schd	Schd	Schd	Schd
Contention window	-	-	-	Sg & Fi	-	Sg & Fi	-	Sg & Fi	Sg & Fi
Clustering	No	No	No	No	Yes	Yes	Yes	No	No
Synchronization	Yes	Yes	Yes	Yes	Yes	Yes	No	Yes	Yes
Channel Technique	-	-	-	-	-	-	FrD	-	LPL&PrS

Table B.10 Hybrid MAC protocols comparison (cont.).

MAC characteristics	RMAC	X-MAC	CMAC	WiseMAC	RATE EST	SP	MFP	ZMAC	MH-MAC
Action against energy waste	Co	Ov&Oh	Oh	Il&Oh	Il&Oh	Il&Oh	Il	Co	
Type of scheduling	-	Local	Dist	Dist	Dist	Dist	Local	Cent	Stag
Roles for the nodes	Fixed	Fixed	Fixed	Fixed	Fixed	Fixed	Fixed	Fixed	Fixed
Node mobility	Static	Static	Static	Static	Static	Static	Static	Static	Static
Node traffic behaviour	Adapt	Adapt	Fixed	Adapt	Adapt	Adapt	Adapt	Adapt	Adapt
Type of Traffic	Sensor	Sensor	Sensor	Sensor	Sensor	Sensor	Sensor	Sensor	Sensor
Node Traffic exchange	Uni	Var	Uni	Var	Var	Var	Var	Var	Var
Traffic load	Hi&Me&Lo	Hi&Me&Lo	Hi&Me&Lo	Hi&Lo	Hi&Lo	Hi&Lo	Hi&Lo	Hi&Lo	Hi&Me&Lo
Slot assignment	Rc&Nghb	Rc&Nghb	Rc&Nghb	-	-	-	-	Rc&Nghb	-
Channel Assignment	-	-	-	-	-	-	-	-	-
Frequency allocation	-	-	-	-	-	-	-	-	-
Frame subinterval Time reduction	-	Sleep&Act	Sched	Act&Sleep	-	-	Act&Sleep	-	-
Duty cycle tunning	-	Tf	Pre	-	-	-	-	Tf	Pre
Preamble Length Reduction	Pigg	Pack & Pigg	-	-	DiS	DiS	-	-	Pack
Channel allocation	Single	Single	Single	Single	Dual	Single	Single	Single	Single
Preamble reception filtering	-	-	-	-	-	-	Micro	-	-
Communication modes	-	-	-	-	-	-	-	-	Be
Multiplex techniques	TDMA	CSMA	CSMA	TDMA/CSMA	TDMA/CSMA	TDMA/CSMA	TDMA/CSMA	TDMA & CSMA	CSMA
Wake-up techniques	-	Schd	Schd	Schd	Radio	Schd	Schd	Schd	Schd
Contention window	Sg & Var	-	Sg & Var	-	-	-	-	Sg & Var	Sg & Var
Clustering	No	No	No	No	No	No	No	Yes	No
Synchronization	Yes	No	Yes	Yes	Yes	Yes	No	Yes	Yes/No
Channel Technique	PrS	PrS	CA	PrS	PrS	PrS	PrS&CCA	PrS	PrS

Table B.11 Hybrid MAC protocols comparison (cont.).

MAC characteristics	IEEE 802.15.4	Funneling MAC	SCP-MAC	CrankShaft	PARMAC	HMAC	GANGS	HyMAC	TMCP
Action against energy waste	Co	Co	Co&Oh&Il	Oh	Il&Co	Il&Co		Co	Co
Type of scheduling	Cent	Cent	Glob&Local	Local	Dist	Local	Dist	Cent	Co
Roles for the nodes	Fixed	Fixed	Fixed	Fixed	Fixed	Fixed	Fixed	Fixed	-
Node mobility	Adp & Sta	Static	Static	Static	Static	Static	Static	Static	Fixed
Node traffic behaviour	Adapt	Adapt	Adapt	Adapt	Fi	Fi	Fi	Fi	Static
Type of Traffic	Sensor &Multi	Conv	Sensor	Sensor	Sensor	Sensor	Sensor	Sensor	Fi
Node Traffic exchange	Uni& Var	Var	Uni& Var	Uni& Var	Uni	Uni	Uni	Uni	Sensor
Traffic load	Lo& Hi	Hi&Me&Lo	Hi&Me&Lo	Hi&Me&Lo	Me	Hi&Lo	Hi&Lo	Hi&Lo	Uni
Slot assignment	On	On	On	Rc	Rc	On	Dist	Rc	Hi&Lo
Channel Assignment	standard	-	-	-	-	-	-	Multi	Rc
Frequency allocation	standard	-	-	-	-	-	-	Multi	Multi
Frame subinterval			Sleep&Sched	Sleep&Sched	-	-	-	-	Multi
Time reduction		Tf	Tf	Tf	-	-	-	-	
Duty cycle tunning			Tf	Tf	-	-	-	-	
Preamble Length Reduction			Pigg	-	-	-	-	-	
Channel allocation		Single	Single	Single	Single	Single	Single	Multi	Multi
Preamble reception filtering			-	-	-	-	-	-	
Communication modes	Both	Be	-	-	-	-	-	-	
Multiplex techniques	TDMA/CSMA	TDMA/CSMA	CSMA	TDMA	TDMA	CSMA&TDMA	CSMA&TDMA	TDMA/FDMA	TDMA/FDMA
Wake-up techniques	Schd	Schd	Schd	Schd	Schd	Schd	Schd	Schd	-
Contention window	Sg & Fi	Sg & Fi	Db & Fi	Sg & Fi	Sg & Fi	Sg & Fi	Sg & Fi	Sg & Fi	-
Clustering	Yes	No	No	Yes	Yes	No	Yes	No	No
Synchronization	Yes	Yes	Yes	Yes	Yes	Yes	Yes*	Yes	Yes
Channel Technique	PrS	ScP&LPL	ScP&LPL	ScP	-	-	-	LPL	CD &CA

Table B.12 QoS MAC protocols comparison.

MAC characteristics	Q-MAC	EQ-MAC	CoCo	MMSPEED
Action against energy waste			Co&Ov	
Type of scheduling	NoB	Dist	Cent	-
Roles for the nodes	Fixed	Fixed	Fixed	Fixed
Node mobility	Static	Static	Static	
Node traffic behaviour	Fixed	Fixed	Fixed	
Type of Traffic	Prio Sensor	Prio Sensor	Prio Sensor	Prio Sensor
Node Traffic exchange	Uni	Uni	Uni	Var
Traffic load	Hi&Me&Lo	Hi&Me&Lo	Hi&Me&Lo	Hi&Me&Lo
Slot assignment	Sd	Sd	-	Sd
Channel Assignment	-	-	-	-
Frequency allocation	-	-	-	-
Frame subinterval Time reduction	-	-	-	-
Duty cycle tunning	Pre	-	-	
Preamble Length Reduction	-	-	-	-
Channel allocation	Single	Single	Single	Single
Preamble reception filtering	-	-	-	-
Communication modes	-	-	-	-
Multiplex techniques	CSMA	TDMA/CSMA	802.11	CSMA
Wake-up techniques	Schd	Schd/-	Schd	Schd
Contention window	Sg & Fi	Sg & Fi	-	Sg & Fi
Clustering	No	No	Yes	No
Synchronization	Yes	Yes	No	Yes
Channel Technique	QoS	QoS	QoS	QoS

Table B.13 Cross-Layer MAC protocols comparison.

MAC characteristics	Chang, A	MACRO	MAC-CROSS	MINA	Yuan, J	Seada, K	Chiang, M	Madan, R	Cui, S
Action against energy waste			Co&Il&Ho&Ov						
Type of scheduling	-	-	Dist	Cent	Dist			Cent	
Roles for the nodes	Fixed	Fixed	Fixed	Fixed	Fixed	Fixed	Fixed	Fixed	Fixed
Node mobility	Static	Static	Static	Adapt	Static	Static	Adapt	Static	Static
Node traffic behaviour	Fixed	Fixed	Fixed	Fixed	Fixed	Fixed	Fixed	Fixed	Fixed
Type of Traffic	Sensor	Sensor	Sensor	Sensor	Sensor	Sensor	Sensor	Sensor	Sensor
Node Traffic exchange	Var	Uni	Uni	Uni	Var	Uni	Var	Uni	Uni
Traffic load	Hi&Me&Lo	Hi&Me&Lo	Hi&Me&Lo	Hi&Me&Lo	Hi&Me&Lo	Lo	Hi&Me&Lo	Hi&Me&Lo	Hi&Me&Lo
Slot assignment	-	-	Rc&Sd	Rc&Sd	-	-	Rc&Sd	Rc&Sd	Rc&Sd
Channel Assignment	-	-	-	Rc&Sd	-	-	-	-	-
Frequency allocation	-	-	-	Multi	-	-	-	Single	-
Frame subinterval Time reduction	-	Sched&sleep	-	-	-	-	-	-	-
Duty cycle tunning	-	-	Pre	-	-	-	-	-	-
Preamble Length Reduction	-	-	-	-	-	-	-	-	-
Channel allocation	Single	Single	Single	Multi	Multi	Single	Multi	Single	Single
Preamble reception filtering	-	-	-	-	-	-	-	-	-
Communication modes	-	-	-	Be	-	Be	-	-	-
Multiplex techniques	CSMA	CSMA	CSMA	TDMA + CDMA/FDMA	CDMA/OFDM		CDMA	TDMA	TDMA
Wake-up techniques	Schd	Schd	Schd	Schd	-	-	-	Schd	Schd
Contention window	-	-	Sg & Fi		-	-	-	Sg & Fi	Sg & Fi
Clustering	No	No	Yes	Yes	No	No	No	No	Yes
Synchronization	Yes	No	Yes	Yes	No	No	Yes/No	No	No

Table B.14 Cross-Layer MAC protocols comparison (cont.).

MAC characteristics	TSMP	Multi-LMAC	RL-MAC	U-MAC	G-MAC	R-MAC2	1-hop MAC	MERLIN	GeRAF
Action against energy waste	Nal	Co	Co&Il&Ho&Ov	Il	Il&Oh&Co		Il	Il	Oh
Type of scheduling	Cent	Dist	Dist	Dist	Cent	Dist	Local	Cent	Dist
Roles for the nodes	Fixed	Fixed	Fixed	Fixed	Rt	Fixed	Fixed	Fixed	Fixed
Node mobility	Static	Static	Static	Static	Static	Static	Static	Static	Adapt
Node traffic behaviour	Fixed	Adapt	Adapt	Adapt	Adapt	Fixed	Fixed	Fixed	Fixed
Type of Traffic	Sensor	Sensor	Sensor	Sensor	Prio Sensor	Sensor	Sensor	Sensor	Sensor
Node Traffic exchange	Var	Uni	Uni	Var	Uni	Uni	Uni	Uni	Uni
Traffic load	Hi&Lo	Hi&Me&Lo	Hi&Me&Lo	Hi&Me&Lo	Hi&Me&Lo	Hi&Me&Lo	Hi&Lo	Hi&Lo	Hi&Me&Lo
Slot assignment	On	Rc&Sd	Sd	-	On	Rc&Sd	-	Rc&Sd	Rc&Nghb
Channel Assignment	Rc	On	-	-	-	-	-	-	-
Frequency allocation	Multi	Multi	-	-	-	-	-	-	-
Frame subinterval Time reduction	Act	-	Act&Sleep	Act&Sleep	Sleep	-	Act&Sleep	-	Sched&sleep
Duty cycle tunning	App	Tf	Stat	Util	Pre	Dt	-	Pre	Pre
Preamble Length Reduction	-	-	-	-	-	Pack	-	-	-
Channel allocation	Multi	Multi	Single	Single	Single	Single	Single	Single	Single
Preamble reception filtering	-	-	-	-	-	-	-	-	-
Communication modes	Be	-	-	-	-	-	-	-	-
Multiplex techniques	TDMA/FDMA	TDMA/FDMA	CSMA	CSMA	TDMA	CSMA	TDMA/CSMA	CSMA	CSMA
Wake-up techniques	Schd	Schd	Schd	Schd	Schd	Schd	Schd	Schd	Schd
Contention window	Sg & Var	Sg & Fi	Sg & Fi	Sg & Fi	Sg & Fi	Sg & Fi	-	Sg & Fi	Sg & Var
Clustering	No	No	Yes	No	Yes	No	No	Yes	No
Synchronization	Yes	Yes	Yes	Yes	Yes	Yes	Yes	Yes	Yes

C

O-QPSK modulation for the IEEE 802.15.4 PHY layer at 2.4 GHz

This Appendix presents the offset O-QPSK modulation applied in the context of the IEEE 802.15.4 PHY layer. The signal spreading and despreading is also addressed, allowing to improve the SNR and reduce the effects of interferers.

C.1 QPSK Modulation

The Quadrature phase-shift keying (QPSK) is a digital modulation used to transmit two different messages over the same frequency band. From a bit stream two bits are taken at a time and mapped into signals as shown in Table C.1. Since each bit occupies one bit interval (T_b), the signals corresponding to the "digits" (or symbols) 00, 01, 11, 10 last a symbol duration of $T_s = 2T_b$. The QPSK modulation exploits the fact that $V cos(2\pi f_c t)$ and $V sin(2\pi f_c t)$ are orthogonal over the interval $(0, T_b)$ when $f_c = k/T_b$, where k is an integer. Figure C.1 presents the block diagram for the QPSK modulator.

The bit stream is first converted to an Non-Return-to-Zero Level (NRZ-L) waveform $i(t)$ with ± 1 levels. The waveform $i(t)$ is then demultiplexed into even, F_i, and odd, $Q(i)$, bit streams (waveforms) where and are mnemonics for the in-phase and quadrature, respectively. The individual bits in each stream modulates the in-phase carrier, $V cos(2\pi f_c t)$, and quadrature carrier,

Table C.1   QPSK signal mapping.

Bit pattern	Message	Signal transmitted	
00	m_1	$s_1(t) = V cos(2\pi f_c t)$	$0 \leqslant t \leqslant T_s = 2T_b$
01	m_2	$s_2(t) = V sin(2\pi f_c t)$	$0 \leqslant t \leqslant T_s = 2T_b$
10	m_3	$s_2(t) = -V sin(2\pi f_c t)$	$0 \leqslant t \leqslant T_s = 2T_b$
11	m_4	$s_2(t) = -V cos(2\pi f_c t)$	$0 \leqslant t \leqslant T_s = 2T_b$

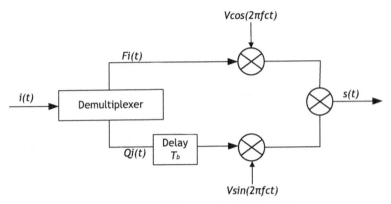

Figure C.1   Block diagram of a QPSK modulator.

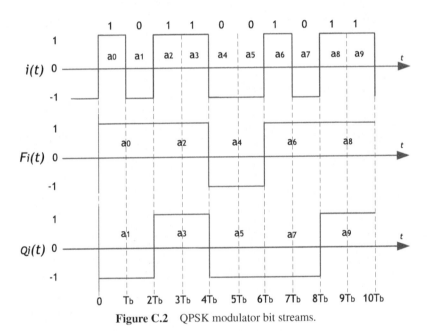

Figure C.2   QPSK modulator bit streams.

$V \sin(2\pi f_c t)$, respectively. The bit sequences for the different branches of a QPSK modulator are presented in Figure C.2.

The transmitted signal is given by:

$$s(t) = F_i(t)V\cos(2\pi f_c t) + Q_i(t)V\sin(2\pi f_c t) \qquad \text{(C.1)}$$

Rewritten the expression, the transmitted signal is given by:

$$s(t) = \sqrt{F_i^2(t) + Q_i^2(t)} V \cos\left(2\pi f_c t - \tan^{-1}\left(\frac{Q_i(t)}{F_i(t)}\right)\right) \quad \text{(C.2)}$$

$$= \sqrt{2} V \cos[2\pi f_c t - \Theta(t)]$$

The phase is $\Theta(t)$ is determined as follows:

$$\Theta(t) = \begin{cases} -3\Pi/4 & if \ F_i = -1, \ Q_i = -1 \ (bits \ are \ 00) \\ 3\Pi/4 & if \ F_i = -1, \ Q_i = +1 \ (bits \ are \ 01) \\ -\Pi/4 & if \ F_i = +1, \ Q_i = -1 \ (bits \ are \ 10) \\ \Pi/4 & if \ F_i = +1, \ Q_i = +1 \ (bits \ are \ 11) \end{cases} \quad \text{(C.3)}$$

Figures C.3 and C.4 presents the in-phase and quadrature components by considering the bit sequence presented in Figure C.2.

Figure C.5 presents the QPSK transmitted waveform.

The idea of using a QPSK signal is to have an envelope that is ideally constant. However, this modulation can be used by different communication

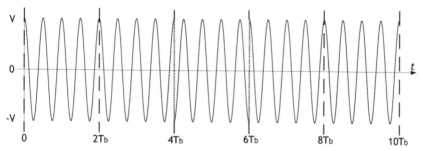

Figure C.3 In-phase component of the QPSK modulator.

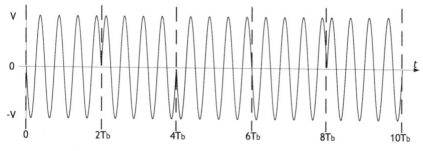

Figure C.4 Quadrature component of the QPSK modulator.

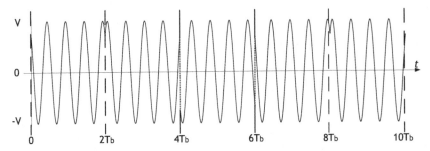

Figure C.5   QPSK waveform.

standards so as the signals must be band-limited by a band-pass filter, in order to fulfil the channel bandwidth requirements. Hence, the filter function will degrade the envelope properties. Moreover, every time a phase shift of Π occurs, the envelope will pass through zero momentary. Therefore, before transmitting, the QPSK signal is typically amplified by using a nonlinear power amplifier, the filter side lobes of the signal spectrum are also recreated. In order to avoiding such situations the of use linear amplifiers is advised. However, as explained in [NS09], these amplifiers have less efficiency when compared with the nonlinear amplifiers. The solution to deal with the Π phase shift problem is to consider an O-QPSK signal. This way, the possible changes are 0 or $\pm\Pi/2$ whilst the phase changes occur more frequently.

C.2 O-QPSK Modulation

The difference between O-QPSK and QPSK consists in offsetting the F_i and Q_i bit streams by T_b. For the sake of simplicity, we assume that $T_b = T_c$, where T_c is the inverse of the chip rate [WPA07a]. Therefore, the phase changes occur more frequently, namely every T_b, and not every $T_s = 2T_b$ like in the QPSK modulation. Figure C.6 presents the block diagram for an O-QPSK modulator. The bit sequences for the different branches of an O-QPSK modulator are presented in Figure C.7.

The O-QPSK modulated signal is given by:

$$s(t) = F_i(t)V cos(2\pi f_c t) + Q_i(t)V sin(2\pi f_c t) \qquad (C.4)$$

The O-QPSK odd and even chips (i.e., 32-streams) are generated in the same way as they would be generated for QPSK, with the exception that for the O-QPSK modulation the $F_i(t)$ and $Q_i(t)$ bit streams are offseted by T_b. By using the offset, the phase changes that occur in the combined

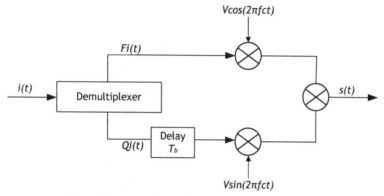

Figure C.6 Block diagram of an O-QPSK modulator.

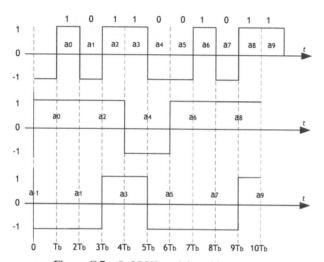

Figure C.7 O-QPSK modulator bit streams.

signal never exceed $\pm\Pi/2$. As shown in Figure C.7, the O-QPSK signal cannot undergo a change of Π like in QPSK since the F_i bit stream has a transition in the middle of the Q_i bit stream. Hence, we conclude that the O-QPSK modulation limits the maximum abrupt phase shift in $s(t)$ whilst having better performance when compared with QPSK by avoiding having $F_i(t)$ and $Q_i(t)$ simultaneously crossing the 0 value, which could cause large amplitude variations in the envelope. Since in linear amplifiers zero crossing needs high power consumption, the use of O-QPSK also decreases the energy spent by the radio transceiver.

Figures C.8 and C.9 present, the in-phase and quadrature components for the bit sequence presented in Figure C.7. Figure C.10 shows the O-QPSK transmitted waveform itself.

The IEEE 802.15.4 radio transceivers operating in the 2.4 GHz band may employ a DSSS spreading technique. As a consequence the transmitter signal takes-up more bandwidth than the required to transmit the information signal being modulated. The name "spread spectrum" derives from the fact that

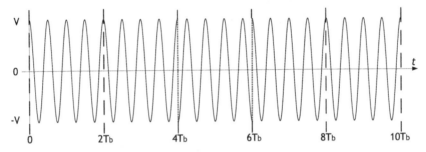

Figure C.8   In-phase component of the O-QPSK modulator.

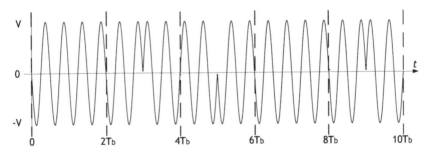

Figure C.9   Quadrature component of the O-QPSK modulator.

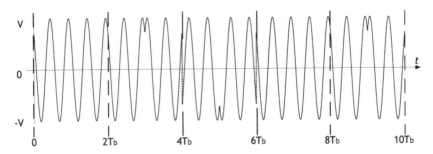

Figure C.10   O-QPSK waveform.

Figure C.11 Block diagram for the spreading and modulation functions.

the carrier signals occur over the full bandwidth (spectrum) of the sensor nodes transmitting frequency. The DSSS technique employs an O-QPSK modulation offering data rates up to 250 kb/s. The modulation and spreading technique uses a 16-ary quasi-orthogonal modulation, where four information bits are used to select one of 16 nearly orthogonal Pseudo-random Noise (PN) sequences to be transmitted.

Next, the PN sequences for the successive data symbols are concatenated and the aggregated chips are modulated onto the carrier by using the O-QPSK modulator with half-sine pulse shaping, as shown in Figure C.11.

Table C.2 presents the mapping between the data symbols and the 32-chip PN sequences. The data symbol bits are read from left to the right instead of right to left in this table. The bit-to-symbol module is responsible for mapping each octet (i.e., a group of eight bits) of the binary data by using the 4 least significant bits (LSBs) (b0, b1, b3, b4) to map one data symbol, and the 4 most significant bits (MSBs) (b4, b5, b6, b7) to map the next data symbol. Each

Table C.2   Symbol-to-chip mapping in the 2.4 GHz band.

Data symbol (decimal)	Data symbol (binary) (b0 b1 b2 b3)	Chip values (c0 c1 ... c30 c31)
0	0 0 0 0	1 1 0 1 1 0 0 1 1 1 0 0 0 0 1 1 0 1 0 1 0 0 1 0 0 0 1 0 1 1 1 0
1	1 0 0 0	1 1 1 0 1 1 0 1 1 0 0 1 1 1 0 0 0 0 1 1 0 1 0 1 0 0 1 0 0 0 1 0
2	0 1 0 0	0 0 1 0 1 1 1 0 1 1 0 1 1 0 0 1 1 1 0 0 0 0 1 1 0 1 0 1 0 0 1 0
3	1 1 0 0	0 0 1 0 0 0 1 0 1 1 1 0 1 1 0 1 1 0 0 1 1 1 0 0 0 0 1 1 0 1 0 1
4	0 0 1 0	0 1 0 1 0 0 1 0 0 0 1 0 1 1 1 0 1 1 0 1 1 0 0 1 1 1 0 0 0 0 1 1
5	1 0 1 0	0 0 1 1 0 1 0 1 0 0 1 0 0 0 1 0 1 1 1 0 1 1 0 1 1 0 0 1 1 1 0 0
6	0 1 1 0	1 1 0 0 0 0 1 1 0 1 0 1 0 0 1 0 0 0 1 0 1 1 1 0 1 1 0 1 1 0 0 1
7	1 1 1 0	1 0 0 1 1 1 0 0 0 0 1 1 0 1 0 1 0 0 1 0 0 0 1 0 1 1 1 0 1 1 0 1
8	0 0 0 1	1 0 0 0 1 1 0 0 1 0 0 1 0 1 1 0 0 0 0 0 0 1 1 1 0 1 1 1 1 0 1 1
9	1 0 0 1	1 0 1 1 1 0 0 0 1 1 0 0 1 0 0 1 0 1 1 0 0 0 0 0 0 1 1 1 0 1 1 1
10	0 1 0 1	0 1 1 1 1 0 1 1 1 0 0 0 1 1 0 0 1 0 0 1 0 1 1 0 0 0 0 0 0 1 1 1
11	1 1 0 1	0 1 1 1 0 1 1 1 1 0 1 1 1 0 0 0 1 1 0 0 1 0 0 1 0 1 1 0 0 0 0 0
12	0 0 1 1	0 0 0 0 0 1 1 1 0 1 1 1 1 0 1 1 1 0 0 0 1 1 0 0 1 0 0 1 0 1 1 0
13	1 0 1 1	0 1 1 0 0 0 0 0 0 1 1 1 0 1 1 1 1 0 1 1 1 0 0 0 1 1 0 0 1 0 0 1
14	0 1 1 1	1 0 0 1 0 1 1 0 0 0 0 0 0 1 1 1 0 1 1 1 1 0 1 1 1 0 0 0 1 1 0 0
15	1 1 1 1	1 1 0 0 1 0 0 1 0 1 1 0 0 0 0 0 0 1 1 1 0 1 1 1 1 0 1 1 1 0 0 0

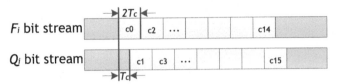

Figure C.12   O-QPSK chip offsets.

symbol will be responsible for specifying one of the 16 nearly orthogonal 32-bit chip PN sequences for the transmission [WPA07a].

Moreover, for the IEEE 802.15.4 standard the O-QPSK even-indexed chips are represented by the bit stream (F_i) onto the in-phase and the odd-indexed chips are represented by the bit stream (Q_i) onto the quadrature phase offset by T_c, as shown in Figure C.12. The data symbol is represented by a 32-chip sequence, as shown in Table 3.3. The chip rate is 2.0 Mchip/s since it is 32 times the symbol rate (i.e, 62.5 ksymbol/s).

C.3 Minimum Shift Keying

If the rectangular pulse shapes used in O-QPSK are replaced by the following sinusoidal pulse shapes using $cos(\pi t/2T_c)$ and $sin(\pi t/2T_c)$ in the I and Q components respectively, the resulting modulation is known as Minimum shift keying (MSK). In [Far11], the authors present a typical half-sin pulse shape filter in order to modify the shape of the binary pulses, as shown in Figure C.13.

Equations (C.5) and (C.6) present the half-sin pulse shape filter functions responsible to replace each bit stream with one half of sinusoidal signal in the in-phase and in quadrature respectively.

$$Filter_F(t) = \begin{cases} cos(\pi t/2T_c), \ 0 \leqslant t \leqslant 2T_c \\ 0, \quad otherwise \end{cases} \tag{C.5}$$

$$Filter_Q(t) = \begin{cases} sin(\pi t/2T_c), \ 0 \leqslant t \leqslant 2T_c \\ 0, \quad otherwise \end{cases} \tag{C.6}$$

where, T_c, represents the width of the bit.

Figure C.13   Half-sin pulse shaping filter.

Figure C.14 presents the block diagram for the MSK modulator.

After applying the half-sine pulse shaping filter, the MSK modulated signal is given by:

$$s(t) = F_i(t)cos\left(\frac{\pi t}{2T_c}\right)Vcos(2\pi f_c t) - Q_i(t)sin\left(\frac{\pi t}{2T_c}\right)Vsin(2\pi f_c t) \quad (C.7)$$

To represent the MSK signal a chip rate of 2.0 Mchip/s (T_c = 500 ns) was considered. Figures C.15 and C.16 present the in-phase and quadrature components for the MSK modulator. Figure C.17 shows the resulting MSK transmitted waveform.

As it was already mentioned, the IEEE 802.15.4 defines that, for a radio transceiver operating in the 2.4 GHz band the PHY symbol rate shall be 62.5 ksymbol/s. Since the 32 chips are transmitted in the same time periods for transmitting 4 bits (one symbol), 32 times the symbol rate (i.e., 32×62.500), gives us a 2 Mchip/s rate.

A summary of all the calculations performed in order to obtain the value for T_c (as explained before, T_c is the inverse of the chip rate) are presented as

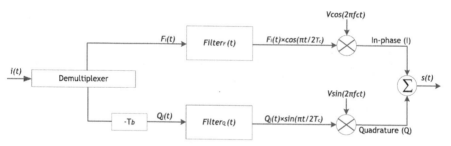

Figure C.14   MSK Modulator.

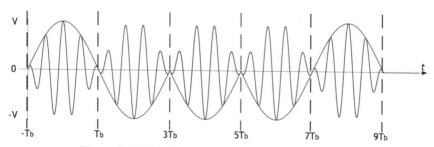

Figure C.15   In-phase component of the MSK modulator.

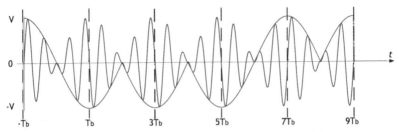

Figure C.16   Quadrature component of the MSK modulator.

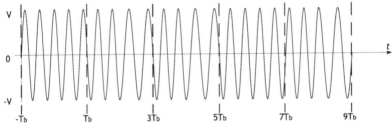

Figure C.17   MSK waveform.

follows:

$$\begin{cases} maximum\ data\ rate = 250\ kbps \\ maximum\ symbol\ data\ rate = 62.5\ ksymbols/s \\ bits\ per\ symbol = 4 \\ symbol\ period = 1/symbol\ frequency \\ bits\ per\ symbol = 250\ kbps/62.5\ ksymbols/s = 4 \\ symbol\ period = 1/62.5\ ksymbols/s = 16\mu s \\ T_c = 1/2\ Mchip/s = 500ns \end{cases} \quad (C.8)$$

C.4 Signal Spreading

As explained before in the context of the O-QPSK modulation, every 4 bits are mapped in a 32-chip PN sequence. As so, there will be an increase of the over-the-air bit rate by a factor of eight. The total bandwidth will also increases by a factor of eight. Since the bit rate is proportional to the signal bandwidth, the signal bandwidth also increases by a factor of eight. The basic concept of signal spreading and despreading is shown in Figure C.18.

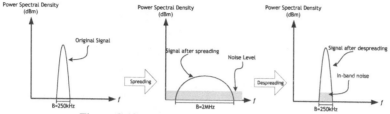

Figure C.18 Signal spreading and despreading.

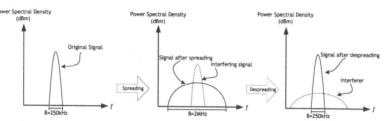

Figure C.19 Signal spreading and despreading by considering the effect of an interferer.

After spreading the bandwidth will be 2 MHz, and the magnitude of the power spectral density will decrease since the signal energy is "spread" over a larger bandwidth. In the despreading side, the signal energy will be concentrated back to the original bandwidth. The 32-chip PN sequence received is then compared with one of the 16 possible chip sequences presented in Table C.2. The most similar will be selected as the receiver chip sequence, being converted in the original data bit sent. Notice that the despreading technique does not affect the noise level in the 250 kHz bandwidth of interest. By increasing the desired band without increasing the noise level, the effective SNR is increased by the signal despreading. This increase in the signal SNR will directly improve the receiver sensitivity.

This improvement in the SNR is referred in the IEEE 802.15.4 as the Spreading gain (SG). The value of the SG is equal to the ratio of the signal bit rate after spreading to the signal bit rate before spreading. For the 2.4 GHz band the SG is given by:

$$SG = 10 \times log_{10} \left(\frac{2 \times 10^6}{250 \times 10^3} \right) \simeq 9 \, dB \tag{C.9}$$

By considering the presence of an interfering signal, when the signal arrives at the destination, the despreading technique will concatenate the signal energy back to the 250 kHz bandwidth. And the interfering signal will be "spread" over a large bandwidth, Figure 3.21. Moreover, in the receiver a filter will be applied to remove the spectral contents outside the 250 kHz bandwidth. The remaining portion of the interferer signal in the frequency band of interest has much lower power compared to the original interferer signal.

The signal-to-interference ratio (SIR) can be defined as the ratio of designer signal power to the total interference power:

$$SIR = \frac{Desired\ Signal\ Power}{Total\ Interference\ Power} \tag{C.10}$$

References

[ADB⁺04] Anish Arora, Prabal Dutta, Sandip Bapat, Vinod Kulathumani, Hongwei Zhang, Vinayak Naik, Vineet Mittal, Hui Cao, Murat Demirbas, Mohamed Gouda, et al. A line in the sand: a wireless sensor network for target detection, classification, and tracking. *Computer Networks*, 46(5):605–634, 2004.

[ASSC02a] I. F. Akyildiz, W. Su, Y. Sankarasubramaniam, and E. Cayirci. Wireless Sensor Networks: A Survey. *Computer Networks*, 38:393–422, 2002.

[ASSC02b] Ian F Akyildiz, Weilian Su, Yogesh Sankarasubramaniam, and Erdal Cayirci. A survey on sensor networks. *IEEE communications magazine*, 40(8):102–114, 2002.

[AT809] *AT86RF231 Datasheet*, 2009. Datasheet.

[BDWL10] Abdelmalik Bachir, Mischa Dohler, Thomas Watteyne, and Kin K. Leung. MAC Essentials for Wireless Sensor Networks. *IEEE Communications Surveys Tutorials*, 12(2):222–248, 2010.

[Ber01] Sklar Bernard. Digital communications fundamentals and applications. *Prentice Hall, New Jersey, USA*, 2001.

[BGdMT06] Gunter Bolch, Stefan Greiner, Hermann de Meer, and Kishor S. Trivedi. *Queueing Networks and Markov Chains: Modeling and Performance Evaluation with Computer Science Applications*. Wiley-Interscience, 2nd edition, April 2006.

[BGLA01] Lichun Bao and J. J. Garcia-Luna-Aceves. A new approach to channel access scheduling for ad hoc networks. In *Proc. of the 7th Annual International Conference on Mobile Computing and Networking (MobiCom' 01)*, pages 210–221, July 2001.

[BMFD06] M.I. Brownfield, K. Mehrjoo, A.S. Fayez, and IV Davis, N.J. Wireless sensor network energy-adaptive mac protocol. In *Proc. of the 3rd IEEE Consumer Communications and Networking Conference (CCNC 2006)*, volume 2, pages 778–782, January 2006.

[BMS⁺08] Ghulam Bhatti, A Mehta, Zafer Sahinoglu, Jinyun Zhang, and R Viswanathan. Modified beacon-enabled ieee 802.15. 4 mac for lower latency. In *Global Telecommunications Conference, 2008. IEEE GLOBECOM 2008. IEEE*, pages 1–5. IEEE, 2008.

[BV⁺09] Chiara Buratti, Roberto Verdone, et al. Performance analysis of ieee 802.15. 4 non beacon-enabled mode. *IEEE Transactions on Vehicular Technology*, 58(7):3480, 2009.

[BVBL14] Luís M Borges, Fernando J Velez, Norberto Barroca, and António S Lebres. Frame capture and reliability based decider implementation in the mixim ieee 802.15. 4 framework. In *Proceedings of the 7th International ICST Conference on Simulation Tools and Techniques*, pages 98–107. ICST (Institute for Computer Sciences, Social-Informatics and . . . , 2014.

[BVL14] L. M. Borges, F. J. Velez, and A. S. Lebres. Survey on the characterization and classification of wireless sensor network applications. *IEEE Communications Surveys Tutorials*, 16(4):1860–1890, Fourthquarter 2014.

[BYAH06] Michael Buettner, Gary V. Yee, Eric Anderson, and Richard Han. X-mac: a short preamble mac protocol for duty-cycled wireless sensor networks. In *Proc. of the 4th International Conference on Embedded Networked Sensor Systems (SenSys '06)*, pages 307–320, New York, NY, USA, 2006. ACM.

[CBF11] R. Verdone C. Buratti, M. Martalò and G. Ferrari. *Sensor Networks with IEEE 802.15.4 Systems*. Springer, New York, NY, USA, 2011.

[CBV04a] P Chatzimisios, AC Boucouvalas, and V Vitsas. Effectiveness of rts/cts handshake in ieee 802.11 a wireless lans. *Electronics letters*, 40(14):915–916, 2004.

[CBV04b] P Chatzimisios, AC Boucouvalas, and V Vitsas. Optimisation of rts/cts handshake in ieee 802.11 wireless lans for maximum performance. In *Global Telecommunications Conference Workshops, 2004. GlobeCom Workshops 2004. IEEE*, pages 270–275. IEEE, 2004.

[CC107] *CC1100 Datasheet*, 2007. Datasheet.

[CL54] H. Chernoff and E.L. Lehmann. The Use of Maximum Likelihood Estimates in χ^2 Tests for Goodness of Fit. *Annals of Mathematical Statistics*, 25(3):579–586, 1954.

[COO12] Coolness, Nov. 2012.

[CUK$^+$10] M Sanaullah Chowdhury, Niamat Ullah, Md Humaun Kabir, Pervez Khan, and Kyung Sup Kwak. Throughput, delay and bandwidth efficiency of ieee 802.15. 4a using css phy. In *Information and Communication Technology Convergence (ICTC), 2010 International Conference on*, pages 158–163. IEEE, 2010.

[DEA$^+$06] Ilker Demirkol, Cem Ersoy, Fatih Alagoz, et al. Mac protocols for wireless sensor networks: a survey. *IEEE Communications Magazine*, 44(4):115–121, 2006.

[DHNH03] S. Dulman, L. Hoesel, T. Nieberg, and P. Havinga. Collaborative communication protocols for wireless sensor networks. In *Proc. of the European Research on Middleware and Architectures for Complex and Embedded Systems Workshop*, Pisa, Italy, April 2003.

[DL03] T. van Dam and K. Langendoen. An adaptive energy-efficient MAC protocol for wireless sensor networks. In *Proc. of the Third ACM International Conference on Embedded Networked Sensor Systems (SenSys'03)*, pages 171–180, Los Angeles, CA, USA, November 2003.

[DZM12] Jingshan Duan, Yiqun Zhuang, and Lixiang Ma. An adaptive rts/cts mechanism in ieee 802.15. 4 for multi-hop networks. In *Computational Problem-Solving (ICCP), 2012 International Conference on*, pages 155–159. IEEE, 2012.

[EH02] A. El-Hoiydi. Aloha with Preamble Sampling for Sporadic Traffic in Ad Hoc Wireless Sensor Networks. In *Proc. of IEEE International Conference on Communications (ICC 2002)*, volume 5, pages 3418–3423, New York, USA, 2002.

[EHD04a] A. El-Hoiydi and J.-D. Decotignie. Wisemac: an ultra low power mac protocol for the downlink of infrastructure wireless sensor networks. In *Proc. of the Ninth International Symposium on Computers and Communications (ISCC 2004)*, pages 244–251, Alexandria, Egypt, June 2004.

[EHD04b] Amre El-Hoiydi and J-D Decotignie. Wisemac: an ultra low power mac protocol for the downlink of infrastructure wireless sensor networks. In *Computers and Communications, 2004. Proceedings. ISCC 2004. Ninth International Symposium on*, volume 1, pages 244–251. IEEE, 2004.

[EV06] S.C. Ergen and P. Varaiya. PEDAMACS: Power Efficient and Delay Aware Medium Access Protocol for Sensor Networks. *IEEE Transactions on Mobile Computing*, 5(7):920–930, July 2006.

[Far08] Shahin Farahani. *ZigBee Wireless Networks and Transceivers*. Newnes, Newton, MA, USA, 2008.

[Far11] Shahin Farahani. *ZigBee wireless networks and transceivers*. Newnes, 2011.

[Fel68] William Feller. *An Introduction to Probability Theory and Its Applications*, volume 1. Wiley, January 1968.

[GCB03] J. A. Gutiérrez, E. H. Callaway, and R. L. Barrett. *Low-Rate Wireless Personal Area Networks: Enabling Wireless Sensors with IEEE 802.15.4*. IEEE Standards. Standards Information Network - IEEE Press, United States of America, 1 edition, 2003.

[GCB04] Jose A Gutierrez, Edgar H Callaway, and Raymond L Barrett. *Low-rate wireless personal area networks: enabling wireless sensors with IEEE 802.15. 4*. IEEE Standards Association, 2004.

[GEWD05] D. Gracanin, M. Eltoweissy, A. Wadaa, and L. A. Dasilva. A service-centric model for wireless sensor networks. *IEEE Journal on Selected Areas in Communications*, 23(6):1159–1166, 2005.

[HHT02] M.J. Handy, M. Haase, and D. Timmermann. Low Energy Adaptive Clustering Hierarchy with Deterministic Cluster-Head Selection. In *Proc. of the 4th International Workshop on Mobile and Wireless Communications Network*, pages 368–372, Stockholm, Sweden, September 2002.

[HL07] Gertjan P. Halkes and Koen Langendoen. Crankshaft: An Energy-Efficient MAC-Protocol for Dense Wireless Sensor Networks. In *Proc. of the fourth European Wireless Sensor Network (EWSN07)*, pages 228–244, Delft, The Netherlands, January 2007.

[HNL07] M Healy, T Newe, and E Lewis. Efficiently securing data on a wireless sensor network. *Journal of Physics: Conference Series*, 76:012063, jul 2007.

[HNS+06a] Seongil Han, Yongsub Nam, Yongho Seok, Taekyoung Kwon, and Yanghee Choi. Two-phase collision avoidance to improve scalability in wireless lans. In *GLOBECOM*, 2006.

[HNS⁺06b] Seongil Han, Yongsub Nam, Yongho Seok, Taekyoung Kwon, and Yanghee Choi. WLC29-4: Two-phase Collision Avoidance to Improve Scalability in Wireless LANs. In *IEEE Globecom 2006*, pages 1–5, Nov 2006.

[HvDL05] Gertjan P. Halkes, Tijs van Dam, and Koen Langendoen. Comparing energy-saving mac protocols for wireless sensor networks. *Mobile Networks and Applications (MONET)*, 10(5):783–791, 2005.

[HXS⁺13] Pei Huang, Li Xiao, Soroor Soltani, Matt W Mutka, and Ning Xi. The evolution of mac protocols in wireless sensor networks: A survey. *IEEE communications surveys & tutorials*, 15(1):101–120, 2013.

[IEE02] IEEE Standard for Telecommunications and Information Exchange Between Systems - LAN/MAN - Specific Requirements - Part 15: Wireless Medium Access Control (MAC) and Physical Layer (PHY) Specifications for Wireless Personal Area Networks (WPANs). *IEEE Std 802.15.1-2002*, pages 1–473, 14 2002.

[IEE03a] IEEE Standard for Information Technology - Telecommunications and Information Exchange Between Systems - Local and Metropolitan Area Networks - Specific Requirements Part 15.3: Wireless Medium Access Control (MAC) and Physical Layer (PHY) Specifications for High Rate Wireless Personal Area Networks (WPANs). *IEEE Std 802.15.3-2003*, pages 0–315, 2003.

[IEE03b] Ieee standard for information technology - telecommunications and information exchange between systems - local and metropolitan area networks specific requirements part 15.4: Wireless medium access control (mac) and physical layer (phy) specifications for low-rate wireless personal area networks (lr-wpans). *IEEE Std 802.15.4-2003*, pages 0–670, 2003.

[IEE03c] IEEE Standard for Information Technology- Telecommunications and Information Exchange Between Systems- Local and Metropolitan Area Networks- Specific Requirements- Part 11: Wireless LAN Medium Access Control (MAC) and Physical Layer (PHY) Specifications. *ANSI/IEEE Std 802.11, 1999 Edition (R2003)*, pages i–513, 2003.

[IEE04] IEEE Standard for Local and Metropolitan Area Networks Part 16: Air Interface for Fixed Broadband Wireless Access Systems. *IEEE Std 802.16-2004 (Revision of IEEE Std 802.16-2001)*, pages 1–857, 2004.

[IEE07] IEEE Standard for Information Technology - Telecommunications and Information Exchange Between Systems - Local and Metropolitan Area Networks - Specific Requirement Part 15.4: Wireless Medium Access Control (MAC) and Physical Layer (PHY) Specifications for Low-Rate Wireless Personal Area Networks (WPANs). *IEEE Std 802.15.4a-2007 (Amendment to IEEE Std 802.15.4-2006)*, pages 1–203, 2007.

[Ins07] Texas Instruments. Cc2420 datasheet-2.4 ghz ieee 802.15. 4/zigbee-ready rf transceiver (rev. b). *Chipcon AS [Revised Feb. 2012]*, 2007.

[IVHJH11] Ozlem Durmaz Incel, Lodewijk Van Hoesel, Pierre Jansen, and Paul Havinga. Mc-lmac: A multi-channel mac protocol for wireless sensor networks. *Ad Hoc Networks*, 9(1):73–94, 2011.

[JBA07] Vivek Jain, Ratnabali Biswas, and Dharma P. Agrawal. Energy-Efficient and Reliable Medium Access in Sensor Networks. In *Proc. of the IEEE International Symposium on a World of Wireless, Mobile and Multimedia Networks (WoW-MoM 2007)*, pages 1–8, June 2007.

[JBT06] Kyle Jamieson, Hari Balakrishnan, and Y.C. Tay. Sift: a MAC Protocol for Event-Driven Wireless Sensor Networks. In *Proc. of the Third European Workshop on Wireless Sensor Networks (EWSN)*, volume 3868, pages 260–275, Zurich, Switzerland, February 2006.

[JDB99] Raj Jain, Arjan Durresi, and Gojko Babic. Throughput fairness index: An explanation. 1999.

[JH09] M.R. Jongerden and B.R. Haverkort. Which battery model to use? *IET Software*, 3(6):445–457, December 2009.

[KA13] Bilal Muhammad Khan and Falah H Ali. Collision free mobility adaptive (cfma) mac for wireless sensor networks. *Telecommunication Systems*, 52(4):2459–2474, 2013.

[KHH05] Mauri Kuorilehto, Marko Hännikäinen, and Timo D. Hämäläinen. A Survey of Application Distribution in Wireless Sensor Networks. *EURASIP Journal Wireless Communications Networking*, 2005(5):774–788, Oct 2005.

[KSC08] Youngmin Kim, Hyojeong Shin, and Hojung Cha. Y-MAC: An Energy-Efficient Multi-channel MAC Protocol for Dense Wireless Sensor Networks. In *Proc. of the International Conference on Information Processing in Sensor Networks (IPSN '08)*, pages 53–63, St. Louis, Missouri, USA, April 2008.

[KSEG+08] A. Kuntz, F. Schmidt-Eisenlohr, O. Graute, H. Hartenstein, and M. Zitterbart. Introducing Probabilistic Radio Propagation Models in OMNeT++ Mobility Framework and Cross Validation Check with NS-2. In *Proceedings of the 1st International Conference on Simulation Tools and Techniques for Communications, Networks and Systems & Workshops*, Simutools '08, Brussels, BEL, 2008. ICST (Institute for Computer Sciences, Social-Informatics and Telecommunications Engineering).

[KSK+08] M. Kuorilehto, J. Suhonen, M. Kohvakka, P. Hämäläinen, M. Hännikäinen, and T.D. Hamalainen. *Ultra-Low Energy Wireless Sensor Networks in Practice: Theory, Realization and Deployment*. John Wiley & Sons, 2008.

[KSW+08] A. Köpke, M. Swigulski, K. Wessel, D. Willkomm, P. T. Klein Haneveld, T. E. V. Parker, O. W. Visser, Hermann S. Lichte, and S. Valentin. Simulating wireless and mobile networks in omnet++ the mixim vision. In *Proc. of the 1st international conference on Simulation tools and techniques for communications, networks and systems & workshops (Simutools '08)*, pages 71–79, Marseille, France, March 2008.

[KT75] Leonard Kleinrock and Fouad Tobagi. Packet switching in radio channels: Part i–carrier sense multiple-access modes and their throughput-delay characteristics. *IEEE transactions on Communications*, 23(12):1400–1416, 1975.

[KW05a] H. Karl and A. Willig. *Protocols and Architectures for Wireless Sensor Networks*. John Wiley & Sons, Chichester, UK, 2005.

[KW05b] Holger Karl and Andreas Willig. *Protocols and Architectures for Wireless Sensor Networks*. John Wiley & Sons, 2005.

[LDCO06] W. L. Lee, A. Datta, and R. Cardell-Oliver. FlexiMAC: A flexible TDMA-based MAC protocol for fault-tolerant and energy-efficient wireless sensor networks. In *Proc. of 14th IEEE International Conference on Networks (ICON '06)*, volume 2, pages 1–6, Singapore, Republic of Singapore, September 2006.

[LDMM+05] Benoît Latré, Pieter De Mil, Ingrid Moerman, Niek Van Dierdonck, Bart Dhoedt, and Piet Demeester. Maximum throughput and minimum delay in ieee 802.15. 4. In *International Conference on Mobile Ad-Hoc and Sensor Networks*, pages 866–876. Springer, 2005.

[LE06] Zhenzhen Liu and I. Elhanany. RL-MAC: A QoS-Aware Reinforcement Learning based MAC Protocol for Wireless Sensor Networks. In *Proc. of the 2006 IEEE International Conference on Networking, Sensing and Control (ICNSC '06)*, pages 768–773, Florida, USA, April 2006.

[LEQ05] Yang Liu, I. Elhanany, and Hairong Qi. An Energy-Efficient QoS-aware Media Access Control Protocol for Wireless Sensor Networks. In *Proc. of the IEEE International Conference on Mobile Adhoc and Sensor Systems Conference (MASS 2005)*, pages 191–194, Washington, DC, USA, November 2005.

[LHA08] Hieu Khac Le, D. Henriksson, and T. Abdelzaher. A Practical Multi-channel Media Access Control Protocol for Wireless Sensor Networks. In *Proc. of the International Conference on Information Processing in Sensor Networks (IPSN '08)*, pages 70–81, St. Louis, Missouri, USA, April 2008.

[LKR04] G. Lu, B. Krishnamachari, and C.S. Raghavendra. An adaptive energy-efficient and low-latency mac for data gathering in wireless sensor networks. In *Proc. of the 18th International Parallel and Distributed Processing Symposium*, page 224, Santa Fe, New Mexico, April 2004.

[LQW04] P. Lin, C. Qiao, and X. Wang. Medium access control with a dynamic duty cycle for sensor networks. In *Proc. of the 2004 IEEE Wireless Communications and Networking Conference (WCNC 2004)*, volume 3, pages 1534–1539, Atlanta, Georgia, USA, March 2004.

[LRW04] E.-Y.A. Lin, J.M. Rabaey, and A. Wolisz. Power-efficient Rendez-Vous Schemes for Dense Wireless Sensor Networks. In *Proc. of the 2004 IEEE International Conference on Communications*, volume 7, pages 3769–3776, Paris, France, June 2004.

[LSR04] Huan Li, P. Shenoy, and K. Ramamritham. Scheduling communication in real-time sensor applications. In *Proc. of the 10th IEEE Real-Time and Embedded Technology and*

Applications Symposium (RTAS 2004), pages 10–18, Toronto, Canada, May 2004.

[Mah06] Nitaigour P. Mahalik. *Sensor Networks and Configuration: Fundamentals, Standards, Platforms, and Applications.* Springer-Verlag New York, Inc., Secaucus, NJ, USA, 2006.

[Mah07] N. P. Mahalik. *Sensor Networks and Configuration: Fundamentals, Standards, Platforms, and Applications.* Springer, New York, NY, USA, 2007.

[MCP⁺02] Alan Mainwaring, David Culler, Joseph Polastre, Robert Szewczyk, and John Anderson. Wireless sensor networks for habitat monitoring. In *Proc. of the 1st ACM international workshop on Wireless sensor networks and applications (WSNA '02)*, pages 88–97, Atlanta, Georgia, USA, September 2002.

[MiX13] Mixim - modeling framework for mobile and fixed wireless networks, Oct 2013.

[MMU11] Junya Mizuguchi, Masashi Murata, and Wataru Uemura. A broadcasting method based on rts/cts for an ad-hoc network. In *Consumer Electronics (ISCE), 2011 IEEE 15th International Symposium on*, pages 104–106. IEEE, 2011.

[NBC20] Luís M. Borges Norberto Barroca, Fernando J. Velez and Periklis Chatzimisios. Performance Enhancement of IEEE 802.15.4 by employing RTS/CTS and Frame Concatenation. *IET Wireless Sensor Systems*, 10(6):308–319, 2020.

[NBC22] Fernando J. Velez Norberto Barroca, Luís M. Borges and Periklis Chatzimisios. Impact of using CCS PHY and RTS/CTS Combined with Frame Concatenation in the IEEE 802.15.4 Performance. *submitted to IET Wireless Sensor Systems*, March 2022.

[NH93] Homayoun Nikookar and Homayoun Hashemi. Statistical modeling of signal amplitude fading of indoor radio propagation channels. In *Universal Personal Communications, 1993. Personal Communications: Gateway to the 21st Century. Conference Record., 2nd International Conference on*, volume 1, pages 84–88. IEEE, 1993.

[Nic05] D. Niculescu. Communication paradigms for sensor networks. *IEEE Communications Magazine*, 43(3):116–122, March 2005.

[NS09] Ha H. Nguyen and Ed Shwedyk. *A First Course in Digital Communications*. Cambridge University Press, Cambridge, UK, 2009.

[OMN13] Omnet++ network simulation framework, Sep 2013.

[oST01] Information Technology Laboratory (National Institute of Standards and Technology). *Announcing the Advanced Encryption Standard (AES)* . Computer Security Division, Information Technology Laboratory, National Institute of Standards and Technology, Gaithersburg, MD :, 2001.

[PCB00] Nissanka B Priyantha, Anit Chakraborty, and Hari Balakrishnan. The cricket location-support system. In *Proceedings of the 6th annual international conference on Mobile computing and networking*, pages 32–43. ACM, 2000.

[PHC04] Joseph Polastre, Jason Hill, and David Culler. Versatile low power media access for wireless sensor networks. In *Proc. of the Second ACM International Conference on Embedded Networked Sensor Systems (SenSys'04)*, pages 95–107, Baltimore, MD, USA, November 2004.

[PJ04] H. Pham and S. Jha. An Adaptive Mobility-Aware MAC Protocol for Sensor Networks (MS-MAC). In *Proc. of the 2004 IEEE International Conference on Mobile Ad-hoc and Sensor Systems*, pages 558–560, Vancouver, Canada, October 2004.

[PTR] Prentice Hall PTR. *Data and Computer Communications, 8 edition*.

[R+02] Theodore S Rappaport et al. *Wireless communications: principles and practice*, volume 2. Prentice Hall PTR, New Jersey, USA, 2002.

[RAdS+00] J.M. Rabaey, M.J. Ammer, Jr. da Silva, J.L., D. Patel, and S. Roundy. Picoradio supports ad hoc ultra-low power wireless networking. *Computer*, 33(7):42–48, July 2000.

[RDA+09] Jérôme Rousselot, Jean-Dominique Decotignie, Marc Aoun, Peter van Der Stok, Ramon Serna Oliver, and Gerhard Fohler. Accurate timeliness simulations for real-time wireless sensor networks. In *Computer Modeling and Simulation, 2009. EMS'09. Third UKSim European Symposium on*, pages 476–481. IEEE, 2009.

[REHD08] J. Rousselot, A. El-Hoiydi, and J.-D. Decotignie. Low power medium access control protocols for wireless sensor networks.

In *Proc. of the 14th European Wireless Conference (EW 2008)*, pages 1–5, Prague, Czech Republic, June 2008.

[RLLK08] Jiho Ryu, Jeongkeun Lee, Sung-Ju Lee, and Taekyoung Kwon. Revamping the ieee 802.11 a phy simulation models. In *Proceedings of the 11th international symposium on Modeling, analysis and simulation of wireless and mobile systems*, pages 28–36. ACM, 2008.

[ROJ08] Antonio G. Ruzzelli, Gregory M. P. O'Hare, and Raja Jurdak. MERLIN: Cross-layer Integration of MAC and Routing for Low Duty-Cycle Sensor Networks. *Ad Hoc Networks*, 6(8):1238–1257, Nov 2008.

[RWA$^+$08] Injong Rhee, A. Warrier, M. Aia, Jeongki Min, and M.L. Sichitiu. Z-MAC: A Hybrid MAC for Wireless Sensor Networks. *IEEE/ACM Transactions on Networking*, 16(3):511–524, June 2008.

[RWR04] Shad Roundy, Paul Kenneth Wright, and Jan M Rabaey. Energy scavenging for wireless sensor networks. 2004.

[SAW12] Sa-wsn, Nov. 2012.

[SK11] Sujesha Sudevalayam and Purushottam Kulkarni. Energy harvesting sensor nodes: survey and implications. *IEEE Communications Surveys and Tutorials*, 13(3):443–461, 2011.

[SKA04] Sameer Sundresh, Wooyoung Kim, and Gul Agha. Sens: A sensor, environment and network simulator. In *Proceedings of the 37th annual symposium on Simulation*, page 221. IEEE Computer Society, 2004.

[SS07] Xiaolei Shi and G. Stromberg. SyncWUF: An Ultra Low-Power MAC Protocol for Wireless Sensor Networks. *IEEE Transactions on Mobile Computing*, 6(1):115–125, January 2007.

[SSZ09] Maoheng Sun, Kaijian Sun, and Youmin Zou. Analysis and improvement for 802.15. 4 multi-hop network. In *Communications and Mobile Computing, 2009. CMC'09. WRI International Conference on*, volume 2, pages 52–56. IEEE, 2009.

[SW04] H. Schwetlick and A. Wolf. PSSS - Parallel Sequence Spread Spectrum a Physical Layer for RF Communication. In *Proc. of 2004 IEEE International Symposium on Consumer Electronics*, pages 262–265, Algarve, Portugal, September 2004.

[SWM⁺07] Wilson So, Jean Walrand, Jeonghoon Mo, et al. Mcmac: A parallel rendezvous multi-channel mac protocol. In *Wireless Communications and Networking Conference, 2007. WCNC 2007. IEEE*, pages 334–339. IEEE, 2007.

[TJB04] Y.C. Tay, K. Jamieson, and H. Balakrishnan. Collision-Minimizing CSMA and its Applications to Wireless Sensor Networks. *IEEE Journal on Selected Areas in Communications*, 22(6):1048–1057, aug. 2004.

[TK85] Hideaki Takagi and Leonard Kleinrock. Throughput analysis for persistent csma systems. *IEEE Transactions on Communications*, 33:627–638, 1985.

[TPS⁺05] Gilman Tolle, Joseph Polastre, Robert Szewczyk, David Culler, Neil Turner, Kevin Tu, Stephen Burgess, Todd Dawson, Phil Buonadonna, David Gay, et al. A macroscope in the redwoods. In *Proceedings of the 3rd international conference on Embedded networked sensor systems*, pages 51–63. ACM, 2005.

[TSGJ11] Lei Tang, Yanjun Sun, Omer Gurewitz, and David B Johnson. Em-mac: a dynamic multichannel energy-efficient mac protocol for wireless sensor networks. In *Proceedings of the Twelfth ACM International Symposium on Mobile Ad Hoc Networking and Computing*, page 23. ACM, 2011.

[Var10] A. Varga. Omnet++. In K. Wehrle, M. Günes, and J. Gross, editors, *Modeling and Tools for Network Simulation*, pages 35–58. Springer Verlag, September 2010.

[VBBC19] Fernando J Velez, Luís M Borges, Norberto Barroca, and Periklis Chatzimisios. Two innovative energy efficient IEEE 802.15. 4 MAC sub-layer protocols with packet concatenation: employing RTS/CTS and multichannel scheduled channel polling. *Wearable Technologies and Wireless Body Sensor Networks for Healthcare*, page 241, 2019.

[vDL03] Tijs van Dam and Koen Langendoen. An Adaptive Energy-Efficient MAC Protocol for Wireless Sensor Networks. In *Proc. of the 1st International Conference on Embedded Networked Sensor Systems*, SenSys '03, pages 171–180, New York, NY, USA, November 2003. ACM.

[VDMC08] R. Verdone, D. Dardari, G. Mazzini, and A. Conti. *Wireless Sensor and Actuator Networks: Technologies, Analysis and Design*. Academic Press, London, UK, 2008.

[VH08] András Varga and Rudolf Hornig. An overview of the omnet++ simulation environment. In *Proc. of the 1st international conference on Simulation tools and techniques for communications, networks and systems & workshops (Simutools '08)*, pages 1–10, ICST, Brussels, Belgium, Belgium, 2008.

[VHH04] L. F. W. Van Hoesel and P. J. M. Havinga. A Lightweight Medium Access Protocol (LMAC) for Wireless Sensor Networks: Reducing Preamble Transmissions and Transceiver State Switches. In *Proc. of the 1st International Workshop on Networked Sensing Systems (INSS)*, pages 205–208, Tokyo, Japan, June 2004. Society of Instrument and Control Engineers (SICE).

[vHH09] L.F.W. van Hoesel and P.J.M. Havinga. Analysis of a self-organizing algorithm for time slot selection in schedule-based medium access. Technical Report TR-CTIT-09-06, University of Twente, Enschede, January 2009.

[VVL+11] Frank Vanheel, Jo Verhaevert, Eric Laermans, Ingrid Moerman, and Piet Demeester. Automated linear regression tools improve rssi wsn localization in multipath indoor environment. *EURASIP Journal on Wireless Communications and Networking*, 2011(1):38, 2011.

[WBD+06] T. Watteyne, A. Bachir, M. Dohler, D. Barthe, and I. Auge-Blum. 1-hopMAC: An Energy-Efficient MAC Protocol for Avoiding 1 -hop Neighborhood Knowledge. In *Proc. of the 3rd Annual IEEE Communications Society Sensor and Ad Hoc Communications and Networks (SECON '06)*, volume 2, pages 639–644, Reston, VA, U.S.A., Sep 2006.

[WCS11] Yik-Chung Wu, Qasim Chaudhari, and Erchin Serpedin. Clock synchronization of wireless sensor networks. *IEEE Signal Processing Magazine*, 28(1):124–138, 2011.

[WIS12] Wisebed project, Nov. 2012.

[WLA05] Part 11: Wireless LAN Medium Access Control (MAC) and Physical Layer (PHY) specifications, Nov 2005.

[WPA07a] Approved IEEE Draft Amendment to IEEE Standard for Information Technology-Telecommunications and Information Exchange Between Systems-Part 15.4:Wireless Medium Access Control (MAC) and Physical Layer (PHY) Specifications for Low-Rate Wireless Personal Area Networks (LR-WPANS): Amendment to Add Alternate Phy (Amendment of IEEE Std 802.15.4), 2007.

[WPA07b] Approved ieee draft amendment to ieee standard for information technology-telecommunications and information exchange between systems-part 15.4:wireless medium access control (mac) and physical layer (phy) specifications for low-rate wireless personal area networks (lr-wpans): Amendment to add alternate phy (amendment of ieee std 802.15.4). *IEEE Approved Std P802.15.4a/D7, Jan 2007*, 2007.

[WPA+07c] Sinam Woo, Woojin Park, Sae Young Ahn, Sunshin An, and Dongho Kim. Knowledge-based exponential backoff scheme in ieee 802.15. 4 mac. In *International Conference on Information Networking*, pages 435–444. Springer, 2007.

[WPA11] Part 15.4: Low-Rate Wireless Personal Area Networks - Revision of IEEE Std 802.15.4-2006, Sep 2011.

[WSKW09] K. Wessel, M. Swigulski, A. Köpke, and D. Willkomm. Mixim: the physical layer an architecture overview. In *Proc. of the 2nd International Conference on Simulation Tools and Techniques (Simutools'09)*, pages 78–86, Rome, Italy, March 2009.

[WWJ+05] Kamin Whitehouse, Alec Woo, Fred Jiang, Joseph Polastre, and David Culler. Exploiting the capture effect for collision detection and recovery. In *Embedded Networked Sensors, 2005. EmNetS-II. The Second IEEE Workshop on*, pages 45–52. IEEE, 2005.

[YBO08] B. Yahya and J. Ben-Othman. An Energy Efficient Hybrid Medium Access Control Scheme for Wireless Sensor Networks with Quality of Service Guarantees. In *Proc. of the IEEE Global Telecommunications Conference 2008 (IEEE GLOBECOM 2008)*, pages 1–5, New Orleans, LA, USA, Dec 2008.

[YBO09] Bashir Yahya and Jalel Ben-Othman. Towards a classification of energy aware mac protocols for wireless sensor networks. *Wireless Communications and Mobile Computing*, 9(12):1572–1607, 2009.

[YHE04a] W Ye, J Heidemann, and D Estrin. Medium Access Control with Coordinated Adaptive Sleeping for Wireless Sensor Networks. *IEEE/ACM Networking Transactions*, 12(3):493–506, June 2004.

[YHE04b] Wei Ye, John Heidemann, and Deborah Estrin. Medium access control with coordinated adaptive sleeping for wireless sensor networks. *IEEE/ACM Transactions on Networking (ToN)*, 12(3):493–506, 2004.

[YSH06a] W Ye, F Silva, and J Heidemann. Ultra-Low Duty Cycle MAC with Scheduled Channel Polling. In *Proc. of the Fourth ACM International Conference on Embedded Networked Sensor Systems (SenSys'06)*, pages 321–333, Boulder, Colorado, USA, Nov 2006.

[YSH06b] Wei Ye, Fabio Silva, and John Heidemann. Ultra-low duty cycle mac with scheduled channel polling. In *Proceedings of the 4th international conference on Embedded networked sensor systems*, pages 321–334. ACM, 2006.

[YV06a] X Yang and NH Vaidya. A Wireless MAC Protocol using Implicit Pipelining. *IEEE Transactions on Mobile Computing*, 5(3):258–273, 2006.

[YV06b] Xue Yang and Nitin H Vaidya. A wireless mac protocol using implicit pipelining. *IEEE Transactions on Mobile Computing*, 5(3):258–273, 2006.

[ZZJ09] J. Zheng, J. Zheng, and A. Jamalipour. *Wireless Sensor Networks: A Networking Perspective*. John Wiley & Sons, 2009.

Index

About the Authors

Fernando J. Velez (M'93–SM'05) received the Licenciado, M.Sc. and Ph.D. degrees in Electrical and Computer Engineering from Instituto Superior Técnico, Technical University of Lisbon in 1993, 1996 and 2001, respectively. Since 1995 he has been with the Department of Electromechanical Engineering of Universidade da Beira Interior, Covilhã, Portugal, where he is Assistant Professor. He is also a senior researcher of Instituto de Telecomunicações. Fernando was an IEF Marie Curie Research Fellow in King's College London (KCL) in 2008/09(OPTIMOBILE IEF) and a Marie Curie ERG fellow at Universidade da Beira Interior from 2010 until March 2013 (PLANOPTI ERG). Fernando is the coordinator of the Instituto de Telecomunicações team in the Marie Skłodowska-Curie ITN Action (TeamUp5G) that started in 2019. He made or makes part of the teams of several European and Portuguese research projects on mobile communications and he was the coordinator of six Portuguese projects. Recently, he was the coordinator of **CONQUEST** (CMU/ECE/0030/2017), an exploratory project from Carnegie Mellon University (CMU) Portugal, a collaboration with the Department of Engineering and Public Policy from CMU. He has authored three books, 14 book chapters, 160 papers and communications in international journals and conferences, plus 39 in national conferences. Fernando is currently the IEEEVTS Region 8 (Europe, Middle East and Africa) Chapter Coordinator (nominated by VTS in 2010) and was the elected IEEE VTS Portugal Chapter coordinator from 2006 until 2014. He is the elected Chair of the IEEE ComSoc Portugal Chapter since Spring 2022. He is also member of the Board of Directors from the Telecommunications Specialization of Ordem dos Engenheiros. Prof. Velez was the coordinator of the WG2 (on Cognitive Radio/Software Defined Radio Co-existence Studies) of COST IC0905 TERRA. His main research areas are cellular planning tools, traffic from mobility, cross-layer design, spectrum sharing/aggregation, and cost/revenue performance of advanced mobile communication systems.

Luís M. Borges is currently a Network and Performance Specialist Engineer at NOKIA. Worked as a RAN Engineer at NetOne (Angola) within the WiMAX and LTE project as well as worked in Ericsson within the Vodafone PT SRAN modernization project. Luís M. Borges received the Licenciatura and Ph.D degrees in Electrical Engineering from Universidade da Beira Interior, Covilhã, Portugal, in 2006 and 2013, respectively. Luís M. Borges was a Post-Doc researcher in CREaTION project. He was also research assistant at Instituto de Telecomunicações, Covilhã. He made or makes part of the team of COST 2100 and COST IC1004 European projects, and he participated in SMART-CLOTHING and PROENERGY-WSN Portuguese projects. Luís M. Borges is a member of the Ordem dos Engenheiros (EUREL). His main research areas are or were Mobile networks and optimization, WSN, medium access protocols, cross-layer design, hardware development, network modelling, and application development, cognitive radio and energy harvesting.

Norberto Barroca received his Licenciado degree in electronics and telecommunications engineering from Polytechnic Institute of Castelo Branco in 2007 and his M.Sc. and Ph.D. degrees in electrical and computer engineering from the University of Beira Interior in 2009 and 2014, respectively. He was a research assistant at Instituto de Telecomunicações, while pursuing research for the Ph.D. degree. His research interests included wireless sensor networks and cross-layer design.

Periklis Chatzimisios serves as Professor in the Department of Information and Electronic Engineering of the International Hellenic University (Greece), where he leads the Communications, Cybersecurity and Internet of Things (CCIoT) Research Group. He has been awarded the title of Researcher Professor by the Department of Electrical and Computer Engineering of the University of New Mexico (USA). He is also a Visiting Fellow in the Faculty of Science & Technology, at Bournemouth University (UK). Prof. Chatzimisios has participated as a researcher and scientific officer in many European and national research and development projects funded by European Commission, Institute of International Education (IIE) and national agencies. He helds various management positions as Coordinator/WG Chair/Vice Chair and Training/Standardization Chair in related research projects such as TeamUp5G, SwiftV2X, COST CA20120 INTERACT.

He is/has been involved in several standardization and IEEE activities under the IEEE Communication Society (ComSoc) serving as the Chair of the ComSoc Young Professionals (YP) Standing Committee and as a

Member of the ComSoc Education Services Board, the ComSoc Standards Program Development Board, the ComSoc Industry Outreach Board and the ComSoc Online Content Board. He is the Chair of the Communications Chapter and Professional Activities for the IEEE Greece Section. In the past he served as Chair/Vice Chair/Secretary of the Technical Committee on Information Infrastructure and Networking (TCIIN), Vice Chair of the Technical Committee on Big Data (TCBD), Secretary of the Technical Committee on Cognitive Networks (TCCN) and he is an active member of the IEEE Future Networks (5G) Initiative. For more than 15 years he has been a member and/or Chairman of the Scientific/Steering/Organizing Committee of hundreds of international conferences and Founder/Organizer/Co-Chair for many Workshops which are co-allocated with major IEEE conferences. He served as Editor-in-Chief of the IEEE Standards E-Magazine (eZine) as well as in editorial board positions for several journals published by IEEE (such as IEEE Network, IEEE Communication Standards Magazine, IEEE Wireless Communication Magazine, IEEE Communications Surveys & Tutorials, IEEE Access, IEEE Communication Letters) and other publishing houses. In the past he was the Director during 2014-2016 (co-Director during 2012-2014) for the E-Letter of the IEEE Technical Committee on Multimedia Communications (MMTC). He is the author/editor of 8 books and more than 160 peer-reviewed papers and book chapters on the topics of performance evaluation and standardization of mobile/wireless communications, Internet of Things, 5G communications, the 4th Industrial Revolution, Smart Cities and vehicular networking. He is in the Top 2% Most Influential Scientists in the Stanford University list for 2021 and 2020 in the area of Networking & Telecommunications.